M.R.C.
BRAIN METABOLISM UNIT

Functions of the Septo-Hippocampal System

The Ciba Foundation for the promotion of international cooperation in medical and chemical research is a scientific and educational charity established by CIBA Limited-now CIBA-GEIGY Limited-of Basle. The Foundation operates independently in London under English trust law.

Ciba Foundation Symposia are published in collaboration with Elsevier Scientific Publishing Company / Excerpta Medica / North-Holland Publishing Company in Amsterdam.

Elsevier / Excerpta Medica / North-Holland, P.O.Box 211, Amsterdam

Functions of the Septo-Hippocampal System

Ciba Foundation Symposium 58 (new series)

Elsevier · Excerpta Medica · North-Holland
Amsterdam · Oxford · New York

© *Copyright 1978 Ciba Foundation*

All rights reserved. No part of this publication may be reproduced or transmitted in any form or by any means, electronic or mechanical, including photocopying and recording, or by any information storage and retrieval system, without permission in writing from the publishers.

ISBN 0-444-90032-2

Published in July 1978 by Elsevier/Excerpta Medica/North-Holland, P.O. 211, Amsterdam and Elsevier/North-Holland, Inc., 52 Vanderbilt Avenue, New York, N.Y. 10017.

Suggested series entry for library catalogues: Ciba Foundation Symposia.
Suggested publisher's entry for library catalogues: Elsevier/Excerpta Medica/North-Holland.

Ciba Foundation Symposium 58 (new series)
446 pages, 95 figures, 8 tables

Library of Congress Cataloging in Publication Data

Symposium on Functions of the Septo-Hippocampal System, London, 1977.
    Functions of the septo-hippocampal system.

  (Ciba Foundation symposium, new ser.: 58)
  Bibliography: p.
  Includes index.
    1. Septum (Brain)–Congresses. 2. Hippocampus (Brain)–Congresses. I. Title. II. Series: Ciba Foundation. Symposium: new ser., 58.
[DNLM: 1. Hippocampus–Physiology–Congresses. 2. Septum pellucidum–Physiology–Congresses. W3 C161F v. 58 1977 / WL314 S989f 1977]
QP383.2.S95 1977   599'.01'88   78-18687
ISBN 0-444-90032-2

Printed in The Netherlands by Casparie, Alkmaar

Contents

L. WEISKRANTZ Chairman's introduction 1

J. A. GRAY Objectives of the meeting 3

G. S. LYNCH, G. ROSE and C. M. GALL Anatomical and functional aspects of the septo-hippocampal projections 5
Discussion 20

L. W. SWANSON The anatomical organization of septo-hippocampal projections 25
Discussion 44

J. STORM-MATHISEN Localization of putative transmitters in the hippocampal formation (with a note on the connections to septum and hypothalamus) 49
Discussion 80

P. ANDERSEN Long-lasting facilitation of synaptic transmission 87
Discussion 102

J. F. DEFRANCE, J. C. STANLEY, J. E. MARCHAND and R. B. CHRONISTER Cholinergic mechanisms and short-term potentiation 109
Discussion 122

General Discussion I: Monoaminergic inputs to the hippocampus 127
Nicotinic transmission in the hippocampus 130
Specific pathways between septum and hippocampus 138

O. S. VINOGRADOVA and E. S. BRAZHNIK Neuronal aspects of septo-hippocampal relations 145
Discussion 171

J. O'KEEFE and A. H. BLACK Single unit and lesion experiments on the sensory inputs to the hippocampal cognitive map 179
Discussion 192

V

C. H. VANDERWOLF, R. KRAMIS and T. E. ROBINSON Hippocampal electrical activity during waking behaviour and sleep: analyses using centrally acting drugs   199
Discussion   221

S. P. GROSSMAN   An experimental dissection of the septal syndrome   227
Discussion   260

J. A. GRAY, J. FELDON, J. N. P. RAWLINS, S. OWEN and N. MCNAUGHTON   The role of the septo-hippocampal system and its noradrenergic afferents in behavioural responses to non-reward   275
Discussion   300

General Discussion II: Hippocampal units and theta activity   309
Mechanism of theta activity   319
Functional significance of theta activity   321

D. S. OLTON   The function of septo-hippocampal connections in spatially organized behaviour   327
Discussion   343

H. URSIN, T. DALLAND, B. ELLERTSEN, T. HERRMANN, T. B. JOHNSEN, P. J. LIVESEY, Z. ZAIDI and H. WAHL   Multivariate analysis of the septal syndrome   351
Discussion   369

L. WEISKRANTZ   A comparison of hippocampal pathology in man and other animals   373
Discussion   388

Final General Discussion: Frequency potentiation and memory   407
Further comments on the hippocampal cognitive map theory   412
Functions of the septo-hippocampal system in man and other animals   417

Index of contributors   429

Subject index   431

Participants

Symposium on Functions of the Septo-Hippocampal System, held at the Ciba Foundation, London, 18th–20th October 1977

Chairman: L. WEISKRANTZ Department of Experimental Psychology, University of Oxford, South Parks Road, Oxford OX1 3UD

P. ANDERSEN The Institute of Neurophysiology, University of Oslo, Karl Johansgt. 47, Oslo 1, Norway

E. AZMITIA Department of Anatomy, University of Cambridge, Downing Street, Cambridge CB2 3DY

A. BJÖRKLUND Department of Histology, University of Lund, Biskopsgatan 5, S-223 62 Lund, Sweden

A. H. BLACK Department of Psychology, McMaster University, 1280 Main St West, Hamilton, Ontario, Canada L8S 4K

J. F. DE FRANCE Department of Neurobiology and Anatomy, Health Sciences Center, University of Texas Medical School, P.O. Box 20708, Houston, Texas 77030, USA

D. GAFFAN Department of Experimental Psychology, University of Oxford, South Parks Road, Oxford OX1 3UD

J. A. GRAY Department of Experimental Psychology, University of Oxford, South Parks Road, Oxford OX1 3UD

S. P. GROSSMAN Department of Behavioral Sciences (Biopsychology), University of Chicago, Green Hall, 5848 South University Avenue, Chicago, Illinois 60637, USA

W. P. KOELLA Section 4 (Neuro- and Psychopharmacology), Pharmacology Research Laboratories, Pharmaceuticals Division, CIBA-GEIGY Limited, K 125.1109, CH-4002 Basle, Switzerland

P. J. LIVESEY Department of Psychology, The University of Western Australia, Nedlands, Western Australia 6009

G. S. LYNCH Department of Psychobiology, University of California, Irvine, California 92717, USA

P. MOLNÁR Department of Physiology, University Medical School, H-7643 Pécs, Hungary

J. O'KEEFE   Department of Anatomy and Embryology, University College (London), Gower Street, London WC1E 6BT

D. S. OLTON   Department of Psychology, The Johns Hopkins University, Baltimore, Maryland 21218, USA

J. B. RANCK, Jr.   Department of Physiology, Downstate Medical Center, State University of New York, Box 131, 450 Clarkson Avenue, Brooklyn, New York 11203, USA

J. N. P. RAWLINS   Department of Experimental Psychology, University of Oxford, South Parks Road, Oxford OX1 3UD

T. E. ROBINSON   Department of Psychology, University of California, Irvine, California 92717, USA

M. SEGAL   Isotope Department, The Weizmann Institute of Science, P.O. Box 26, Rehovot, Israel

B. SREBRO   Institute of Physiology, University of Bergen, Arstadveien 19, N-500 Bergen, Norway

J. STORM-MATHISEN   Anatomical Institute, University of Oslo, Karl Johansgt. 47, Oslo 1, Norway

L. W. SWANSON   Department of Anatomy and Neurobiology, Washington University School of Medicine, 660 South Euclid Avenue, St Louis, Missouri 63110, USA

H. URSIN   Department of Physiological Psychology, Institute of Psychology, University of Bergen, Arstadveien 21, 5000 Bergen, Norway

C. H. VANDERWOLF   Department of Psychology, Faculty of Social Science, University of Western Ontario, London 72, Ontario, Canada

OLGA S. VINOGRADOVA   Department of Memory Problems, Institute of Biophysics, Academy Biological Centre, Puschino-on-Oka, Moscow Distr. 142292, USSR

G. WINOCUR   Department of Psychology, Trent University, Peterborough, Ontario, K91 7B8, Canada

J. ZIMMER   Institute of Anatomy, University of Aarhus, DK 8000 Aarhus C, Denmark

Editors: KATHERINE ELLIOTT *(Organizer)* and JULIE WHELAN

Chairman's introduction

L. WEISKRANTZ

Department of Experimental Psychology, University of Oxford

We have a very exciting topic for this symposium: we also have a great number of different disciplines and approaches. We hope that out of the entire discussion some kind of synthesis will emerge, or at least some agreement about where and why we disagree. I personally hope that we can keep certain broad issues in mind during the symposium. The striking aspect of the hippocampus is the anatomical elegance of its structure, revealed in detail in the past few years. In contrast there is really appalling ignorance about what this elegance means. I am not sure that the septal area is quite so elegant in appearance but the hippocampus certainly is. Anatomical information will be presented to us. In recent years certain favourite anatomical connections have been stressed in discussions of the septo-hippocampal system, but there are older connections too: anatomists keep on adding connections; they very rarely subtract any! These 'older' connections, between for example the hippocampus and thalamus, are rarely stressed now and it may be worth reminding ourselves that we are dealing with a rather more complex anatomical system than just the pathways that we happen to be interested in at the moment.

We shall be given information on neurotransmitters, and it is worth asking ourselves *why* God made so many different transmitters when presumably it would be possible for all synapses to work on *one* transmitter, if the only function is to get signals from one neuron to another. The question here is what are the functional implications of a dependence on a particular transmitter, if there is one, in this particular anatomical system.

We shall hear electrophysiological information also, and I hope in discussion we can ask ourselves how we go from the single cell, which is the favourite unit of analysis these days, to populations of cells and their organizational structure. That is a standard problem but by no means a simple one to overcome.

We shall also be discussing function. There are a great many different approaches to the question of function. Historically, the suggested role of the hippocampus in memory is one with which we are all familiar. We are also familiar with its suggested role in emotional behaviour—in frustration and punishment—and relevant to that will be information on the endocrinological aspects of the hippocampus. We know now about its possible role in spatial functions. We know that it seems to have some connection with voluntary movement. We know about its suggested role in various kinds of disinhibitory and perseverative behaviour, and in habituation.

I hope we can direct our attention to three questions: first, is it possible to subsume all these different kinds of functions under a single heading? Is there some fundamental scheme in the Platonic sense which we see the shadows of, from time to time, in the various hypotheses that have been put forward? It would be interesting and important to see whether we can get agreement on that. The second question is: what are the ways in which the different suggested functions can be studied independently? Is an animal, for example, who is impaired on spatial function going to be impaired on various other aspects of behaviour *because* he has a spatial deficit? We must know the answer to this before we can decide whether we are dealing with several independent functions. Finally, the third question is related to the second but is not quite the same: the title of the symposium contains the words 'septo-hippocampal system'. By the end, we may want to know whether we are in fact dealing with systems in the plural, rather than one single system.

Objectives of the meeting

J. A. GRAY

Department of Experimental Psychology, University of Oxford

The sciences of brain and behaviour are beginning to knit together. So far, they have produced little more than a fringe, outlining the way in to the brain (perception) and the way out (motor processes). But it is now becoming possible to weave at least a coarse fabric with which to clothe the middle; and the septo-hippocampal system looks to be a favourable point at which to get on with the weaving.

The anatomy of the septo-hippocampal system is relatively well worked out, by comparison at least with other parts of the limbic system. It is, furthermore, an anatomy which is above all *orderly*; so orderly, in fact, that guessing from its structure to its function has become a favourite pastime of arm-chair theorists. All too often, indeed, such theorists have relied solely on anatomy (bolstered perhaps by computer analogies) in their speculations. But this is not because other data are lacking. On the contrary, since the early 1960s there has been a growing flood of reports of the behavioural effects of various kinds of damage to, or stimulation of, the different parts of the septal area and the hippocampal formation; and of the behavioural correlates of various kinds of electrical recording from these structures. Thus there is plenty of opportunity for *controlled* speculation as to the functions of the septo-hippocampal system; and that, I believe, is the chief business of this symposium.

Of course, such speculation has been going on non-stop for a long time. But it is an odd feature of research in this field that workers who use a particular technique take account only of the results of other workers using similar methods to their own. This has led to quite divergent views all becoming firmly accepted, each by the adherents of only one technique. For example, students of the behavioural correlates of the hippocampal theta rhythm for the most part believe this to be related to motor behaviour; those who study the effects of lesions in animals have generally been persuaded that the septal area and the hippocampus exercise some sort of inhibitory function; and those who study lesions in man have usually credited the hippocampus with a role in memory. These different groups of workers may refer to each other's results; indeed, they often publish within the covers of the same volumes: but one gets the strong impression that they never really *talk* to each other. We hope that they will do just that during the next three days.

The situation I have described has been going on for quite some time: why, then, is now a better time to talk than hitherto? There are several reasons. First, there have recently appeared a couple of volumes which have brought together many of the data about the hippocampus (Isaacson & Pribram 1975) and the septal area (DeFrance 1976) taken separately; though, oddly, this is the first symposium to consider them together, in spite of the well-known close connections between the two structures. Thus we can concern ourselves principally with understanding the significance of these data, rather than with recording them. Second, in the past five or six years there have been very important advances—and some dramatic changes—in our knowledge of the basic anatomy and physiology of the septo-hippocampal system; but the implications of these advances have not yet been fully worked out. Third, there have developed, in the same period, a number of radically new approaches, often based on new experimental methods, to the behavioural functions of the septo-hippocampal system; but these new ideas have not yet been fully brought up against each other, or against the data which supported older ideas.

We hope, therefore, that the next three days will force us to look carefully at *all* of the data, old *and* new, anatomical, physiological *and* behavioural, dealing with the septal area *and* with the hippocampal formation, in an effort to get to grips with the question: what are the functions of the septo-hippocampal system? At each step in our scrutiny, we should ask: What do we know? What ought to be done to resolve those points about which we are ignorant? Which are the issues which really divide us, and which are the ones which we merely describe differently?

References

DeFrance, J. (ed.) (1976) *The Septal Nuclei*, Plenum Press, New York
Isaacson, R. L. & Pribram, K. H. (eds.) (1975) *The Hippocampus*, 2 vols., Plenum Press, New York

Anatomical and functional aspects of the septo-hippocampal projections

GARY LYNCH*, GREG ROSE[+] and CHRISTINE GALL*

Department of Psychobiology and *School of Social Sciences*[+], *University of California, Irvine, California*

Abstract The origins, distribution, and cellular targets of the septo-hippocampal projections are reviewed. It appears that the distribution of acetylcholinesterase-positive neurons in the medial septum and diagonal bands and those cells labelled after injections of horseradish peroxidase into the hippocampus coincide; however, the possibility of a non-acetylcholinesterase septal projection remains. Good agreement is found between the distribution of hippocampal acetylcholinesterase and the patterning of silver grains after injection of [^{3}H]leucine into the medial septum. A major target of septal efferents to the hippocampus is the interneuron population; the possibility of septal mediation of intrahippocampal circuitry via this anatomical arrangement is discussed.

The hippocampus of mammals receives two major projections from the telencephalon, and these are in several important respects quite different from one another. By far the dominant afferent of the hippocampus is that from the entorhinal cortex; this projection generates a tightly packed, extremely well developed terminal field which covers nearly 70% of the dendritic tree of the granule cells of the dentate gyrus as well as a significant portion of the apical dendrites of the pyramidal cells (Blackstad 1958; Raisman *et al.* 1965; Hjorth-Simonsen 1972; Hjorth-Simonsen & Jeune 1972). Quantitative electron microscopic studies have indicated that in some regions the entorhinal terminals generate as much as 90% of the total synaptic population (Matthews *et al.* 1976; Lee *et al.* 1977). In marked contrast to this is the septo-hippocampal afferent system. These fibres are few in number and rather than restricting themselves to specific laminae they are found scattered throughout several sites in hippocampus (Raisman 1966; Mosko *et al.* 1973). The contrasting anatomical properties of the two input systems would imply that the septum exerts a quite different type of influence on the physiology of the hippocampal formation than does the entorhinal cortex. It is this idea which serves as the theme of this review.

The understanding of the entorhinal-hippocampal connections has benefited greatly from recent neuroanatomical and electrophysiological investigations. Stated simply, entorhinal fibres terminate on the outer portions of the dendrites of the granule and pyramidal cells (Nafstad 1967; Steward 1976) and when activated produce robust extracellular postsynaptic potentials (EPSPs) in those cells (Andersen *et al.* 1966; Dudek *et al.* 1976). It seems evident that this projection provides a powerful relay between thalamic and cortical sites (Van Hoesen & Pandya 1975) and the primary cells of the hippocampus (i.e. the granule cells of the dentate gyrus).

Unfortunately the system of septo-hippocampal connections has not proved as amenable a subject for either anatomical or electrophysiological work. The region from which the system is generated is small and is traversed by fibre bundles travelling to and from the hippocampus, confounding analysis of the origin of the projections which actually arise from within the septum. Furthermore, the sparse and scattered nature of the distribution of the septal axons within the hippocampus provides any number of difficulties for tracing experiments (Mosko *et al.* 1973). However, the application of recently developed neuroanatomical methods has provided valuable new information, with the result that a clearer picture of the organization of the septal afferents of the hippocampus has begun to emerge. In this paper we shall make use of these data to reexamine the origins, distribution, and cellular targets of the septo-hippocampal projections. Following this, brief consideration will be given to the question of what role the septal afferents play in the operation of the hippocampus and an attempt will be made to link the data from the anatomical studies with the results of neurobehavioural investigations.

THE ORIGINS OF THE SEPTAL PROJECTION INTO HIPPOCAMPUS

Until quite recently, information on the location of neurons which generate the septo-hippocampal projection came primarily from studies using the anterograde degeneration method and acetylcholinesterase histochemistry (AChE; EC 3.1.1.7). With regard to the former, Raisman (1966) reported that the medial but not lateral septum projected into the hippocampus and dentate gyrus. These results were somewhat compromised by the fibre-of-passage problem; that is, it was difficult to be certain that the observed degeneration originated from neurons located in the septum rather than from fibres interrupted by the lesion which pass through the septal area. Studies by Lewis & Shute (1967) using the AChE histochemical method demonstrated

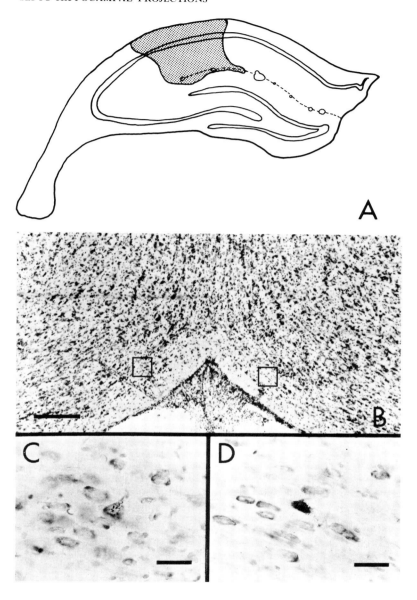

Fig. 1. After injection of horseradish peroxidase (30% w/v) into subfield CA1 of the rostral hippocampus of the rat (illustrated schematically in A by the stippled area) retrogradely labelled neurons were observed in the nucleus of the diagonal bands both ipsilateral and contralateral to the injection site. B shows a coronal section through the septal area which demonstrated such a labelling pattern. The small boxes in the lower portion of the figure identify the location of two neurons shown at higher power in C and D. Calibration: B, bar = 200 μm; C,D, bar = 25 μm.

that the medial septum–diagonal band area contained a collection of densely staining (i.e. AChE-positive) cells and by means of a series of clever lesion experiments the authors were able to conclude that these cells generated a septo-hippocampal projection.

The experiments of Segal & Landis (1974) and, more recently, Meibach & Siegel (1976) using the method of retrograde transport of horseradish peroxidase (La Vail et al. 1973) have shed new light on the question of the origins of the septal afferents to the hippocampus. Their results largely confirm the conclusions of the earlier work but add the additional and important information that the septo-hippocampal projection exhibits a topographical relationship. It appears that the most medial aspects of the septum innervate the rostral hippocampus while more lateral portions of the medial nucleus and adjacent area send fibres to progressively more caudal and ventral regions of the hippocampal formation. Whether this organization holds true for the entire length of the medial septal nucleus–diagonal band complex is not clear.

From these studies it appears that hippocampal afferents from the septum are exclusively ipsilateral in origin. However, our own experiments using horseradish peroxidase histochemistry demonstrate that a small but significant number of cells which send their axons to the hippocampus are located in the contralateral medial septum–diagonal band complex of the rat (Fig. 1). This result supports an earlier observation (Mellgren & Srebro 1973) of a bilateral loss of hippocampal acetylcholinesterase after discrete unilateral lesions of the medial septum. That others using horseradish peroxidase were unable to identify this bilateral origin for the septo-hippocampal projection may have been the result of using a less sensitive chromagen in their development procedure (Mesulam 1976).

Taken together the evidence from the horseradish peroxidase and acetylcholinesterase studies is quite good that a septo-hippocampal system exists and originates within the medial and ventral aspects of the septal complex; however, it is not clear that the projections described by the two methods are the same. The possibility remains that there exist septal afferents of the hippocampus which do not contain AChE. A necessary first step in investigating this question is a comparison of the distribution and appearance of the AChE-positive cells with those cells which are labelled after large injections of horseradish peroxidase into hippocampus.

Fig. 2 summarizes the location of the AChE-positive neurons in the septum and nucleus of the diagonal bands. The animal used for this figure was treated with diisopropylfluorophosphate (DFP), an irreversible inhibitor of AChE, and then allowed 24 hours to synthesize new enzyme before sacrifice.

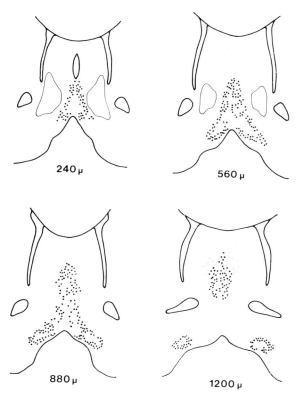

FIG. 2. Schematic representation of acetylcholinesterase-positive neurons located in the septal region after pretreatment with diisopropylfluorophosphate. The number of μm below each drawing indicates the position of the section from the anterior end of the nucleus of the diagonal bands.

Since this enzyme is produced only in the cell soma, at this short survival interval AChE-containing neurons stain heavily while the dendritic and axonal processes (which have not yet received the enzyme by somatofugal transport) remain unstained. The result is that the location and features of the AChE-positive cells can be seen unobscured by neuropil staining (Lynch *et al.* 1972). A comparison of material treated in this way with control (i.e. no DFP) rats provides a more complete picture than is found with AChE histochemistry alone.

In general, AChE-positive cells are found throughout the medial septal nucleus and diagonal bands region. These cells do not, however, constitute a homogeneous population. On the basis of differences in morphology and

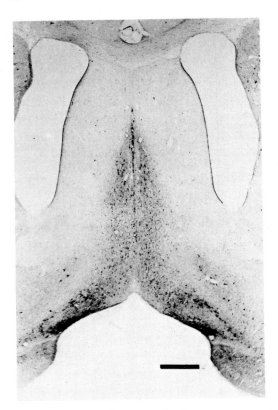

FIG. 3. Low power photomicrograph of the septum stained by AChE histochemistry in a rat previously treated with diisopropylfluorophosphate. At this coronal level the various AChE-positive cell types appear in separate fields. The cells of the central portion of the medial septal nucleus (MSN) are noticeably more faintly stained than those of the dorsal MSN 'cap'. The cells of the diagonal band region, seen in an ovoid configuration near the ventral face of the section, are similarly distinct from central MSN cells on the basis of staining intensity and gross morphology. A small population of intensely stained 'midline cells' comprise the fourth distinct AChE-positive subgroup. Bar = 500 μm.

staining intensity several distinct cell groups can be described. This heterogeneity is most apparent at mid-septal levels where the various groups occupy distinct regions of the complex (Fig. 3). At this level the majority of the histochemically identified somata in the medial septal nucleus belong to small, lightly staining bipolar cells. The few visible processes of these neurons are aligned almost exclusively dorsoventrally. In contrast to the medial septal nucleus cells, AChE-positive cells of the diagonal band region are larger and much more intensely stained, with branching dendritic and occasional axonal processes visible. As can be seen in Figs. 3 and 4c, these

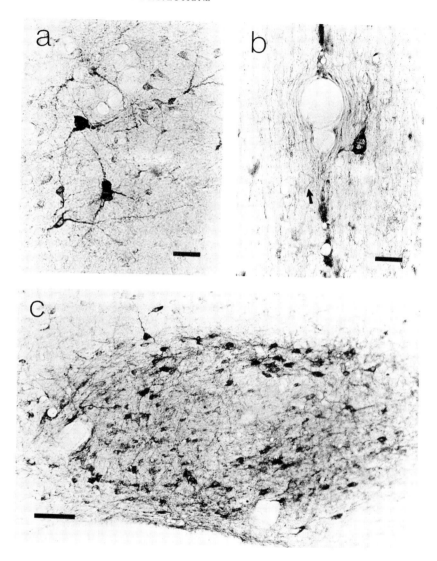

FIG. 4. Photomicrographs of AChE-positive cells in the medial septum-diagonal band complex in a rat pre-treated with diisopropylfluorophosphate. The densely staining cells of micrograph *(a)* found at the periphery of the diagonal bands 'ovoid group' illustrate the excellent delineation of basic morphology possible with this technique. Micrograph *(b)*, taken at the midline of the medial septal nucleus, shows the process (arrow) of a cell on the right half of that nucleus cross the midline to bifurcate on the contralateral side. Although not frequent, such 'crossing' processes are regularly observed. The ovoid configuration of the diagonal band AChE-positive group is apparent in micrograph *(c)*. The disposition of cell bodies as well as the alignment of their processes creates an ovoid AChE-positive ring. Calibration: a,b, bar = 25 μm; c, bar = 100 μm.

neurons are arranged in an ovoid cluster ventral and medial to the nucleus accumbens septi. The dendritic processes appear to conform to the curve of the ovoid ring with the centre remaining cell- and process-poor. Two smaller but distinct AChE-positive cell populations are found along the midline and at the dorsal tip of the medial septal nucleus. The midline neurons stain intensely and have an oval, elongated soma; occasionally one thick process aligned with the midline is evident. The cells which 'cap' the medial septal nucleus seem to represent yet another distinct group in that they are somewhat larger and much more intensely stained than the neurons of the medial septal nucleus and more frequently display dendritic processes out of the dorsoventral alignment.

The distribution of the neurons which were labelled after a large injection of horseradish peroxidase in the rostral hippocampus is shown in Fig. 5. The labelled cells are found along the entire anterior-posterior extent of the medial septal nucleus and nucleus of the diagonal band and in this sense are co-extensive with the AChE-positive cells. In the case illustrated the neurons are restricted to a medial position but this may be due to the fact that the enzyme injection was limited to the rostral hippocampus. (As previously mentioned, Meibach & Siegel (1977) found more laterally placed neurons after injections of horseradish peroxidase into the 'temporal' aspects of the hippocampus). The retrograde horseradish peroxidase method provides only a minimum of cytological detail but it was evident from these experiments that several classes of cells corresponding approximately to those described using the AChE histochemical method were labelled.

The observation that AChE- and horseradish peroxidase-positive cells are co-extensive in their distribution suggests that the septo-hippocampal system may be composed entirely of cholinesterase-positive elements. As will be seen, this conclusion would also be in agreement with the results of studies comparing the organization of the septal projections in hippocampus as demonstrated by autoradiography and AChE histochemistry. Despite this, a caveat is in order—experiments in which AChE and horseradish peroxidase histochemistries have been combined have indicated the presence of numerous retrogradely labelled cells which do not stain for acetylcholinesterase (Rose & Lynch, unpublished data; Mesulam et al. 1977). Whether this is due to inadequate staining with the AChE method or provides evidence of non-AChE septal projections to hippocampus remains a subject for further work. At the present, it can be concluded that: (1) the distribution of acetylcholinesterase-containing neurons corresponds quite well with the origins of the septo-hippocampal system and (2) it is likely that several morphologically distinct types of AChE-positive cells project to the hippocampus.

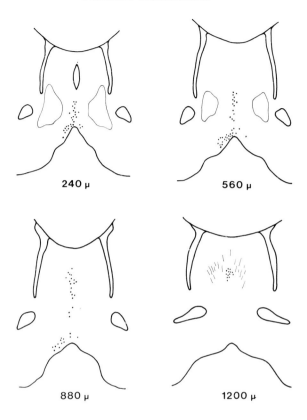

FIG. 5. Schematic representation of the distribution of retrogradely labelled neurons after an injection of 30% horseradish peroxidase into the rostral hippocampus. Anterior–posterior levels are indicated as in Fig. 2.

THE DISTRIBUTION AND TARGETS OF THE SEPTO-HIPPOCAMPAL SYSTEM

Again work using the anterograde degeneration method and AChE histochemistry has provided useful descriptions of the distribution of the septal projections in the hippocampus; furthermore, studies comparing the projections as shown by the two methods have found good although not complete agreement. Fig. 6 illustrates the distribution of the acetylcholinesterase staining in a cross-section of hippocampus and compares this with the pattern seen after a large injection of concentrated [^{3}H]leucine into the medial septal area. In the dentate gyrus the two patterns are quite similar in that the densest innervation is in the infragranular region while a thin band is detected immediately above the granule cells. AChE histochemistry provides evidence

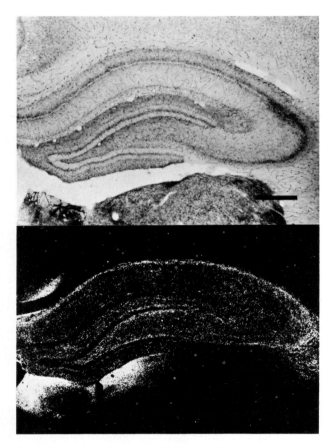

FIG. 6. Coronal sections through the rostral hippocampus stained for acetylcholinesterase *(top)* and after a large injection of [^{3}H]leucine into the medial septum *(bottom)*. In general, there is good correspondence between AChE-positive sites in the rostral hippocampus and those areas demonstrating radioactive labelling after injection of the tritiated amino acid into the septum. For further details see text. Bar = 500 μm.

for a modest projection in the middle molecular layer which has been difficult to establish satisfactorily with the anterograde degeneration method (Mosko *et al.* 1973) but is somewhat better illustrated by autoradiography. In the regio inferior pyramidal cell fields, the heaviest innervation as revealed by both methods is in the stratum oriens. The projections of the septum to the apical dendritic fields of the pyramidal cells is sparse and comparisons between methods are accordingly difficult; the stratum moleculare of CA1 is notably devoid of input while a very thin zone between this region and the stratum radiatum can be seen to receive a projection by both AChE histochemistry and autoradiographic tracing. A band of moderate intensity of AChE activity is found along the basal axis of the CA1 pyramidal cells but this has no apparent counterpart with either autoradiographic or Fink-Heimer tracing methods.

These results support earlier conclusions (Mosko et al. 1973; Mellgren & Srebro 1973) that the pattern of the acetylcholinesterase staining in hippocampus corresponds quite closely to the distribution of the septo-hippocampal projection. The AChE pattern is the slightly more extensive of the two and this may indicate that the hippocampus receives or generates AChE-positive afferents other than those from the septum (Storm-Mathisen & Blackstad 1964). Alternatively, it is possible that these discrepancies are due to the sparse nature of the projections in question—as the number of fibres and terminals which contain radiolabel decreases it becomes increasingly more difficult to distinguish them with certainty from background.

In a positive sense, it is evident that the septum does not send any major projections into hippocampus which terminate outside the zones receiving the AChE-positive fibres. In light of the earlier discussion of possible non-AChE components of the septo-hippocampal system it seems likely that if two types of septal inputs exist, they are intertwined and have the same regions for targets.

The distribution of the septo-hippocampal projections as described above is highly suggestive with regard to the cellular targets of these axons. The infra-granular region contains a heterogeneous population of interneurons and is essentially devoid of dendrites from either the granule or pyramidal neurons (Ramon y Cajal 1911). Therefore, the observation that this area contains a dense concentration of septal axons and terminals strongly suggests that these fibres are targeted for one or more types of interneurons (Mosko et al. 1973). The distribution of septal projections in the regio inferior also suggests this conclusion, in that the location of the terminals and a previously described group of interneurons coincides.

That this interneuron population is the target of septal afferents has been partially confirmed by autoradiographic studies using [^{3}H]adenosine. This nucleoside readily crosses synaptic junctions, thereby providing a means of identifying the target cells of neurons which have transported it down their axons (Schubert & Kreutzberg 1975). After the injection of [^{3}H]adenosine into the medial septum, the compound (or one of its derivatives) is transported into the hippocampus and with appropriate survival periods is found in a number of postsynaptic targets (Rose & Schubert 1977). Fig. 7 shows an example of a densely labelled cell in the infragranular region of the dentate

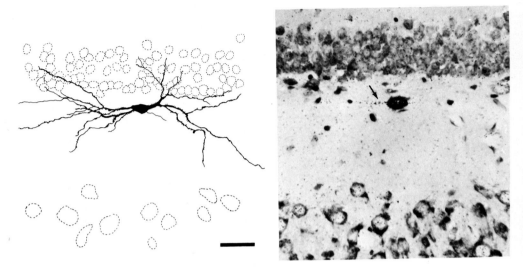

FIG. 7. *Left:* Camera-lucida drawing of a Golgi-stained cell in the hilus of the dentate fascia generally described as an interneuron. *Right:* Intrahilar cell labelled two days after the injection of [³H]adenosine into the septum. Radioactive label is seen to extend into the proximal portions of the cell dendrites (arrow). Both the horizontal orientation of its dendritic process and its position suggest that this neuron is of similar type to the cell drawn on the left. Bar = 50 μm.

gyrus which resembles a type of interneuron commonly encountered in Golgi material. Cajal (1911) discussed this class of cells and noted that their axons ramified in the dendritic field of the granule cells. Other interneuron 'types' have been labelled in stratum oriens of CA3, but identification of these is less certain because of possible confusion with the displaced pyramidal cells also occasionally found in this layer. The granule cells are also labelled after septal injections of [³H]adenosine, although not as heavily as are the interneurons. This may suggest that although septal axons terminate on both these hippocampal cell types, a greater percentage of the innervation is targeted for the interneuron population.

From these observations it is not possible to decide if any selectivity exists in the septal innervation of interneurons; of particular interest is the question of whether those cells which generate the presumably inhibitory basket plexuses on the granule cells are involved. In any event, the relationship between septal fibres and interneurons provides a mechanism whereby the numerically inferior septal projections to the dentate could exert potent physiological effects in the face of the relatively massive perforant path input. Specifically, the interneurons generate extensive axonal arborizations and in some cases these make synaptic contacts with hundreds of granule cells (Ramon y Cajal 1911). Furthermore, the location and inhibitory character of the basket endings is such as to exert a profound control over the physiology of the granule cells (Andersen *et al.* 1964, 1966). Thus by innervating the interneuron population the numerically inferior septal inputs to the hippocampus could exert an influence which would be far greater than could be achieved by their direct termination on hippocampal pyramidal or granule cell dendrites.

These anatomical observations provide a suggestion about the type of role that the septum might perform in the operation of the hippocampus. If, as argued above, these projections are primarily targeted for interneurons, then activation of the septum could be expected to alter the inhibitory control that these local cells exert over the granule and pyramidal cells. In this context, a recent electrophysiological study (Alvarez-Leefmans & Gardner-Medwin 1975) demonstrated that septal stimulation could regulate to a remarkable degree the efficacy of perforant path volleys in eliciting granule cell discharge. An explanation for this result could well be supplied by the hypothesis that septal contacts on interneurons modulate their activity and thus allow the 'biasing', in effect, of the response of hippocampal neurons to their excitatory inputs.

SOME FUNCTIONAL IMPLICATIONS OF THE ANATOMICAL DATA

The anatomical data discussed above suggest that the septal projections may serve to modulate the response of the hippocampus to the activation of its massive afferent input from the entorhinal cortex. Ultimately, evaluation of this and similar hypotheses must come from experiments in which the activity of hippocampal neurons after the activation of entorhinal projections is studied over time and the contributions of the septal inputs are carefully analysed. Neurobehavioural studies done in collaboration with Dr S. A. Deadwyler have been conducted which, we hope, will be appropriate for these purposes, and the initial results have been encouraging. Obviously, the first condition that had to be met was the development of a paradigm which led to the reliable discharge of the entorhinal projections. After considerable experimentation, it was found that a tone cue would generate a robust evoked response in the outer molecular layer of the dentate gyrus as a rat learned that operant responses performed within seconds of the signal would be reinforced.

Since the laminar distribution of the tone-evoked response (as recorded with a movable microelectrode) was nearly identical to that produced by electrical stimulation of the entorhinal cortex it is very likely that the recorded potentials are due to a synchronous discharge of the perforant path. Surprisingly, the evoked response produced under these circumstances did not elicit reliable discharge of the granule cells. However, when a second, non-rewarded tone was added to the situation and the rat was required to discriminate between tones, the evoked responses in the outer molecular layer were followed by prolonged high frequency discharges of the granule cells, although addition of the second tone caused no discernible increase in the amplitude of the evoked potential to either tone. Eventually very different patterns of granule cell discharge were established to two tones, the prolonged burst of cell firing being seen in association with the positive (i.e. rewarded) tone, and an initial burst followed by a rapid return to background levels after the negative (non-rewarded) tone.

From these results it appears that activation of the entorhinal projections, even in a synchronous fashion, is not by itself an adequate condition to discharge the granule cells and hence initiate activity in the first segment of the intrahippocampal circuit. This strongly implies that the 'responsivity' (threshold) of the granule cells is under the control of an agency other than the entorhinal cortex and it is likely that this control is exerted at the level of the cell body. A regulatory mechanism of this type resembles the function ascribed to the septo-hippocampal system in the hypothesis discussed above. Studies in which the firing behaviour of the neurons in the medial septal nucleus and diagonal bands complex are followed in the one- versus two-tone behavioural situation (i.e. in two circumstances in which the response of the granule cells to an entorhinal input is very different) could provide a partial test of these ideas.

That the primary hippocampal circuit is not triggered even in a behavioural situation which results in the synchronous discharge of the entorhinal afferents to the dentate gyrus serves as a reminder that even the suggestively 'simple' organization of the hippocampus does not make this structure an easy target for functional analysis. Afferent strength is normally predicted by terminal number; however, from these experiments the conclusion must be drawn that in a functional sense this may not always hold true. In the dentate gyrus afferent projection fibres from the perforant path constitute by far the dominant input to this structure. It could be, however, that in normal (i.e. behavioural) circumstances even this massive innervation is unable to trigger discharge of the postsynaptic granule cells without some kind of 'cooperation' from other inputs. Appreciation of the circumstances which lead to the appropriate interactions of these very different afferent systems may provide some of the clues needed to understand the physiological and behavioural processes in which the hippocampus participates.

ACKNOWLEDGEMENTS

We wish to acknowledge the skilful technical assistance of Ms Brigitta Flick and Ms Jo Ann Hendrickson and the perseverance of Ms Darlene Thompson in typing the manuscript. This work was supported in part by grant BNS 76-17370 from the National Science Foundation as well as a Career Development Award to G.L. C.M.G. was supported by a predoctoral fellowship from the National Institutes of Health.

References

ALVAREZ-LEEFMANS, F. J. & GARDNER-MEDWIN, A. R. (1975) Influences of the septum on the hippocampal dentate area which are unaccompanied by field potentials. *J. Physiol. (Lond.)* 249, 14-16

ANDERSEN, P., ECCLES, J. C. & LOYNING, Y. (1964) Pathway of postsynaptic inhibition in the hippocampus. *J. Neurophysiol.* 27, 608-619

ANDERSEN, P., HOLMQVIST, B. & VOORHOEVE, P. E. (1966) Entorhinal activation of dentate granule cells. *Acta Physiol. Scand.* 66, 448-460

BLACKSTAD, T. W. (1958) On the termination of some afferents to the hippocampus and the fascia dentata. *Acta Anat.* 35, 202-214

DUDEK, F. E., DEADWYLER, S. A., COTMAN, C. W. & LYNCH, G. (1976) Intracellular responses from granule cell layer in slices of rat hippocampus: perforant path synapse. *J. Neurophysiol.* 39, 384-393.

HJORTH-SIMONSEN, A. (1972) Projection of the lateral part of the entorhinal area to the hippocampus and fascia dentata. *J. Comp. Neurol.* 146, 219-232

HJORTH-SIMONSEN, A. & JEUNE, B. (1972) Origin and termination of the hippocampal perforant path in the rat studied by silver impregnation. *J. Comp. Neurol.* 144, 215-232

LA VAIL, J. H., WINSTON, K. R. & TISH, A. (1973) A method based on retrograde intra-axonal transport of protein for identification of cell bodies of origin of axons terminating within the CNS. *Brain Res.* 58, 470-477

LEE, K. S., STANFORD, E. J., COTMAN, C. & LYNCH, G. (1977) Ultrastructural evidence for bouton proliferation in the partially deafferented dentate gyrus of the adult rat. *Exp. Brain Res.* 29, 475-485

LEWIS, P. R. & SHUTE, C. C. D. (1967) The cholinergic limbic system: projections to hippocampal formation, medial cortex, nuclei of the ascending cholinergic reticular system and the subfornical organ and supra-optic crest. *Brain* 90, 521-537

LYNCH, G. S., LUCAS, P. A. & DEADWYLER, S. A. (1972) The demonstration of AChE containing neurons within the caudate nucleus of the rat. *Brain Res.* 45, 617-621

MATTHEWS, D. A., COTMAN, C. & LYNCH, G. (1976) An electron microscopic study of lesion-induced synaptogenesis in the dentate gyrus of the adult rat. I. Magnitude and time course of degeneration. *Brain Res.* 115, 1-21

MEIBACH, R. C. & SIEGEL, A. (1977) Efferent connections of the septal area in the rat: an analysis utilizing retrograde and anterograde transport methods. *Brain Res.* 119, 1-20

MELLGREN, S. I. & SREBRO, B. (1973) Changes in acetylcholinesterase and distribution of degenerating fibres in the hippocampal region after septal lesion in the rat. *Brain Res.* 52, 19-35

MESULAM, M. (1976) The blue reaction product in horseradish peroxidase neurohistochemistry: incubation parameters and visibility. *J. Histochem. Cytochem.* 24, 1273-1280

MESULAM, M., VAN HOESEN, G. & ROSENE, D. G. (1977) Substantia innominata, septal area and nuclei of the diagonal band in rhesus monkey: organization of efferents and their acetylcholinesterase histochemistry. *Neurosci. Abstr.* 3, 202

MOSKO, S., LYNCH, G. & COTMAN, C. W. (1973) The distribution of septal projections to the hippocampus of the rat. *J. Comp. Neurol.* 152, 163-174

NAFSTAD, P. H. J. (1967) An electron microscopic study of the termination of the perforant path fibers in the hippocampus and the fascia dentata. *Z. Zellforsch. Mikrosk. Anat. 76*, 532–542

RAISMAN, G. (1966) The connexions of the septum. *Brain 89*, 317–348

RAISMAN, G., COWAN, W. M. & POWELL, T. P. S. (1965) The extrinsic afferent, commissural and association fibres of the hippocampus. *Brain 88*, 963–996

RAMON Y CAJAL, S. (1911) *Histologie du Système Nerveux de l'Homme et des Vertèbres*, vol. II, Maloine, Paris

ROSE, G. & SCHUBERT, P. (1977) Release and transfer of [^{3}H]adenosine derivatives in the cholinergic septal system. *Brain Res. 121*, 353–357

SCHUBERT, P. & KREUTZBERG, G. W. (1975) [^{3}H]adenosine, a tracer for neuronal connectivity. *Brain Res. 85*, 317–319

SEGAL, M. & LANDIS, S. (1974) Afferents to the hippocampus of the rat studied with the method of retrograde transport of horseradish peroxidase. *Brain Res. 78*, 1–15

STEWARD, O. (1976) Topographic organization of the projections from the entorhinal area to the hippocampal formation of the rat. *J. Comp. Neurol. 167*, 285–314

STORM-MATHISEN, J. & BLACKSTAD, T. W. (1964) Cholinesterase in the hippocampal region. Distribution and relation of architectonics and afferent systems. *Acta Anat. 56*, 216–253

VAN HOESEN, G. W. & PANDYA, D. N. (1975) Some connections of the entorhinal (area 28) and perirhinal (area 35) cortices of the rhesus monkey. I. Temporal lobe afferents. *Brain Res. 95*, 1–24

Discussion

Gray: In your neurobehavioural experiment I didn't understand whether the changed unit response of the granule cells occurred first to the unreinforced second tone and only then started occurring to the reinforced tone, or whether (after you introduced the unreinforced tone) it first occurred to the reinforced tone. That is critical for the interpretation of the experiment.

Lynch: Because of the random pattern in which the tones are presented and the averaging procedures we used, it is not possible to give a satisfactory answer to your question.

Gray: If you keep going until the discrimination is learnt perfectly, does the unit response to the correct tone start dying out in the granule cells or does it stay permanently?

Lynch: The unit response appears to be permanent.

Winocur: Does that pattern entirely reverse when the reward value of the stimuli reverses?

Lynch: When the value of the tone is reversed, yes, the pattern of unit responses also reverses.

Winocur: Have you done this over a series of reversals?

Lynch: We did two reversals—we can't go much further than that because of the difficulty of maintaining electrodes over several days.

Weiskrantz: To what extent is this pattern dependent on the actual emission of a response?

Lynch: The rat has to respond with a nose-poke. The rats are maintained at 85% of body weight on water deprivation and obtain a lick of water when they break a photocell with the nose-poke response. There is an interesting feature that we sometimes see, which may relate to your point. Occasionally the rat generates a long burst to the incorrect tone, and on these occasions, he usually makes a response.

Ursin: Could the response you get be due to anything like a feedback based on the animal making a response, by moving the vibrissa, or just moving his head?

Lynch: Yes, it could be, but remember that the onset of the unit discharges is very rapid. It happens within 15-20 ms of the time the medial geniculate cells discharge—that is, the beginning of the evoked response can be measured within 15-20 ms of the time the medial geniculate fires. When we look at the animal during the time we are getting the electrical activity we do not see any movement.

Weiskrantz: So it could be feed-forward as well as feedback?

Lynch: Yes.

Andersen: Do you really want us to believe that you can modify the discharge pattern of cells without changing the corresponding field potentials?

Lynch: We do not detect any changes in the size of the averaged evoked response when the tones begin to elicit the unit discharges. A number of things could be going on such that a constant perforant path volley could be more efficient in terms of driving its targets. As has been shown, the septal projections terminate in part on the interneurons below the granule cells and some of these interneurons generate the inhibitory basket complexes. It is also possible that the septal projections have the same targets in the CA3 field. The suggestion is a simple one; the septal inputs via the interneurons modulate the response of the granule cells to the perforant path and may also allow the CA3 to dentate granule cell feedback system to begin operation. When we add the second tone perhaps we are creating conditions under which the tone brings the septal inputs into play.

Andersen: But how can the field potential be unchanged in that situation? The membrane resistance will also fall, and the field potential will drop, not increase. The discharges will also decrease, not increase as your data show.

Lynch: But could we detect such changes? The perforant path field responses are being produced by 'natural' cues and are more variable than those produced by direct stimulation of the perforant path. We also don't have any idea of how large a change of cell body potential we need to allow the granule neurons to discharge.

Andersen: You referred to Cajal. He showed different types of cells, and their axonal ramification, but little about connectivity. We don't know whether the two cells which you are talking about are connected.

Lynch: I agree with you that we can't make statements about connectivity with the Golgi technique, as we can't with any of the traditional light microscopic tracing methods.

Weiskrantz: Dr Lynch, is your suggestion that this modulation is produced by the septal nucleus based on circumstantial evidence, or is there any crucial evidence that in the absence of the septal nucleus the same kind of modulation occurs?

Lynch: No, there is no direct evidence. The interneuronal population is visited by inputs originating in the raphe and locus coeruleus, but, as I said, our strategy was to see if we could find circumstances in which the response of the granule cells to a naturally elicited perforant path volley would change. These are conditions which we feel are optimal for detecting the operation of the septo-hippocampal system.

Ranck: An observation perhaps related to that of Dr Lynch is that of Winson & Abzug (1977). They find that the response in dentate to stimulation of the perforant path changes depending on the behavioural state of the animal. One of the curious things about this is that they only see changes in field potentials at relatively high intensities of stimulation. One might think that the greatest sensitivity of the system would be at relatively small field potentials when the input–output relations had the steepest slope. So Winson and Abzug, like Dr Lynch, do not find a field potential change in a situation where we know the system has been changed and where a change would be expected.

Weiskrantz: Two separate issues arise here, firstly, the nature of the conditioning procedure and the changes that occur at a gross level; and, secondly, a much more detailed question at a different level as to what actual action field potentials are involved and the internal fine-grained organization at the electrical level.

Andersen: Whatever the complexity in pathways of the volley finally hitting the target cell, the mechanism when it hits that cell is simple. The size of the slow field potential is linearly related to the injected synaptic current. There is also a direct relation between the size of the field potential, and the latency and number of cell discharges. Whenever you have an increase or decrease in the number of cell discharges, you must also have a change in the field potential with correctly placed electrodes. Therefore, if you have a changed discharge pattern, I cannot see how you cannot have seen a changed field potential. Could it be drowned in an irregular background? I think your work is fascinating. It is only the explanation that we disagree on.

Lynch: The question is why the granule cells are discharging to a tone they normally ignore when circumstances are changed. We suggest that other inputs are added to the situation and that events at the cell body level are changed. However, if membrane changes are produced in the soma they should have an effect on field responses found in dendrites, as Dr Andersen says. We are restricted by the variance in chronic recording to natural stimuli and the fact that relatively small membrane events may be involved. But the specific hypothesis that the septo-hippocampal inputs are responsible predicts certain experiments and fits the anatomy.

Vinogradova: I understand that you do not see granule cell responses to stimuli outside the conditioning paradigm. That seems strange to me, because we, with Dr Bragin, have done a lot of work on the dentate (Vinogradova & Bragin 1975; Bragin *et al.* 1976). We worked essentially with unreinforced sensory stimuli, and especially with tonal stimuli. We observed strong effects of tonal stimuli on granule cell discharges outside the conditioning paradigm. Some of the granule cell responses are just short bursts, 'on'-effects, but many cells have responses consisting of on-effects, then about 50 ms pauses, and then phasic bursts equal in duration to a tonal stimulus. About 70% of responsive cells respond actively to the stimuli. Other cells responded by prolonged tonic inhibition during and after the tone. We moved the recording microelectrode along the longitudinal axis of the hippocampus, while giving a click, and we observed the active segments where the cells respond to a click by a burst, as a whole population. These segments were surrounded by areas where cells respond to click by inhibition of their activity. We think that such a spatial distribution of responses has some relation to the organization of the perforant path terminations in the dentate. Probably in your experiments you did not record in such an active zone, or perhaps there is a species difference? We use rabbits, not rats.

Lynch: I did not mean to say that we did not see granule cell discharges in response to tones. What I intended to say was that without appropriate conditioning we did not see cell discharges that were correlated with the evoked response in the perforant path zone.

Grossman: Dr Lynch, what do the granule cells do when the animal makes a response in the absence of a tone?

Lynch: We have many situations with the rats making this particular response in the absence of the tone. We haven't analysed it specifically, but from what I can see, the cells aren't doing anything.

Koella: On the neuroanatomical side, have you any evidence about feedback from the granule cells which are being influenced from the septum, back to the very same septal cells?

And secondly, on function, you said there is some evidence of a spatial organization in the sense that there is a dorso-medial to ventro-lateral pattern in the septum which projects as such to the hippocampus. What do you think is the functional basis of this pattern?

Lynch: On the question of feedback to the septum, I have no data, though it is possible that the granule cells fire the CA3 and trans-synaptically the CA1 zones and one of these perhaps goes into the septal zone which generates an input to the granule cells.

On the second question, we are talking essentially about a system which is organized such that the medial aspect of the septum is going to the rostral hippocampus and the lateral septum is going to the temporal hippocampus. The point I would suggest is that this organization extends through the entire medial septal–diagonal band complex, from very different cell types and different subnuclei of the septum. That is all I can tell you, except that if structure predicts function, then I suspect that this complex is doing different things to the same population of granule cells.

References

BRAGIN, A. G., VINOGRADOVA, O. S. & EMELYANOV, V. V. (1976) Spatial organization of neuronal responses to stimulation of dentate fascia in the field CA_3 of hippocampus. *Zhurnal Vysshei Nervnoi Deyatel'nosti* 26, 105–111

VINOGRADOVA, O. S. & BRAGIN, A. G. (1975) Sensory characteristics of the hippocampal cortical input. Dentate fascia. *Zhurnal Vysshei Nervnoi Deyatel'nosti* 25, 410–420 (in Russian)

WINSON, J. & ABZUG, C. (1977) Gating of neural transmission in the hippocampus: efficacy of transmission varies with behavioral state. *Science (Wash. D.C.)* 196, 1223–1225

The anatomical organization of septo-hippocampal projections

L. W. SWANSON

Department of Anatomy and Neurobiology, Washington University School of Medicine, St Louis

Abstract Since the time of Elliot Smith (1910) it has been recognized that the septal complex occupies a pivotal position within the mammalian telencephalon, being strategically placed between the hippocampal formation on the one hand and the basal forebrain and diencephalon on the other. However, it is only in the last few years that the detailed interrelationships between the different nuclear groups within the septum and the various subfields of the hippocampus have been studied. We have recently re-examined the connections of both the septum and the hippocampal formation using the techniques based on the anterograde transport of isotopically labelled proteins and the retrograde transport of the enzyme marker, horseradish peroxidase. Our findings may be summarized as follows. Field CA1 of Ammon's horn and the adjoining subiculum project through the fimbria and pre-commissural fornix upon the lateral septal nucleus of the same side in a topographically ordered manner. Field CA3, on the other hand, projects *bilaterally* upon the lateral septum. The lateral septal nucleus in turn, projects partly upon the medial septal nucleus and nucleus of the diagonal band, and partly to the lateral hypothalamus and the mamillary complex. The medial septal–diagonal band complex projects back, through the fimbria and dorsal fornix, to fields CA3 and CA4 of the hippocampus, to the dentate gyrus, to the subicular complex, and to the entorhinal area. The subicular complex projects through the post-commissural fornix to the anterior thalamic group, the mamillary complex, and the ventromedial and arcuate nuclei of the hypothalamus. Ammon's horn and the subiculum also project to the posterior septal nuclei (triangular and septofimbrial), which in turn send their output to the habenular and interpeduncular nuclei. The significance of these projections is analysed in a review of the major known afferent and efferent connections of the septum and hippocampus, and the cell groups to which they project directly.

It has been appreciated for many years, due mainly to the influence of Elliot Smith (1910), that the septum occupies a critical position in the forebrain. Since it develops within the dorsal part of the lamina terminalis, just rostral to the interventricular foramen, it is uniquely placed between certain parts of the telencephalon—in particular the hippocampus and the amygdala—and at least two parts of the diencephalon—the hypothalamus and the habenular complex. Thus, despite its relatively unimpressive appearance in the primate brain, the septal region has been subjected to intensive study with virtually every available neuroanatomical tool, and most of its major connectional relationships are now well understood.

The basic relationship between the septum and the hippocampus has been known since the time of Honneger (1892), and recent studies have in essence simply defined, in progressively more detail, the afferent and efferent connections of each of the many subdivisions of these two structures (see Powell & Cowan 1955; Nauta 1958; Guillery 1956; Raisman 1966; Meibach & Siegel 1977). It is not my intention here to review in detail the literature on the connections of the septum and the hippocampus; this has been adequately done in the reports just cited, and in the monograph edited by Isaacson & Pribram (1975). Rather, I shall summarize the results of an ongoing series of experiments carried out in our own laboratories over the past years using mainly the autoradiographic and certain histochemical methods. Most of this work has focused on a single species, the albino rat, partly because so many current behavioural and physiological studies use this animal, and partly because the size of its brain makes it possible to process, in a reasonable period of time, the large amount of material necessary to resolve certain anatomical issues. It is important to emphasize this, for while a detailed knowledge of the connections of these regions may be especially useful at this time, it is clear that there are important species differences in this circuitry (e.g., Valenstein & Nauta 1959; Geneser-Jensen 1972; Wyss *et al.* 1977). Despite this caveat, it seems likely that most of the major afferent, intrinsic, and efferent connections of the septum and hippocampus are common to most mammals, and that a clear understanding of their organization in any one species should shed light on some of the still enigmatic problems about their function.

TOPOGRAPHIC CONSIDERATIONS

In acallosal mammals, such as the marsupial opossum, the close topographic relationship between the septum and hippocampus is obvious since they are more or less contiguous (Fig. 1). However, in the majority of mammals, the development of the corpus callosum has had the effect of separating the two regions, although they remain connected by the fibres of the fornix system, and there are no other nuclear masses interposed between them (Fig. 2).

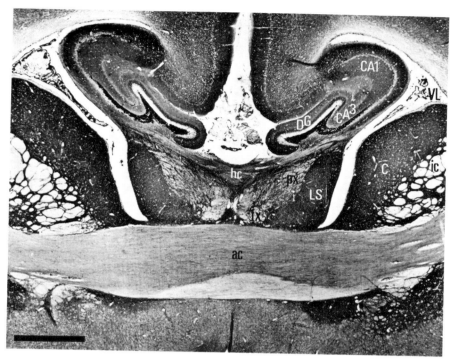

FIG. 1. The septum and hippocampus as seen in a frontal section through the anterior commissure in the acallosal opossum. Note the contiguity between the two at this level. Timm's stain; scale: 1.0 mm. *Abbreviations for all figures:* ACB, n. accumbens; AH, Ammon's horn; AON, anterior olfactory n.; ATN, anterior thalamic n.; AV, anteroventral n.; BST, bed n.; C, caudate n.; CTF, central tegmental field; DBB, diagonal band n.; DG, dentate gyrus; ENT, entorhinal area; H(HAB), habenula; HYP, hypothalamus; IPN, interpeduncular n.; LC, locus coeruleus; LS, lateral septal n.; M(MAM), mamillary body; MePO, median preoptic n.; MPO, medial preoptic area; MS, medial septal n.; OT, olfactory tubercle; PS, posterior septal n.; PT, parataenial n.; PVT, paraventricular n. of the thalamus; SB, subicular complex; SI, substantia innominata; TT, taenia tecta; VL, lateral ventricle; VMH, ventromedial n. of the hypothalamus; VTA, ventral tegmental area; 25, infralimbic area; ac, anterior commissure; cc, corpus callosum; dhc, dorsal hippocampal commissure; fi, fimbria; fr, fasciculus retroflexus; fx, fornix; hc, hippocampal commissure; ic, internal capsule; mt, mamillo-thalamic tract; px, pre-commissural fornix; sm, stria medullaris.

This relationship is especially clear during the ontogenetic development of the forebrain, since at early stages the anlagen of the septum and the hippocampus occupy adjoining regions of the telencephalic vesicles (Hines 1922). Despite their early proximity, the subsequent cytoarchitectonic organization of the two regions is quite different: the septal region is a complex of subcortical nuclei, whereas the hippocampal formation is a beautifully laminated series of cortical fields.

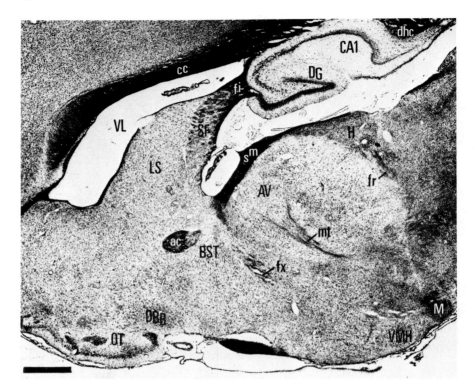

Fig. 2. A parasagittal section through the forebrain of the rat to show the close relationship between the septum and hippocampus, which are reciprocally interconnected by fibres running in the fimbria. This section also shows the major sites in direct receipt of septo-hippocampal projections, including the habenula, anterior thalamic nuclei, mamillary body, and ventromedial nucleus of the hypothalamus. Klüver-Barrera stain; scale: 1.0 mm.

Several different parcellations of the septal region (Young 1936; Fox 1940; Andy & Stephan 1964) have been suggested. On developmental, connectional, and cytoarchitectonic grounds, it may best be regarded as consisting of four major divisions (Swanson & Cowan 1976) (Fig. 3). Taken together, the lateral and medial divisions of the septum are reciprocally interconnected with the hippocampal formation; the ventral division, comprising the bed nucleus of the stria terminalis or simply the bed nucleus, is similarly related to the amygdala; and the posterior division appears to bring the habenular and interpeduncular nuclei under the influence of the hippocampus.

The lateral division contains primarily medium-sized neurons and has been subdivided into dorsal, intermediate, and ventral parts on the basis of neuronal size and packing density, and on certain connectional differences. All parts

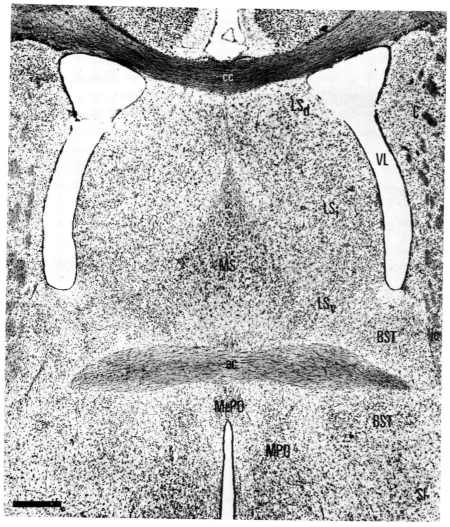

FIG. 3. A frontal section through the anterior commissure of the rat to show the major divisions of the septal region, except for the posterior group, which lies more caudally in the vicinity of the ventral hippocampal commissure. Klüver-Barrera stain; scale: 0.5 mm.

of the medial division contain a mixture of very large to small neurons; traditionally, it is divided (somewhat arbitrarily) into the medial septal nucleus dorsally and the nucleus of the diagonal band ventrally. As discussed below, it appears that both parts of the medial division share basically the same pattern of afferent, intrinsic, and efferent connections. The posterior division consists of two distinct cell groups—the septofimbrial and the triangular septal nuclei. The former lies among the fibres of the pre-commissural division of the fornix, caudal to, and continuous with, the lateral septal nucleus; its cells are, on the average, somewhat larger and more scattered than those in the lateral septal nucleus. The triangular nucleus consists of smaller, densely packed neurons within the more medial fibres of the pre-commissural fornix, and among the fibres of the ventral hippocampal commissure. The bed nucleus of the stria terminalis is a large, heterogeneous region associated, as its name implies, with the stria terminalis, and is distinguished most clearly by its massive input from the amygdala and the ventral part of the subiculum. Although there is no consensus on the parcellation of this cell group, connectional evidence suggests that there are distinct medial, lateral, posterior, sub-commissural, and preoptic components (Swanson & Hartman 1975; Swanson 1976; Swanson & Cowan 1976; Krettek & Price 1977). The bed nucleus separates the septal region from the preoptic region ventrally and the caudate nucleus laterally.

Although the terminology used to describe the fields of the hippocampal formation has undergone a long and complex evolution, the general scheme proposed by Blackstad in 1956 has gained wide acceptance. It is perhaps easiest to consider the hippocampal formation as a series of adjacent cortical fields extending from the entorhinal area to the dentate gyrus. On the basis of their connections, these fields fall into four groups corresponding to the entorhinal area, the subicular complex, Ammon's horn, and the dentate gyrus (Swanson & Cowan 1976; see their Fig. 3). The entorhinal area and dentate gyrus apparently give rise almost exclusively to intrahippocampal connections, in the rat at least, while Ammon's horn and the subicular complex project through the pre- and post-commissural parts of the fornix system. The morphology of the dentate gyrus, Ammon's horn, and entorhinal area in the rat are well known; the subicular complex, however, has until recently received little attention, and is still relatively poorly understood. It consists of the subiculum, presubiculum, parasubiculum, and postsubiculum, the latter being a differentiation of the dorsal part of the presubiculum. The 'retrosplenial e' field (see Haug 1976) also appears to form part of the dorsal presubiculum as defined by Swanson & Cowan (1977). Although each field in the subicular complex is quite distinct cytoarchitectonically, the subiculum itself resembles Ammon's horn more than the pre- and parasubiculum, which in turn more closely resemble the entorhinal area. This cytoarchitectonic 'ambivalence' of the subiculum proper is reflected in the fact that it projects through both the pre- and post-commissural divisions of the fornix.

CONNECTIONS BETWEEN THE SEPTUM AND HIPPOCAMPUS

There is now clear autoradiographic evidence about the fields of origin and the topographic organization of the well-known hippocampal input to the septal region (Swanson & Cowan 1977). Such fibres arise from all septotemporal levels of fields CA3 and CA1 of Ammon's horn and the adjacent subiculum, and course rostrally in a topographically organized manner such that progressively more temporal (ventral) neurons project through progressively more lateral parts of the fimbria. This arrangement is preserved in the septum, where fibres from septal parts of fields CA3, CA1, and the subiculum end in the dorsal part of the lateral septal nucleus, and most probably in the adjacent septofimbrial nucleus as well (Fig. 4). Similarly, fibres from intermediate (occipital) parts of the same fields end primarily in the intermediate part of the lateral septal nucleus and the posterior nuclei, and fibres from the temporal (ventral) pole project to the ventral part of the lateral septal nucleus and the posterior nuclei, as well as to the bed nucleus of the stria terminalis. Interestingly, field CA3 projects bilaterally to the septum, as first suggested by Raisman *et al.* (1966), while field CA1 and the subiculum project ipsilaterally. Furthermore, field CA1 and the subiculum project throughout the rostrocaudal extent of the lateral septal nucleus; field CA3, on the other hand, appears to innervate only the caudal two-thirds of the nucleus. Recent Fink-Heimer evidence suggesting a medial to lateral organization of hippocampal inputs to the septum (Siegel *et al.* 1974; Siegel & Edinger 1976), does not appear to be correct since fibres of passage caudal to the lesions were interrupted; other evidence (Meibach & Siegel 1977) suggesting that hippocampal input to the septum arises primarily in the

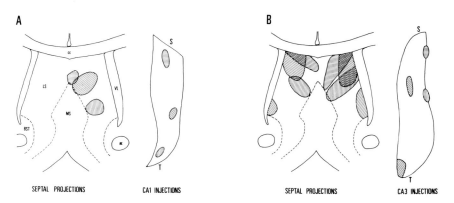

FIG. 4. A summary of the topographic organization of hippocampal inputs from fields CA1 (A) and CA3 (B) to the lateral septum; S, septal (dorsal); T, temporal (ventral).

subiculum, but is not clearly related to cytoarchitectonically defined cortical fields, is sometimes difficult to evaluate critically as presented.

The lateral septal nucleus, then, receives a massive, topographically organized input from Ammon's horn and the subiculum; it in turn projects heavily to the adjacent medial septal–diagonal band complex (Raisman 1966; Swanson & Cowan 1976). This is a particularly important connection since the latter projects back to the hippocampal formation by way of the fornix system (Daitz & Powell 1954), and most probably through the cingulum to a limited extent (Swanson & Cowan 1976). The cholinergic medial septal–diagonal band input is remarkably extensive (Mellgren & Srebro 1973; Swanson & Cowan 1976): virtually all parts of the hippocampal formation appear to be innervated, although the polymorph layer of the dentate gyrus and the parasubiculum receive the most input, and field CA1 and the subiculum receive the least. There is some evidence (Segal & Landis 1974a) for a crude topographic organization in the septo-hippocampal projection, but its significance is unclear in view of other evidence for extensive short interconnections between all parts of the medial septal–diagonal band complex (Fox 1940; Swanson & Cowan 1976).

In summary, it seems clear that Ammon's horn and the subiculum, when considered together, project topographically and bilaterally upon the septal region, which in turn projects rather diffusely back to virtually the entire hippocampal formation. It should also be pointed out that, in a similar way, the bed nucleus receives a massive input from certain parts of the amygdala (De Olmos & Ingram 1972; Krettek & Price 1978), and projects in turn back upon the amygdala (Swanson & Cowan 1976).

EFFERENT CONNECTIONS OF THE SEPTUM AND HIPPOCAMPUS

The number of sites in receipt of direct septal and hippocampal projections is relatively small. Aside from an input to the septal region, the efferent projections of the hippocampal formation appear to arise almost exclusively within the subicular complex. The dorsal part of the subiculum, and the full dorso-ventral extent of the pre- and parasubiculum considered together, project through the post-commissural fornix to the mamillary body (Swanson & Cowan 1975, 1977; Chronister et al. 1975, 1976; Meibach & Siegel 1975, 1977). In the rat, the dorsal part of the subiculum projects bilaterally to all divisions of the medial mamillary nucleus, while the pre- and parasubiculum considered together project bilaterally to the medial nucleus as well as ipsilaterally to the lateral nucleus (Swanson & Cowan 1975, 1977). There is evidence that this projection is topographically organized such that more dorsal parts of the subicular complex project to more dorsal parts of the medial mamillary nucleus (Swanson & Cowan 1975, 1977; Meibach & Siegel 1977).

It is also clear that the pre- and parasubiculum considered together project through the post-commissural part of the fornix to the anterior thalamic nuclei (Swanson & Cowan 1975, 1977; Chronister et al. 1975, 1976). The majority of such fibres end in the ipsilateral anteroventral nucleus, although some also end in the same nucleus on the contralateral side, and perhaps in the ipsilateral anterodorsal and anteromedial nuclei as well (Swanson & Cowan 1977).

The ventral part of the subicular complex (Raisman et al. 1966; Meibach & Siegel 1977), and more specifically the subiculum itself (Swanson & Cowan 1975), gives rise to the medial cortico-hypothalamic tract, which distributes fibres through, and perhaps to, the anterior hypothalamic area to the relatively cell-free zone surrounding the ventromedial and arcuate nuclei of the hypothalamus. In addition, neurons in the ventral part of the subiculum and immediately adjacent parts of field CA1 project through the pre-commissural fornix to the nucleus accumbens, taenia tecta, medial part of the anterior olfactory nucleus, and infralimbic area (Swanson & Cowan 1975, 1977). These projections of the ventral subiculum to the basal telencephalon (including the bed nucleus of the stria terminalis) and the ventromedial nucleus are similar to projections to the same regions from the amygdala (Heimer & Nauta 1969; De Olmos 1972; Krettek & Price 1978), and suggest that it may be more closely related functionally to the amygdala than to the hippocampal formation.

The efferent connections of the septal region, excluding those from the medial division to the hippocampal formation which have already been considered, are relatively well defined. The output from the posterior division is the simplest. The septofimbrial nucleus projects in a highly topographic way through the stria medullaris to the medial habenular nucleus on the same side; and the triangular nucleus sends fibres to the medial and lateral habenular nuclei on both sides of the brain, and to the ipsilateral part of the interpeduncular nucleus through the fasciculus retroflexus on the same side (Swanson & Cowan 1976). An input to the habenula from the posterior septal nuclei has been confirmed in a horseradish peroxidase study (Herkenham & Nauta 1977a). The bed nucleus of the stria terminalis also projects bilaterally to the lateral habenular nuclei and, rostral to this, to the paraventricular and parataenial nuclei of the thalamus on the same side (Swanson & Cowan 1976). The latter two nuclei also receive an input from the ventral part of the lateral septal nucleus (Swanson & Cowan 1976). Thus, cell groups in the dorsomedial part of the thalamus, including the habenular, parataenial, and paraventricular nuclei, receive inputs from the posterior, lateral, and ventral divisions of the septum.

The mamillary body receives a massive and topographically organized input from the septum (Swanson & Cowan 1976), in addition to that from the subicular complex. In particular, the posterior part of the bed nucleus projects to the premamillary nucleus, the anterior part of the bed nucleus and the ventral lateral septal nucleus project to the supramamillary area and the ventral part of the fibre capsule of the mamillary body, the intermediate part of the lateral septal nucleus projects to the region surrounding the lateral mamillary nucleus, and the medial septal–diagonal band complex projects to the medial mamillary nucleus. These septal fibres to the mamillary body travel through the medial forebrain bundle in the lateral preoptic and lateral hypothalamic areas, and it is likely (Raisman 1966), though not yet conclusively shown, that some of them may be in synaptic contact with neurons along the course of the pathway. This is particularly evident in the pathway from the ventral part of the lateral septal nucleus which courses initially through the medial preoptic and anterior hypothalamic areas (Swanson & Cowan 1976).

Of the septal nuclei, the bed nucleus has the most widespread descending connections. It projects through the stria terminalis to the central and medial nuclei of the amygdala, and extensively through the medial forebrain bundle to the nucleus accumbens, to virtually all parts of the preoptic region and the hypothalamus, with the exception of the cellular core of the ventromedial nucleus, to the ventral tegmental area, reticular formation, and central gray of the midbrain, and to the locus coeruleus (Swanson 1976; Swanson & Cowan 1976).

AFFERENT CONNECTIONS OF THE SEPTUM AND HIPPOCAMPUS

A survey of known inputs to the hippocampal formation makes it clear that, in the primate at least, the cerebral cortex is the major source of such fibres, and that they end primarily in the entorhinal area and in the presubiculum. The evidence suggests that 'polysensory' information derived from the primary visual, auditory, and somatosensory regions gains access to the entorhinal area by way of relays in association cortex of the temporal and frontal lobes (Jones & Powell 1970; Van Hoesen & Pandya 1975; Van Hoesen *et al.* 1975; Leichnetz & Astruc 1977). Furthermore, olfactory information appears to gain relatively direct access to the lateral entorhinal area from the olfactory bulb itself (White 1965; Scalia 1965; Heimer 1968; Price 1973), and from the piriform cortex (Cragg 1961; Powell *et al.* 1965; Price 1973; Krettek & Price 1977). The cingulate gyrus was long considered to project to the hippocampal formation, in particular to the entorhinal area (e.g. Gerebtzoff 1939; White 1959; Raisman *et al.* 1965), until Domesick (1969) suggested that this conclusion may have been based on the inadvertent interruption of caudally directed fibres of passage. However, the preliminary results of an autoradiographic study of this problem in our laboratory suggest that wide parts of the cingulate gyrus in the rat do indeed project to the dorsal part of the presubiculum and the postsubiculum, and perhaps to the subiculum and entorhinal area. There is as yet little evidence for direct cortical inputs to the septal region in the rat, although fibres from the proreal gyrus to the lateral septal nucleus have been reported in the cat (Voneida & Royce 1974), and the prefrontal cortex is said to project to the same nucleus in the monkey (e.g., Tanaka & Goldman 1976).

The only other telencephalic input to the hippocampus originates in the amygdala. According to Krettek & Price (1977), the lateral nucleus of the amygdala projects to the lateral entorhinal area and the posterior division of the basolateral nucleus projects to the ventral part of the subiculum and the parasubiculum. As mentioned above, the amygdala also projects massively and topographically to the septal region, in particular to the bed nucleus and the ventral part of the lateral septal nucleus (see Krettek & Price 1978). Thus, recent evidence makes it clear that the septum, hippocampal formation, and amygdala are richly interconnected, and it seems most likely that neural activity in any one region will have a direct influence on activity in the other two.

Three diencephalic sources of input to the hippocampal formation have been identified. First, the anterior thalamic nuclei have been said to project to the subicular complex and entorhinal area, on the basis of degeneration studies (Krieg 1947; Domesick 1973). Second, the nucleus reuniens has recently been shown to project widely to field CA1, the subicular complex, and the entorhinal area (Herkenham 1978). And third, it has been shown that injections of horseradish peroxidase into the hippocampus result in the retrograde labelling of a cell group centred in the supramamillary region (Segal & Landis 1974*a*; Pasquier & Reinoso-Suarez 1976), although the precise intrahippocampal distribution of these fibres has not been established. The septal region also receives a variety of diencephalic inputs, none of which are particularly massive. The lateral preoptic and lateral hypothalamic areas (along with the magnocellular preoptic nucleus) project to the medial septal–diagonal band complex (see Raisman 1966; Swanson 1976) and the ventromedial nucleus of the hypothalamus projects to the bed nucleus (Saper *et al.* 1976). Evidence based on horseradish peroxidase injections (Segal & Landis 1974*b*) suggests that the medial habenular nucleus also projects to the septal region.

Considerable evidence has accumulated over the years for the existence of brainstem inputs to the septum and hippocampus, although their precise cells of origin are only now beginning to be identified with certainty. Early histochemical fluorescence evidence provided an important insight into the problem by suggesting that the septum and hippocampus receive catecholamine- and indoleamine-containing projections from the brainstem (Fuxe 1965). The locus coeruleus has been shown by retrograde (Segal & Landis 1974a) and anterograde (Pickel et al. 1974) transport methods to project to the hippocampus, and the laminar distribution of these noradrenergic fibres was described with dopamine-β-hydroxylase immunohistochemistry (EC 1.14.17.1) (Swanson & Hartman 1975). A moderate number of noradrenergic fibres are found throughout the stratum lacunosum–moleculare of Ammon's horn; many of them appear to enter from layer I of the caudal part of the cingulate gyrus and from the cingulum. The fibres in Ammon's horn enter the hilar area of the dentate gyrus, which appears to constitute the densest noradrenergic terminal field (Koda & Bloom 1977) in the hippocampus. Some hilar fibres appear to continue into the stratum lucidum, accompanying the mossy fibre system. Smaller numbers of fibres innervate the molecular layer of the dentate gyrus and the stratum radiatum and stratum oriens of Ammon's horn. Fibres containing dopamine-β-hydroxylase are also found throughout the septal region (Swanson & Hartman 1975), although many of them appear to continue rostrally into the cingulate gyrus. However, the ventral part of the bed nucleus, just below the anterior commissure, contains a remarkably dense and circumscribed noradrenergic input which does not appear to arise in the locus coeruleus (Ungerstedt 1971).

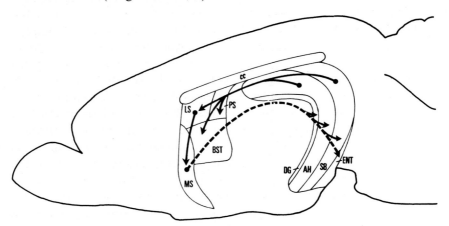

FIG. 5. The major fibre systems interrelating the septum and hippocampal formation.

There is now evidence for a dopaminergic input to the hippocampal formation, particularly to the entorhinal area (Fallon & Moore 1976) and to the lateral septal nucleus (Lindvall 1974). Serotonergic inputs to the hippocampus (e.g., Segal 1975; Moore & Halaris 1976) and septal region (Fuxe 1965) are also known but are still poorly understood; they arise in the dorsal and median raphe nuclei. The dopaminergic innervation presumably comes from the region of the ventral tegmental area (Lindvall 1974). It is also likely that there are other, non-aminergic, brainstem afferents to the septum and hippocampus that remain to be elucidated. For example, Segal & Landis (1974b) report the retrograde labelling of neurons in the dorsal tegmental nucleus after injections of horseradish peroxidase into the septal region.

DISCUSSION

The septum and hippocampus are reciprocally interconnected to such an extent that the function of one clearly cannot be considered without reference to the other. The medial septal complex plays a particularly conspicuous role in this circuitry since it projects to virtually the entire hippocampal formation, and in turn receives a massive input from the hippocampus by way of the lateral septum (Fig. 5). One approach to the problem of clarifying the neural mechanisms underlying the associative, cognitive and visceral functions of the septo-hippocampal complex taken as a whole involves an examination of the centres to which it projects. As outlined above, these include (leaving aside for the moment those receiving inputs from the ventral subiculum and the bed nucleus of the stria terminalis) the mamillary body, the anterior thalamic nuclei, and the habenular and interpeduncular nuclei (Fig. 6). It is clear that for this approach to be fruitful, the projections from these primary terminal fields must next be considered. The mamillary body projects to the anterior thalamic nuclei (Cowan & Powell 1954), to the tegmental nuclei of Gudden (Nauta 1958), and to the tegmental reticular nucleus (Cruce 1977). While the tegmental nuclei of Gudden project back to the mamillary body (Cowan *et al.* 1964), and perhaps to the septum (Segal & Landis 1974b), the tegmental reticular nucleus projects to the cerebellum (Jansen & Brodal 1954) and may thus provide one rather indirect route for hippocampal information to reach the cerebellum.

In addition to its mamillary input, the anterior thalamic nuclei receive a direct input from the septo-hippocampal complex; this information is in turn relayed to the full extent of the cingulate gyrus and to parts of the hippocampal formation itself (see Cowan & Powell 1954). With this in mind, it is reasonable

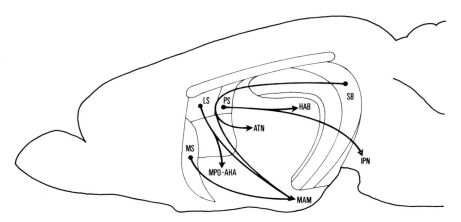

FIG. 6. The major sites in receipt of direct septo-hippocampal projections. The efferent connections of the ventral part of the subiculum and the bed nucleus of the stria terminalis are shown in the next figure.

to suggest that the septum and hippocampus play a major role in controlling the output of the cingulate gyrus. This is particularly important because it provides a route by which information leaving the septo-hippocampal complex can gain access to widespread parts of the brain, including the striatum, the pretectal region, the mediodorsal nucleus, the midbrain reticular formation, the central gray, and the deep pontine nuclei (Domesick 1969). Through these pathways, the septum and hippocampus appear able to influence indirectly neural activity: *(a)* in the striato-thalamo-motor cortical system, *(b)* in the cerebellum by way of the pontine gray, *(c)* in the prefrontal cortex and amygdala through the mediodorsal nucleus, and *(d)* in the complex and still vaguely understood—though undoubtedly important—central gray and reticular formation of the midbrain.

The habenular nuclei receive a massive input from the posterior septal nuclei, which in turn receive their major input from Ammon's horn and/or the subiculum. Recent evidence (Herkenham & Nauta 1977b) suggests that, like the cingulate gyrus, the habenular nuclei project rather widely to regions including the interpeduncular nucleus, the dorsal and median nuclei of the raphe, the superior colliculus, the pars compacta of the substantia nigra, the central gray, lateral hypothalamic and lateral preoptic areas, and the ventromedial nucleus of the thalamus. The significance of these projections is unclear because of the paucity of evidence for behavioural or physiological deficits after habenular ablation; it seems likely, nevertheless, that the habenula can also influence the striato-thalamo-motor cortical system (by way of the substantia nigra) and the central gray, as well as the hypothalamus. It is also clear that the extensive serotonergic projections of the dorsal and median nuclei of the raphe (Fuxe 1965; Conrad *et al.* 1974) may be influenced indirectly by inputs from the septum and hippocampus relayed through the habenula. And finally, projections from the cingulate cortex and the habenula to the ventromedial nucleus of the thalamus may influence the relay of taste and visceral sensory information to the cerebral cortex (see Norgren & Leonard 1973).

As alluded to earlier, it is useful to consider the ventral subiculum and bed nucleus of the stria terminalis together, since their connections are rather unlike those of the rest of the septo-hippocampal complex, and more like those of the amygdala (Fig. 7). Like adjacent parts of the amygdala, the ventral subiculum projects to the ventromedial nucleus of the hypothalamus, and to the bed nucleus, nucleus accumbens, anterior olfactory nucleus, and infralimbic area. In addition, both the ventral subiculum and the bed nucleus receive massive inputs from the amygdala (Krettek & Price 1977, 1978). Considering the magnitude of this input, it would appear that the bed nucleus can reasonably be regarded as an amygdalar relay nucleus, along with the ventromedial nucleus. This assumes greater significance when the efferent connections of both nuclei are considered. In addition to reciprocal pathways back to the amygdala, both have extensive inputs to nuclear groups along the length of the medial forebrain bundle, some fibres reaching as far as the locus coeruleus (Swanson & Cowan 1976; Saper *et al.* 1976).

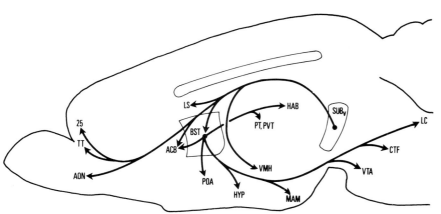

FIG. 7. The efferent connections of the ventral part of the subiculum and the bed nucleus of the stria terminalis, two parts of the septo-hippocampal complex which are closely related to the amygdala on both topological and connectional grounds.

In summary then, the anatomical evidence suggests that the septum and hippocampus, acting primarily through the cingulate cortex, the habenular nuclei, the ventromedial nucleus of the hypothalamus, and the bed nucleus of the stria terminalis, have ample opportunity to influence activity in most of the major neural systems of the brain. The evidence also suggests that the septum and hippocampus receive a wide variety of inputs, either directly or indirectly, from most parts of the neocortex and the amygdala, from many parts of the hypothalamus, and from adrenergic, dopaminergic, and serotonergic cell groups in the brainstem. Rather broadly, then, and leaving aside a variety of feedback loops, the septo-hippocampal complex may be thought of: *(a)* as processing a wide variety of sensory (including visceral) information, *(b)* as being modulated by aminergic inputs from the brainstem, and *(c)* as projecting, more or less directly, to the hypothalamus, the midbrain reticular formation and the central gray, and to the striatum, the substantia nigra, and the pontine gray. Unfortunately, the precise functional significance of many of these pathways is unclear because their synaptic relationships are unknown. The next level of analysis of these complex neural circuits must involve determining the subpopulations of neurons innervated by, and giving rise to, particular pathways, the electrophysiological nature of the synaptic contacts, and the neurotransmitters used.

Whatever neural mechanisms the septo-hippocampal complex is eventually found to be involved in, it is becoming increasingly clear that a great deal of information-processing takes place within the hippocampus itself. Recent evidence (Swanson & Cowan 1977) strongly suggests that the axons of a large fraction of hippocampal formation neurons, including pyramidal cells, form intrahippocampal association and commissural pathways, and do not enter the pre- or post-commissural parts of the fornix system. These intrinsic pathways, most of which appear to be excitatory, seem to interrelate all parts of the hippocampal formation on both sides of the brain. Furthermore, the evidence suggests that each intrahippocampal pathway is topographically organized with respect to the longitudinal axis of its field of origin and the distribution of its terminal field. The greatest challenge in the neurobiology of the hippocampus will be to relate the anatomical organization of its connections to the associative, cognitive and other functions it is thought to subserve.

ACKNOWLEDGEMENTS

This work was supported in part by Grants NS-13267 and NS-10943 from the National Institutes of Health, United States Public Health Service.

References

ANDY, O. J. & STEPHAN, H. (1964) *The Septum of the Cat*, Thomas, Springfield, Ill.

BLACKSTAD, T. W. (1956) Commissural connections of the hippocampal region in the rat with special reference to their mode of termination. *J. Comp. Neurol. 105*, 417–538

CHRONISTER, R. B., SIKES, R. W. & WHITE, L. E. (1975) Postcommissural fornix: origin and distribution in the rodent. *Neurosci. Lett. 1*, 199–202

CHRONISTER, R. B., SIKES, R. W. & WHITE, L. E. (1976) The septo-hippocampal system: significance of the subiculum, in *The Septal Nuclei* (DeFrance, J. F., ed.), pp. 115–132, Plenum Press, New York

CONRAD, L. C. A., LEONARD, C. M. & PFAFF, D. W. (1974) Connections of the median and dorsal raphe nuclei in the rat: an autoradiographic and degeneration study. *J. Comp. Neurol. 156*, 179–206

COWAN, W. M. & POWELL, T. P. S. (1954) An experimental study of the relation between the medial mammillary nucleus and the cingulate cortex. *Proc. R. Soc. Lond. B Biol. Sci. 143*, 114–125

COWAN, W. M., GUILLERY, R. W. & POWELL, T. P. S. (1964) The origin of the mamillary peduncle and other hypothalamic connexions from the midbrain. *J. Anat. 98*, 345–363

CRAGG, B. G. (1961) Olfactory and other afferent connections of the hippocampus in the rabbit, rat, and cat. *Exp. Neurol. 3*, 588–600

CRUCE, J. A. F. (1977) An autoradiographic study of the descending connections of the mammillary nuclei of the rat. *J. Comp. Neurol. 176*, 631–644

DAITZ, H. M. & POWELL, T. P. S. (1954) Studies of the connections of the fornix system. *J. Neurol. Neurosurg. Psychiat. 17*, 75–82

DE OLMOS, J. S. (1972) The amygdaloid projection field in the rat as studied with the cupric-silver method, in *The Neurobiology of the Amygdala* (Eleftherion, B. E., ed.), pp. 145–204, Plenum Press, New York

DE OLMOS, J. S. & INGRAM, W. R. (1972) The projection field of the stria terminalis in the rat brain. An experimental study. *J. Comp. Neurol. 146*, 303–334

DOMESICK, V. B. (1969) Projections from the cingulate cortex in the rat. *Brain Res. 12*, 296–320

DOMESICK, V. B. (1973) Thalamic projections in the cingulum bundle to the parahippocampal cortex of the rat. *Anat. Rec. 175*, 308

ELLIOT SMITH, G. E. (1910) Some problems relating to the evolution of the brain. Arris and Gale lectures. *Lancet 106*, 147–221

FALLON, J. H. & MOORE, R. Y. (1976) Dopamine innervation of some basal forebrain areas in the rat. *Neurosci. Abstr. 2*, 486

FOX, C. A. (1940) Certain basal telencephalic centers in the cat. *J. Comp. Neurol. 72*, 1–62

FUXE, K. (1965) Evidence for the existence of monoamine-containing neurons in the central nervous system. IV. The distribution of monoamine terminals in the central nervous system. *Acta Physiol. Scand.*, Suppl. 247

GENESER-JENSEN, F. A. (1972) Distribution of acetyl cholinesterase in the hippocampal region of the guinea pig. III. The dentale area. *Z. Zellforsch. Mikrosk. Anat. 131*, 481–495

GEREBTZOFF, M. A. (1939) Sur quelques voies d'association de l'écorce cérébrale (recherches anatomo-expérimentales). *J. Belg. Neurol. 39*, 205–221

GUILLERY, R. W. (1956) Degeneration in the post-commissural fornix and in the mamillary peduncle of the rat. *J. Anat. 91*, 350–370

HAUG, F. M. S. (1976) Sulfide silver pattern and cytoarchitectonics of parahippocampal areas in the rat. *Adv. Anat. Embryol. Cell Biol. 52*, 1–73

HEIMER, L. (1968) Synaptic distribution of centripetal and centrifugal nerve fibers in the olfactory system of the rat. An experimental anatomical study. *J. Anat. 103*, 413–432

HEIMER, L. & NAUTA, W. J. H. (1969) The hypothalamic distribution of the stria terminalis in the rat. *Brain Res. 13*, 284–297

HERKENHAM, M. (1978) The connections of the nucleus reuniens thalami: evidence for a direct thalamo-hippocampal pathway in the rat. *J. Comp. Neurol. 177*, 589–609

HERKENHAM, M. & NAUTA, W. J. H. (1977a) Afferent connections of the habenular nuclei in the rat. A horseradish peroxidase study, with a note on the fiber-of-passage problem. *J. Comp. Neurol. 173*, 123–146

HERKENHAM, M. & NAUTA, W. J. H. (1977b) Projections of the habenular nuclei in the rat. *Anat. Rec. 187*, 603

HINES, M. (1922) Studies on the growth and differentiation of the telencephalon in man. The fissure hippocampi. *J. Comp. Neurol. 34*, 73–171

HONNEGER, J. (1892) Vergleichend-anatomische Untersuchungen über den Fornix. *Rec. Zoo. Suisse 5*, 201–434

ISAACSON, R. L. & PRIBRAM, K. H. (eds.) (1975) *The Hippocampus*, vol. 1: *Structure and Development*, Plenum Press, New York

JANSEN, J. & BRODAL, A. (1954) *Aspects of Cerebellar Anatomy*, Gundersen, Oslo

JONES, E. G. & POWELL, T. P. S. (1970) An anatomical study of converging sensory pathways within the cerebral cortex of the monkey. *Brain 93*, 793–826

KODA, L. Y. & BLOOM, F. E. (1977) A light and electron microscopic study of noradrenergic terminals in the rat dentate gyrus. *Brain Res. 120*, 327–335

KRETTEK, J. E. & PRICE, J. L. (1977) Projections from the amygdaloid complex and adjacent olfactory structures to the entorhinal cortex and to the subiculum in the rat and cat. *J. Comp. Neurol. 172*, 723–752

KRETTEK, J. E. & PRICE, J. T. (1978) Amygdaloid projections to subcortical structures within the basal forebrain and brainstem in the rat and cat. *J. Comp. Neurol. 178*, 225–253

KRIEG, W. J. S. (1947) Connections of the cerebral cortex. I. The albino rat. c. Extrinsic connections. *J. Comp. Neurol. 86*, 267–394

LEICHNETZ, G. R. & ASTRUC, J. (1977) Further observations on the efferent connections of the medial granular frontal cortex in macaque monkeys. *Anat. Rec. 187*, 636

LINDVALL, O. (1974) Mesencephalic dopaminergic afferents to the lateral septal nucleus of the rat. *Brain Res. 87*, 89–95

MEIBACH, R. C. & SIEGEL, A. (1975) The origin of fornix fibers which project to the mamillary bodies in the rat: a horseradish peroxidase study. *Brain Res. 88*, 508–512

MEIBACH, R. C. & SIEGEL, A. (1977) Efferent connections of the hippocampal formation in the rat. *Brain Res. 124*, 197–224

MELLGREN, S. I. & SREBRO, B. (1973) Changes in acetylcholinesterase and distribution of degenerating fibers in the hippocampal region after septal lesions in the rat. *Brain Res. 52*, 19–36

MOORE, R. E. & HALARIS, A. (1976) Hippocampal innervation by serotonin neurons of the midbrain raphe in the rat. *J. Comp. Neurol. 164*, 171–184

NAUTA, W. J. H. (1958) Hippocampal projections and related neural pathways to the midbrain in the cat. *Brain 81*, 319–340

NORGREN, R. & LEONARD, C. M. (1973) Ascending central gustatory connections. *J. Comp. Neurol. 150*, 217–238

PASQUIER, D. A. & REINOSO-SUAREZ, F. (1976) Direct projections from hypothalamus to hippocampus in the rat demonstrated by retrograde transport of horseradish peroxidase. *Brain Res. 108*, 165–169

PICKEL, V. M., SEGAL, M. & BLOOM, F. E. (1974) A radioautographic study of the efferent pathways of the nucleus locus coeruleus. *J. Comp. Neurol. 155*, 15–42

POWELL, T. P. S. & COWAN, W. M. (1955) An experimental study of the efferent connections of the hippocampus. *Brain 78*, 115–135

POWELL, T. P. S., COWAN, W. M. & RAISMAN, G. (1965) The central olfactory connections. *J. Anat. 99*, 791–813

PRICE, J. L. (1973) An autoradiographic study of complementary laminar pattern of termination of afferent fibers to the olfactory cortex. *J. Comp. Neurol. 150*, 87–108

RAISMAN, G. (1966) The connections of the septum. *Brain 89*, 317–348

RAISMAN, G., COWAN, W. M. & POWELL, T. P. S. (1965) The extrinsic afferent, commissural and association fibres of the hippocampus. *Brain 88*, 963–996

RAISMAN, G., COWAN, W. M. & POWELL, T. P. S. (1966) An experimental analysis of the efferent projection of the hippocampus. *Brain 89*, 83–108

SAPER, C. B., SWANSON, L. W. & COWAN, W. M. (1976) The efferent connections of the ventromedial nucleus of the hypothalamus of the rat. *J. Comp. Neurol. 169*, 409–442

SCALIA, F. (1965) Some olfactory pathways in the rabbit brain. *J. Comp. Neurol. 126*, 285–310

SEGAL, M. (1975) Physiological and pharmacological evidence for a serotonergic projection to the hippocampus. *Brain Res. 94*, 115–131

SEGAL, M. & LANDIS, S. (1974a) Afferents in the hippocampus of the rat studied with the method of retrograde transport of horseradish peroxidase. *Brain Res. 78*, 1–15

SEGAL, M. & LANDIS, S. (1974b) Afferents to the septal area of the rat studied with the method of retrograde axonal transport of horseradish peroxidase. *Brain Res. 82*, 263–268

SIEGEL, A. & EDINGER, H. (1976) Organization of the hippocampal-septal axis, in *The Septal Nuclei* (DeFrance, J., ed.), pp. 79–113, Plenum Press, New York

SIEGEL, A., EDINGER, H. & OHGAMI, S. (1974) The topological organization of the hippocampal projection to the septal area. A comparative neuroanatomical analysis in the gerbil, rabbit and cat. *J. Comp. Neurol. 157*, 359–378

SWANSON, L. W. (1976) An autoradiographic study of the efferent connections of the preoptic region in the rat. *J. Comp. Neurol. 167*, 227–256

SWANSON, L. W. & COWAN, W. M. (1975) Hippocampo-hypothalamic connections, origin in subicular cortex not Ammon's horn. *Science (Wash. D.C.) 189*, 303–304

SWANSON, L. W. & COWAN, W. M. (1976) Autoradiographic studies of the development and connections of the septal area in the rat, in *The Septal Nuclei* (DeFrance, J., ed.), pp. 37–64, Plenum Press, New York

SWANSON, L. W. & COWAN, W. M. (1977) An autoradiographic study of the organization of the efferent connections of the hippocampal formation in the rat. *J. Comp. Neurol. 172*, 49–84

SWANSON, L. W. & HARTMAN, B. K. (1975) The central adrenergic system. An immunofluorescence study of the location of cell bodies and their efferent connections in the rat utilizing dopamine-B-hydroxylase as a marker. *J. Comp. Neurol. 163*, 487–506

TANAKA, D. & GOLDMAN, P. (1976) Silver degeneration and autoradiographic evidence for a projection from the principal sulcus to the septum in the rhesus monkey. *Brain Res. 103*, 535–540

UNGERSTEDT, U. (1971) Stereotaxic mapping of the monoamine pathways in the rat brain. *Acta Physiol. Scand.*, Suppl. 367

VALENSTEIN, E. S. & NAUTA, W. J. H. (1959) A comparison of the distribution of the fornix system in the rat, guinea pig, cat and monkey. *J. Comp. Neurol. 113*, 337–363

VAN HOESEN, G. & PANDYA, D. (1975) Some connections of the entorhinal (area 28) and perirhinal (area 35) cortices of the rhesus monkey. I. Temporal lobe afferents. *Brain Res. 95*, 1–24

VAN HOESEN, G. W., PANDYA, D. N. & BUTTERS, N. (1975) Some connections of the entorhinal (area 28) and perirhinal (area 35) cortices of the rhesus monkey. II. Frontal lobe afferents. *Brain Res. 95*, 25–38

VONEIDA, T. J. & ROYCE, G. J. (1974) Ipsilateral connections of the gyrus proreus in the cat. *Brain Res. 76*, 393–400

WHITE, L. E. (1959) Ipsilateral afferents to the hippocampal formation in the albino rat. I. Cingulum projections. *J. Comp. Neurol. 113*, 1–32

WHITE, L. E. (1965) Olfactory bulb projections of the rat. *Anat. Rec. 152*, 465–480

WYSS, J. M., SWANSON, L. W. & COWAN, W. M. (1977) Species differences in the projection of Ammon's horn to the ipsilateral and contralateral dentate gyrus. *Anat. Rec. 187*, 753

YOUNG, M. W. (1936) The nuclear pattern and fiber connections of the non-cortical centers of the telencephalon of the rabbit *(Lepus cuniculus)*. *J. Comp. Neurol. 65*, 295–401

Discussion

Weiskrantz: In the older literature (in Nauta's work) there was a projection from hippocampus via fornix to the medial dorsal nucleus of thalamus, and that has some importance in relation to the understanding of amnesia and Korsakoff's syndrome in man.

Swanson: Such a connection has not been confirmed autoradiographically in the rat. There are, however, several possible indirect routes that inputs from the hippocampal formation could take to the mediodorsal nucleus, especially from the amygdala, prefrontal and olfactory cortex.

Gray: Does the ventral subiculum (which you connect with the amygdala inputs and outputs) receive projections from the anterior septal nuclei that you discussed in connection with the rest of the subiculum, or is it unconnected with the medial and lateral septal nuclei?

Swanson: The ventral part of the subiculum receives an input from the medial septal complex, but it is not particularly heavy. The densest input to the hippocampal formation from the medial septal complex is to the hilar part of the dentate gyrus and to the parasubiculum.

Gray: So the ventral subiculum gets as much from the septal area as any other part of the subiculum. Secondly, could you say more about the particular hypothalamic nuclei to which the septo-hippocampal complex projects?

Swanson: Leaving aside the mamillary body, the major direct input from the hippocampal formation is to the relatively cell-free zone surrounding the ventromedial nucleus, from the ventral part of the subiculum. From the work of Ramon y Cajal (1911), Millhouse (1973) and Szentágothai *et al.* (1968) we know that the dendrites of many neurons in the ventromedial and arcuate nuclei project into this cell-free zone. So, at the light microscopic level at least, it would seem that the ventral part of the subiculum projects mainly to these two nuclei in the hypothalamus.

The medial and lateral septal nuclei, and the bed nucleus of the stria terminalis, all appear to project through the medial forebrain bundle to the mamillary bodies. It is still not certain, however, if there are synapses along the medial forebrain bundle, although the degeneration work of Raisman (1966) suggests that there are. This must be confirmed by electron microscopy. The best I can say is that the autoradiographic evidence suggests that there is a diffuse input to the lateral hypothalamic area from the medial and lateral divisions of the septum.

The bed nucleus of the stria terminalis receives a major amygdalar input and projects to virtually the entire hypothalamus; every nucleus in the hypothalamus, with the exception of the cellular core of the ventromedial nucleus, is traversed by fibres sweeping through. Since the density of silver grains decreases gradually as the path descends, it appears that labelled fibres are dropping out along the way. The bed nucleus of the stria terminalis therefore is in a position to have a profound effect on the hypothalamus, which is interesting because it receives inputs from both the hippocampus and the amygdala. So, while it is obvious that the function of the septum cannot be considered without reference to the hippocampus, it is equally obvious that the septum and hippocampus cannot be considered without bringing in the amygdala. This is particularly clear from the work of Krettek & Price (1977), who have shown massive inputs to the entorhinal area and the subicular complex from the amygdala. In fact, an extension of this line of reasoning makes it clear that more and more neural systems have to be added to this framework when all of the secondary and tertiary connections of the so-called septo-hippocampal system are considered. On the other hand, it is difficult to break the hippocampal formation down into functional subunits, because of the tremendous complexity of interconnections between the different cortical fields which comprise it.

Gray: In the older literature, particularly in primates, it is said that there is a direct frontal cortical projection into the hippocampus (e.g. Gardner & Fox 1948; see also a recent report by Leichnetz & Astruc 1976). Is that no longer tenable?

Swanson: It still is, as Van Hoesen and colleague's (1975) work shows. We have also preliminary autoradiographic evidence that wide parts of the cingulate gyrus project directly to the hippocampal formation (J. M. Wyss, L. W. Swanson & W. M. Cowan, unpublished work 1977).

Ursin: Can you say something about the angular bundle?

Swanson: The angular bundle carried the primary output of the entorhinal area to the dentate gyrus and to Ammon's horn.

Andersen: I want to raise the question of homology and ask whether this symposium is in fact about septo-hippocampal function in the rat, or is it about this system in any animal with a hippocampus? Secondly, can your radiolabelled, injected amino açids be taken up by axons as well as cell bodies? How specific is the technique?

Swanson: Valenstein & Nauta (1959) in particular have emphasized that there are species differences in hippocampal connections, and there are certainly species differences in the histochemistry of the hippocampus. We are beginning to reexamine these connections in the primate with the autoradiographic method and there seem, indeed, to be some differences from the rat, although we do not as yet have enough data to make any general statements.

On the question of axon uptake, I can say that this has been tested in a wide variety of sites, and there is no evidence yet that the results of axonal uptake of amino acids can be detected with the sensitivity of our method.

Andersen: Your injection into CA1 seemed to give quite a lot of radioactivity in CA3. However, I think people would agree that there are no such connections from CA1 to CA3.

Swanson: We have interpreted the silver grains you refer to as labelled fibres going through and to the alveus. We have not described any direct connections from field CA3 to field CA1.

Rawlins: As I understand you, you suggest that the projection to the subiculum appears to be larger than the CA1 projection to the medial septum. Secondly, you report all the hippocampal efferent fibres as terminating in the lateral septum, whereas Siegel's group have reported that the efferents from all of the dorsal hippocampus terminate in the medial septum (Siegel *et al.* 1974). Could you suggest why this might be?

Swanson: We have shown light projections from Ammon's horn to the medial septal nucleus. The results of older degeneration studies cannot be explained simply, except perhaps on the basis of additional trans-synaptic degeneration effects.

Regarding the first question, the projection from Ammon's horn back to the subiculum is massive, and the descending output is relatively light; furthermore, this appears to apply to the rat, rabbit, guinea pig, cat and monkey (J. M. Wyss, L. W. Swanson & W. M. Cowan, unpublished work 1977). Fibre counts done several years ago (Powell *et al.* 1957) allow one to estimate that the pre-commissural fornix in the rat contains only 50 000 to 100 000 axons. Since there are about 450 000 pyramidal cells in Ammon's horn, in addition to an unknown number in the subiculum, it would seem that on the order of 20% or less of the pyramidal cells in Ammon's horn and the subiculum leave the hippocampal formation itself, leaving a large number of pyramidal cells that give rise only to intrahippocampal connections.

Björklund: To return to the species difference problem, what are the connections of the entorhinal cortex in other species? Van Hoesen has reported abundant connections between neocortex and the entorhinal area. Are there differences in this respect between rodents and primates?

Swanson: This is an important point. In the primate, Van Hoesen and collaborators have described with degeneration methods a complex series of inputs from association areas of the neocortex to the entorhinal region (Van Hoesen *et al.* 1975; Van Hoesen & Pandya 1975) as well as projections back into the neocortex, especially from the peri- or prorhinal parts of the region (G. W. Van Hoesen, personal communication). More recently, they have described inputs from the subiculum to prefrontal areas FL and FF, to retrosplenial cortex, and to the perirhinal area and a limited part of the adjacent inferotemporal cortex with autoradiographic and horseradish peroxidase methods (Rosene & Van Hoesen 1977). In the rodent, we have described inputs to the perirhinal and prefrontal cortex (primarily to the infralimbic area) from the subiculum, to the perirhinal area from field CA1, and to the retrosplenial and cingulate areas from field CA3 and the postsubiculum (Swanson & Cowan 1977). The cortical connections of the perirhinal area in the rat have not as yet been examined with the new tracing methods, a task which will be difficult to do because of its narrow width. It will be of critical importance to establish whether or not a major route of information leaving the hippocampal formation enters the neocortex via the perirhinal area in all mammalian species, or whether most of it is funnelled out through the fornix system in lower species such as the rat.

Vinogradova: I am worried about the post-commissural fornix. It seems that it comes entirely from the subiculum: that doesn't agree with physiological data, where the pyramidal cells of the hippocampus are usually identified by antidromic excitation from the fornix system and it is difficult to confuse antidromic and synaptic excitation of these cells.

Andersen: There is a discrepancy here, namely that with physiological methods we have not found much evidence for a connection from CA1 through the alveus into the fimbria; a very large proportion of CA1 axons go back to the subiculum—in fact so large that we cannot see any connection to the fimbria, whereas you have found one and you now estimate it to be about 10%. That is the discrepancy.

Zimmer: Dr Swanson, would you summarize the differences between the projections of the dorsal and ventral part of the hippocampal formation, if there are any? Most physiological studies have been done on the dorsal part because that is technically easier.

Swanson: The septal and temporal parts of Ammon's horn appear to have similar descending connections; however, the septal part of field CA3 projects to the cingulate gyrus and the presubiculum, while more temporal parts appear to project to the parasubiculum and entorhinal area. Furthermore, the dorsal part of the subiculum projects to the mamillary body, while its ventral part projects to the ventromedial and arcuate nuclei, and to the nucleus accumbens and certain structures associated with the olfactory system. Experimentally, the major problem is that electrical stimulation or lesions of the dorsal part of the hippocampus invariably involve fibres coming up from more ventral regions.

References

GARDNER, W. D. & FOX, C. A. (1948) Degeneration of the cingulum in the monkey. *Anat. Rec. 100*, 663–664

KRETTEK, J. E. & PRICE, J. E. (1977) Projections from the amygdaloid complex and adjacent olfactory structures to the entorhinal cortex and to the subiculum in the rat and cat. *J. Comp. Neurol. 172*, 723–752

LEICHNETZ, G. R. & ASTRUC, J. (1976) The efferent projections of the medial prefrontal cortex in the squirrel monkey *(Saimiri sciureus)*. *Brain Res. 109*, 455–472

MILLHOUSE, O. E. (1973) The organization of the ventromedial hypothalamic nucleus. *Brain Res. 55*, 71–87

POWELL, T. P. S., GUILLERY, R. W. & COWAN, W. M. (1957) A quantitative study of the fornix-mammillo-thalamic system. *J. Anat. 91*, 419–437

RAISMAN, G. (1966) The connections of the septum. *Brain 89*, 317–348

RAMON Y CAJAL, S. (1911) *Histologie du Système Nerveux de l'Homme et des Vertèbres*, vol. II, Maloine, Paris

ROSENE, D. L. & VAN HOESEN, G. W. (1977) Hippocampal efferents reach widespread areas of cerebral cortex and amygdala in the rhesus monkey. *Science (Wash. D.C.) 198*, 315–317

SIEGEL, A., EDINGER, H. & OHGAMI, S. (1974) The topographical organization of the hippocampal projection to the septal area. A comparative analysis in the gerbil, rat, rabbit and cat. *J. Comp. Neurol. 157*, 359–378

SWANSON, L. W. & COWAN, W. M. (1977) An autoradiographic study of the organization of the efferent connections of the hippocampal formation in the rat. *J. Comp. Neurol. 172*, 49–84

SZENTÁGOTHAI, J., FLERKÓ, B., MESS, B. & HALÁSZ, B. (1968) *Hypothalamic Control of the Anterior Pituitary. An Experimental-Morphological Study*, 3rd edn, Akadémiai Kiadó, Budapest

VALENSTEIN, E. S. & NAUTA, W. J. H. (1959) A comparison of the distribution of the fornix system in the rat, guinea pig, cat and monkey. *J. Comp. Neurol. 113*, 337–363

VAN HOESEN, G. W. & PANDYA, D. N. (1975) Some connections of the entorhinal (area 28) and perirhinal (area 35) cortices of the rhesus monkey. I. Temporal lobe afferents. *Brain Res. 95*, 1–24

VAN HOESEN, G. W., PANDYA, D. N. & BUTTERS, N. (1975) Some connections of the entorhinal (area 28) and perirhinal (area 35) cortices of the rhesus monkey. II. Frontal lobe afferents. *Brain Res. 95*, 25–38

Localization of putative transmitters in the hippocampal formation
With a note on the connections to septum and hypothalamus

JON STORM-MATHISEN*

Norwegian Defence Research Establishment, Division for Toxicology, Kjeller, Norway

Abstract Biochemical assays on microdissected samples, denervation studies, subcellular fractionation, and light and electron microscopic autoradiography of high affinity uptake have been performed to study the cellular localization of transmitter candidates in the rat hippocampal formation.

High affinity uptake of glutamate and aspartate is localized in the terminals of several excitatory systems, such as the entorhino-dentate fibres (perforant path), mossy fibres (from granular cells) and pyramidal cell axons. Thus, in stratum radiatum and oriens of CA1, 85% of glutamate and asparate uptake and 40% of glutamate and aspartate content are lost after lesions of ipsilateral plus commissural fibres from CA3/CA4. Hippocampal efferents also take up aspartate and glutamate, since these activities are heavily reduced in the lateral septum and mamillary bodies after transection of fimbria and the dorsal fornix.

The synthesis (by glutamic acid decarboxylase), content and high affinity uptake of γ-aminobutyrate (GABA) are not reduced after lesions of these or other projection fibre systems. A localization in intrinsic neurons is confirmed by a selective loss of glutamic acid decarboxylase after local injections of kainic acid. Peak concentrations of the enzyme occur near the pyramidal and granular cell bodies, corresponding to the site of the inhibitory basket cell terminals, and in the outer parts of the molecular layers. Some 85% of glutamic acid decarboxylase is situated in 'nerve ending particles'.

Acetylcholine synthesis (by choline acetyltransferase) disappears after lesions of septo-hippocampal fibres. Since 80% of the hippocampal choline acetyltransferase is in 'nerve ending particles', the characteristic topographical distribution of this enzyme should reflect the distribution of cholinergic septo-hippocampal afferents.

Serotonin, noradrenaline, dopamine and histamine are located/synthesized in afferent fibre systems. Some monoamine-containing afferents to the hippocampal formation pass via the septal area, others via the amygdala. The hippocampal formation also contains nerve elements reacting with antibodies against neuroactive peptides, such as enkephalin, substance P, somatostatin and gastrin/cholecystokinin.

---

* *Present address:* Anatomical Institute, University of Oslo, Karl Johansgt. 47, Oslo 1, Norway

The tracing of neurons which use a particular chemical transmitter substance depends on the localization of various 'markers' associated with the alleged role of this substance as a synaptic transmitter. In general, rate-limiting synthesizing enzymes, high affinity uptake and the endogenous transmitter have proved to be suitable as markers in biochemical and histochemical studies, whereas catabolizing enzymes are usually not reliable. Depolarization-induced, Ca^{2+}-dependent release of transmitters from *in vitro* preparations, particularly in combination with precursor studies, should be the tracing method most directly akin to synaptic transmission *in vivo*. This has turned out to be very useful in the case of acidic amino acids, where synthesizing enzymes suitable as markers are not known (Nadler *et al*. 1976, 1977; Hamberger *et al*. 1978; Cotman & Hamberger 1978). However, pitfalls exist, since transmitters may be released non-synaptically (Weinreich & Hammerschlag 1975; Minchin & Iversen 1974). Tracing of transmitter receptors by ligand binding is important for demonstrating the basis for postsynaptic transmitter action, but the binding activities are not necessarily restricted to the synapses and are not suitable as markers in the present sense. Immunohistochemical methods, by virtue of their versatility, sensitivity and specificity, are currently the basis for major advances in the visualization of transmitter markers (Hökfelt *et al*. 1978). Nonetheless, it is necessary to emphasize that the methods are strictly dependent on the purity of the antigen used to raise the antibody (Rossier 1975) and on the extent of cross-reactivity of the antibody. A way of reducing the latter problem may be to apply sequential antibody techniques (Larsson & Rehfeld 1977).

Histochemical visualization and biochemical microassays on dissected samples are suitably combined with denervation experiments. In addition to enabling loss of markers distal to the lesion to be observed as the axons and terminals degenerate, the method shows accumulation of intra-axonal constituents proximal to the lesion as proof of interrupted axonal transport. The cessation of impulse flow combined with continued synthesis is likely to be the basis for the accumulation of the putative transmitters seen in several pathways shortly after axotomy. Demonstration of the latter two phenomena strengthens the evidence obtained from denervation experiments. The most specific lesions are induced by injecting selectively acting neurotoxic substances such as 6-hydroxydopamine, dihydroxylated tryptamines or kainic acid (Schwarcz & Coyle 1977). An interesting method of identifying neurons using a particular transmitter is by inducing a piling-up of the transmitter-synthesizing enzyme in the perikarya by injecting colchicine (Ribak *et al.* 1978).

Aspects of localizing transmitter-specific pathways in general, and the application to the hippocampal formation, are more thoroughly dealt with

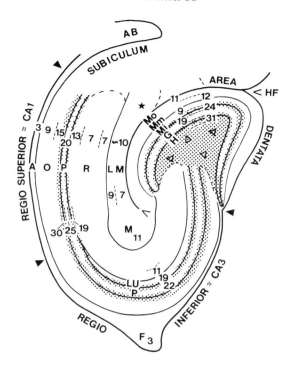

FIG. 1. Schematic drawing of a section transverse to the septo-temporal axis of the rat hippocampal formation. In area dentata and hippocampus stippling indicates the zones with the highest intensities of staining, corresponding to activity per volume, for acetylcholinesterase (Storm-Mathisen 1970) and choline acetyltransferase (P. Kása, personal communication 1975). The activities are high also in subiculum, but the distribution pattern here is more complicated (Storm-Mathisen & Blackstad 1964). The numbers indicate the choline acetyltransferase activity as measured biochemically in microdissected samples (Fonnum 1970; Storm-Mathisen 1972), the unit being 1/10 of the activity in strips of tissue cut at right angle through all layers of CA1—about 20 μmol h^{-1} g^{-1} dry weight. The figures agree very well with the staining results when the variations in dry weight per volume are taken into account (Storm-Mathisen 1970, 1977a). ★, part of molecular layer in CA1/subiculum ('site 31', Storm-Mathisen & Blackstad 1964) having relatively high acetylcholinesterase and choline acetyltransferase activity, partially resistant to lesions of afferent pathways (see text). ▼, limits between the cortical subfields: area dentata, hippocampus regio inferior ($\approx$ CA3), hippocampus regio superior ($\approx$ CA1), and subiculum; > <, bottom and orifice of the hippocampal fissure (HF). Other symbols: A, alveus; AB, angular bundle (psalterium dorsale); F, fimbria; G, granular cell layer of area dentata; H, hilus fasciae dentatae; LM, lacunosal and molecular parts of stratum lacunosum-moleculare of hippocampus CA1; LU, stratum lucidum (mossy fibre layer) of CA3; M, stratum lacunosum-moleculare of hippocampus; Mi, Mm and Mo, inner, middle and outer zones of stratum moleculare of area dentata; O, P and R, stratum oriens, pyramidale and radiatum of hippocampus.

elsewhere (Storm-Mathisen 1977a). The latter paper should be consulted for extensive references to the points made above. This paper briefly reviews previous results and treats more recent results in some detail, particularly those on amino acids. The data referred to are mostly obtained in rats.

ACETYLCHOLINE

The probably cholinergic septo-hippocampal afferents are of primary importance in this symposium, and cholinergic markers were the first to be studied in the hippocampal formation (Shute & Lewis 1961; Lewis et al. 1967). Recent histochemical results on the topography and origin of the septo-hippocampal system are presented elsewhere by G. Lynch & G. Rose (this volume). Briefly, acetylcholinesterase (AChE; EC 3.1.1.7) was found to be concentrated in the vicinity of the pyramidal and granular cell bodies and in the hilus of area dentata. The other layers had relatively minor activities, but a distinct distribution pattern (Storm-Mathisen & Blackstad 1964). Microdissection studies of choline acetyltransferase (EC 2.3.1.6) and AChE have shown that the two enzymes have essentially the same distribution (Fig. 1), suggesting that AChE, which is easily visualized histochemically, in this case is useful as a marker for nerve elements synthesizing acetylcholine. This notion is substantiated by the lesion experiments referred to below,

TABLE 1

Estimated relative importance of the different sources of various 'aminergic' nerve endings in the hippocampal formation of the rat (percentages of total)

	Noradrenaline or dopamine (%)	Serotonin (%)	Histamine (%)	Acetylcholine (%)	γ-Aminobutyrate (GABA) (%)
Dorsal routes	50	80	60	>90	<10
Ventral route	50	20	40	<10	<10
Indigenous[a]	<5	<10	<10	<10	>80

Estimates are based on measurements of noradrenaline uptake, serotonin uptake, choline acetyltransferase and glutamate decarboxylase (Storm-Mathisen & Guldberg 1974), and of histidine decarboxylase (Barbin et al. 1976) after transections of afferent routes.

[a]Glutamic acid decarboxylase, but not choline acetyltransferase or aromatic amino acid decarboxylase, is heavily reduced after intrahippocampal injections of kainic acid (Fonnum & Walaas 1978). Glutamic acid decarboxylase activity does not decrease in the target areas of several fibre projections in the hippocampal formation after appropriate lesions. These systems include ipsilateral and commissural fibres from CA3/4 pyramidal cells, mossy fibres from dentate granular cells, and perforant path fibres from area entorhinalis (see text for references).

but it should be borne in mind that the lesion-resistant activity is not quite as low for AChE (15–20%) as for choline acetyltransferase (<10%). It is worth noting that although the activities per amount of tissue in the low-activity layers are only about one-third of the activities in the high-activity pericellular zones, the latter account for a smaller part of the total amounts of choline acetyltransferase and AChE. A histochemical method for visualizing choline acetyltransferase (Kása 1975) has a fairly low sensitivity and displays the activity only in the juxtacellular layers (P. Kása, personal communication 1975). In sucrose homogenates about 80% of choline acetyltransferase is particulate and this enzyme as well as AChE is concentrated in nerve terminal particles (Fonnum 1970). The prevalence of cholinergic nerve endings is indicated by the report that in homogenates of rat hippocampus about 6% of the synaptosomes stain for AChE (Kuhar & Rommelspacher 1974).

After lesions of the septo-hippocampal pathway less than 10% of choline acetyltransferase (Table 1), and less than 20% of AChE, remain in the hippocampal formation (see Storm-Mathisen 1977*a* for references on lesion experiments). There is a pile-up of enzymes on the septal side of the transections, and only lesions in the medial parts of the septum are effective. Similar results have been obtained for endogenous acetylcholine and high affinity choline uptake. The transient increase in 'bound' (nerve terminal) acetylcholine occurring shortly after lesions may be ascribed to continued synthesis in the absence of impulse flow and release. Lateral septal lesions can lead to similar acute changes (+200% at 20–30 min) without subsequent loss of choline acetyltransferase (Potempska *et al.* 1977). In this case impulse flow could be low in medial septal neurons, because of absence of excitation from the lateral septum.

Choline acetyltransferase in the hippocampal formation is not reduced (93±4% of uninjected controls) and the AChE staining pattern is preserved when intrinsic neurons are destroyed by local injections of kainic acid (Fonnum & Walaas 1978, see below). After the septal afferents have been severed, activities of choline acetyltransferase and AChE higher than 10 and 20% of control remain only in a part of the molecular layer of subiculum/CA1 ('site 31' of Storm-Mathisen & Blackstad 1964). This activity may possibly be due to intrinsic neurons, since it has proved refractory to lesions of various afferent pathways. The existence of terminals of intrinsic cholinergic neurons at this site would be in line with the views of Phillis (1974), who advocates the existence of intrinsic cholinergic inhibitory neurons in superficial layers of the neocortex. Most of the AChE-positive perikarya found in various layers of the hippocampus are probably non-cholinergic (some of them might be glutamic acid decarboxylase-containing interneurons, see below) but could well be cholinoceptive. G. Lynch and G. Rose (this volume) find septo-hippocampal AChE-positive axons synapsing with interneurons.

The data above suggest that the pattern of distribution of choline acetyltransferase and AChE in the hippocampal formation reflects the distribution of terminals of acetylcholine-synthesizing afferents originating in the medial parts of the septum. Other data suggest that these terminals do indeed release acetylcholine on septal stimulation (Smith 1974; Dudar 1975). The actions of this pathway and of acetylcholine in the hippocampal formation are dealt with by J. F. DeFrance et al. (this volume).

5-HYDROXYTRYPTAMINE (SEROTONIN)

Fluorescence histochemistry has shown serotonin-containing fibres and terminals to be concentrated in stratum lacunosum-moleculare of hippocampus and subiculum, and to course through the septal area in the fimbria/fornix and in the cingulum bundle (Fuxe & Jonsson 1974; Björklund et al. 1973). Autoradiographic tracing of axonal protein transport from the raphe nuclei (abolished by 5,6-dihydroxytryptamine) showed similar but more detailed results, and revealed a concentration of labelled terminals in an infragranular zone in area dentata (Moore & Halaris 1975). Using the same approach Azmitia & Segal (1978) have found that raphe-derived fibres, sensitive to 5,7-dihydroxytryptamine, terminate in various parts of the hippocampal formation in an ordered manner. Thus fibres from the median raphe reach stratum oriens and radiatum of CA2–4 through fimbria, subiculum through fornix superior, and hippocampus CA1 and hilus fasciae dentatae via supracallosal bundles. At ventral levels, fibres from the dorsal raphe project via the amygdaloid area to the subiculum and the molecular layer of area dentata. In view of the differences along the dorsoventral axis of the hippocampus with respect to fibre connections (Meibach & Siegel 1977; Swanson & Cowan 1977; Swanson, this symposium) and function (Stevens & Cowey 1973; MacLean 1975) it is noteworthy that the ventral part of the hippocampal formation has a higher serotonin uptake activity than the dorsal part (Storm-Mathisen & Guldberg 1974). The same is true to a lesser extent for noradrenaline uptake, and contents of histamine, histidine decarboxylase and some peptides (see below). It may be relevant in this connection that the ventral part also has the highest oestrogen binding (Pfaff & Keiner 1974).

Raphe lesions abolish more than 80% of endogenous serotonin, serotonin uptake and tryptophan-5-hydroxylase in the hippocampal formation (Kuhar et al. 1972). The median raphe nucleus appears to be the chief source of serotonin in the hippocampal formation (Lorens & Guldberg 1974; Azmitia & Segal 1978), although Palkovits *et al.* (1977) have reported a 73% loss of tryptophan hydroxylase in hippocampus after dorsal raphe lesions. After parasagittal transections interrupting fibres entering through the dorsal 'pathways'—that is, the fimbria, fornix superior, cingulum bundle and supracallosal stria—serotonin uptake in the middle and ventral parts of the hippocampal formation was reduced by 75-85%; there was a 85-90% reduction when the connections with the amygdaloid region, as well as the dorsal pathways, were interrupted by a frontal hemitransection of the brain (Table 1). Transection of the cingulum and other supracallosal structures slightly behind the genu corporis callosi reduced serotonin uptake by 50% in the dorsal third of the hippocampal formation, and by about 20% in the ventral two-thirds (Storm-Mathisen & Guldberg 1974). This agrees with the data on raphe-hippocampal fibres (Moore & Halaris 1975; Halaris *et al.* 1976; Azmitia & Segal 1978), and suggests that the fimbria/fornix is the quantitatively more important of the serotonin-containing afferent fibre tracts. There is no evidence for serotonin-containing perikarya in the hippocampal formation.

There is electrophysiological evidence that the raphe–hippocampal pathway produces inhibitory responses in hippocampal neurons by releasing serotonin (Segal 1975).

CATECHOLAMINES

Catecholamine-fluorescent fibres and terminals have been described as relatively diffusely distributed in the hippocampal formation, with the highest densities of fluorescent structures in hilus fasciae dentatae, and in stratum radiatum of CA3 and stratum moleculare of CA1 and subiculum (Blackstad *et al.* 1967). The catecholamine-containing terminals probably constitute a rather small proportion of all terminals (less than 1%), in agreement with data for monoamines in cortex (see Storm-Mathisen 1977*a* for references). Uptake of noradrenaline, like serotonin uptake, is higher in the ventral than in the dorsal parts of the hippocampal formation (Storm-Mathisen & Guldberg 1974), suggesting that catecholamine nerve endings and axons are more abundant ventrally. Catecholamine-fluorescent fibres seem to enter the hippocampal formation through fimbria and fornix superior, and through supracallosal bundles, as well as through the amygdaloid area (Ungerstedt 1971; Lindvall & Björklund 1974).

In the histochemical studies, most of the fluorescence was interpreted as being due to noradrenaline. However, Swanson & Hartman (1975) found a different distribution using immunohistochemical visualization of dopamine-β-hydroxylase (EC 1.14.17.1), the marker enzyme for noradrenaline-synthesizing structures. Also, dopamine-β-hydroxylase-like reactivity was strongly concentrated in the hilus, but in CA3 there was a conspicuous band in the mossy fibre layer, rather than in stratum radiatum. There was a dopamine-β-hydroxylase-containing axon plexus in stratum lacunosum-moleculare in all parts of hippocampus and subiculum. This was continuous with a plexus in the superficial layers of the retrosplenial cortex and appeared to be the only pathway of entry of fibres. After traversing the septal area, dopamine-β-hydroxylase fibres coursed caudally in a dense bundle in the supracallosal structures medial to the cingulum, some fibres piercing the rostral part of the corpus callosum. No fibres were observed in the fimbria. The differences between catecholamine fluorescence and dopamine-β-hydroxylase immunohistochemistry may be more apparent than real, since both methods are rather difficult and have a limited sensitivity. A problem here is that the antibodies were prepared against dopamine-β-hydroxylase purified from bovine adrenal glands rather than rat brain (Swanson & Hartman 1975). There might also be differences in the relative concentrations of noradrenaline and dopamine-β-hydroxylase between different subpopulations of axons and terminals. However, part of the catecholamine fluorescence could be due to dopamine rather than noradrenaline (Swanson & Hartman 1975), although direct evidence has not been provided for the presence of dopamine fibres in the hippocampus.

Lesions of the locus coeruleus or the ascending noradrenaline-containing pathways induce large reductions in noradrenaline content (64–87%, Thierry *et al.* 1973) and dopamine-β-hydroxylase activity (80%, Ross & Reis 1974) in the hippocampal formation. After transections of the 'dorsal pathways' there was about 50% reduction in noradrenaline uptake, and hemitransections, severing the hippocampal formation from the amygdala in addition, reduced the noradrenaline uptake by more than 90% (Table 1). This suggests that a substantial proportion of the structures accumulating noradrenaline invade the hippocampus through the 'ventral' route. These could be dopamine fibres rather than noradrenaline fibres, since noradrenaline is also transported by the catecholamine transport mechanism of dopamine neurons (Snyder *et al.* 1970). The fact that lesions of the dorsal pathways reduced the content of noradrenaline more than its uptake (by 65–75% compared to 50%), would argue in the same way, although this could possibly be explained by 'compensatory' hyperactivity (Storm-Mathisen & Guldberg 1974). The lack of change in hippocampal noradrenaline uptake after transections of the supracallosal bundles just behind the genu corporis callosi could also be explained by assuming that a substantial proportion of the hippocampus catecholamine fibres are actually dopamine fibres entering through other pathways. However, the latter result is somewhat uncertain because of the existence of fibres piercing the corpus callosum from below at more caudal levels (Swanson & Hartman 1975), and because the survival time (10 days) may have been slightly too short to allow complete degeneration of noradrenaline fibres (Reis et al. 1974). In view of the existing controversies it seems necessary to investigate the catecholamine inputs to the hippocampal formation more closely.

Segal and Bloom have done elegant studies providing evidence that the coeruleo-hippocampal pathway is functionally important in the hippocampal formation and probably acts by releasing noradrenaline on to cyclic AMP-connected β-adrenergic receptors (Segal & Bloom 1974a, b, 1976a, b). Dopamine is also active when applied by iontophoresis to hippocampal neurons (Biscoe & Straughan 1966).

In conclusion, catecholamine fibres to the hippocampus seem to enter via dorsal as well as ventral routes, and may possibly include dopamine fibres in addition to noradrenaline fibres.

HISTAMINE

There is so far no suitable method for visualizing structures containing histamine in the brain. However, largely due to the work of J. C. Schwartz and his collaborators at the Unité de Neurobiologie INSERM in Paris, histamine has been established as a strong transmitter candidate (Schwartz et al. 1976, 1978). Histidine decarboxylase (EC 4.1.1.22), the enzyme catalysing the formation of histamine, now seems very likely to be located in neurons in the mesencephalic reticular formation and hypothalamus, having ascending and descending axons (Barbin et al. 1977). The axons destined for cortical regions derive from the mesencephalic reticular formation and course to the vicinity of the mamillary body before ascending with the other monoamine-containing tracts through the lateral hypothalamic area. Lesions in this area caused the loss of 60% of histidine decarboxylase in the hippocampus (Garbarg et al. 1974). In collaboration with J. C. Schwartz's group, we observed an essentially complete disappearance of this enzyme in hippocampus after a parasagittal cut, severing the dorsal afferents, plus an oblique cut from a lateral approach, separating the hippocampus from the amygdala (Barbin et al. 1976). The reduction in histidine decarboxylase was 80% already at three days and at 6–116 days it was more than 95%. Selective transections of the dorsal and ventral pathways produced about 60% and

30% reductions in the enzyme, respectively, suggesting that both pathways are quantitatively important. Two days after the combined lesions, when histidine decarboxylase was still nearly normal, the histamine content was increased by 140%. A similar but smaller increase in histamine has been observed after lesions of afferents to neocortex (Garbarg et al. 1976). Such an increase shortly after the cessation of impulse flow may be taken as partial evidence that histamine is normally released from nerve endings synthesizing it (see Introduction, p. 50).

Histidine decarboxylase is less concentrated in nerve ending particles than enzymes such as choline acetyltransferase and glutamic acid decarboxylase, but since about 50% was in the crude nerve terminal fraction in hippocampus, a substantial proportion is likely to be situated in nerve endings. In contrast to nerve elements containing serotonin, catecholamines, or peptides, those containing histidine decarboxylase appear to be less concentrated in area dentata than in the other subfields but, like the former, they may be somewhat more concentrated in the ventral than in the dorsal parts of the hippocampal formation (Barbin et al. 1976).

Histamine depresses neurons in the hippocampus, and this depression, as well as some of the inhibitory effect of fornix stimulation, is antagonized by the H_2-receptor antagonist metiamide (Haas & Wolf 1977). In slices of guinea-pig hippocampus cyclic AMP formation is greatly stimulated by H_1- as well as H_2-receptor agonists, and the response is strongly potentiated by noradrenaline (Schwartz et al. 1978).

PEPTIDES

Several peptides have been shown to have behavioural and electrophysiological effects in the brain (De Wied 1977; Phillis & Limacher 1974; Otsuka et al. 1976), and immunohistochemical techniques have revealed the existence of a variety of peptide-containing neurons in the central nervous system (Hökfelt et al. 1978). In the hippocampal formation, particularly in the hilus fasciae dentatae, there are nerve elements reacting with antibodies against several different peptides (T. Hökfelt, personal communication 1977). Thus, perikarya situated in the hilus or between the granular cells, as well as axon plexuses, contain a gastrin-like (or cholecystokinin-like) peptide. Gastrin-like immunoreactive terminals are also seen around pyramidal cells. Somatostatin-positive perikarya occur in the hilus and in oriens bordering alveus. Antibodies against substance P and enkephalin react with terminal plexuses in the stratum radiatum in ventral parts of the hippocampal formation.

Somatostatin-like immunoreactivity (technique different from that of T. Hökfelt) has been described in bouton-like structures densely packed around pyramidal cells in CA1 and 2 in the *dorsal* hippocampus (Petrusz *et al.* 1977). Since somatostatin is inhibitory in its action on these neurons (Rezek *et al.* 1976), it was suggested that the peptide could be localized in the inhibitory basket cell terminals. This suggestion awaits substantiation.

γ-AMINOBUTYRATE (GABA)

Of all the putative transmitters in brain γ-aminobutyric acid (GABA) is perhaps the one whose candidacy rests on the strongest evidence (Roberts *et al.* 1976). Glutamic acid decarboxylase (GAD; glutamate decarboxylase, EC 4.1.1.15), the enzyme catalysing the formation of GABA, appears to be restricted to alleged GABAergic neurons, and is therefore a suitable marker for such neurons. This is emphasized by inhibition studies suggesting that the activity of GAD is of primary importance for the functioning of possible GABAergic synapses (Wood & Peesker 1974; Tapia *et al.* 1975). In the hippocampus GAD is concentrated between the perikarya of the pyramidal and granular cells and, in addition, in the most superficial layers of the cortex (Fig. 2). We observed this in the rat by assaying microdissected samples biochemically (Storm-Mathisen & Fonnum 1971; Storm-Mathisen 1972) in studies prompted by the observation that the pyramidal and granular cells are inhibited by synapses mainly located on their somata, and originating from short-axoned neurons, basket cells (Andersen *et al.* 1964, 1966a). Later electropharmacological studies have provided evidence that this inhibition may operate via GABA receptors (Curtis *et al.* 1970; Segal *et al.* 1975; Andersen *et al.* 1978). The high concentration corresponding to the peripheral branches of the apical dendrites was an unexpected finding, since no inhibition had been reported to occur at this site. However, results with the hippocampal slice preparation, where studies of the dendrites are facilitated, suggest that GABA applied outside the pyramidal cell layer can produce excitatory as well as inhibitory responses, the mechanism for the latter apparently differing from the one operating at the soma (Andersen *et al.* 1978).

The notion that GABA is situated in intrinsic neurons was corroborated by the observation that the GAD activity remained normal after extensive lesions of all known fibre pathways to and within the hippocampal formation (Storm-Mathisen 1972, 1974, 1977b; Storm-Mathisen & Guldberg 1974). Only local lesions designed to interrupt basket cell axons produced a substantial loss of GAD (up to 50%), but not of lactate dehydrogenase (Storm-Mathisen 1977a). When kainic acid was injected locally to destroy intrinsic neurons

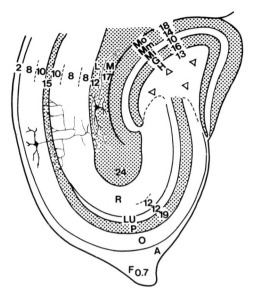

FIG. 2. Stippling indicates zones with higher than average uptake of [³H]GABA or [³H]diaminobutyrate, as determined autoradiographically (T. Taxt & J. Storm-Mathisen, in preparation), which agrees very well with the content of glutamic acid decarboxylase, as visualized immunohistochemically (Barber & Saito 1976). The intensities are somewhat lower in the molecular than in the cellular layers. Numbers indicate the activity of glutamic acid decarboxylase measured biochemically in dissected samples (Storm-Mathisen & Fonnum 1971; Storm-Mathisen 1972), the unit being 1/10 of the activity of strips cut at right angles through all layers of CA1—about 50 μmol h^{-1} g^{-1} dry weight. Note that both stippling and numbers show peaks in the cellular as well as in the outer molecular layers. Unlike the stippling, the numbers are higher in the molecular than in the cellular layers. This may be due to technical reasons (for discussion, see Storm-Mathisen 1977a, 1978). Further note that the 'light' zones account for a substantial proportion of the total amounts of enzyme and [³H]GABA uptake. Also indicated are a basket cell and another short-axoned neuron, based on the drawings of Ramón y Cajal (1893) and Lorente de Nó (1934). See Fig. 1 for explanation of symbols.

(Schwarcz & Coyle 1977), a 64% reduction was observed in GAD without effect on choline acetyltransferase or aromatic amino acid decarboxylase (Fonnum & Walaas 1978).

The evidence that GAD is a neuronal enzyme rests above all on data on the cerebellum and basal ganglia, where up to 90% of its activity has been observed to disappear after lesion of long-axoned alleged GABAergic neurons (Fonnum et al. 1970, 1974; McGeer et al. 1973; Kataoka et al. 1974). Furthermore, about 80–85% of this enzyme is particulate in sucrose homogenates of hippocampus, probably representing nerve terminal particles (Fonnum 1972; Storm-Mathisen 1975). In the present context it should be emphasized that the so-called 'GAD II', once proposed to be a non-neuronal enzyme (Haber *et al.* 1970), appears to be an artifactual phenomenon (Miller & Martin 1973). This term should therefore not be used, in order to avoid confusion with the enzyme extracted from peripheral organs, which appears to differ from GAD purified from brain (Wu 1977). It is not known whether such non-neuronal GAD contributes to the enzyme activity measured in brain samples.

The microchemical results on GAD in rat hippocampus have been confirmed and extended by immunohistochemical studies with antibodies against GAD purified from mouse brain (Barber & Saito 1976). Light microscope preparations (rat hippocampus) showed the enzyme to be distributed in the neuropil with peak concentrations between the cell bodies in the pyramidal and granular layers, as well as in the most superficial molecular layers. The histochemical reaction product occurs in puncta suggestive of nerve terminals. Electron microscopic observations confirmed this and showed GAD-positive nerve endings to form only symmetrical synapses on dendrites and cell bodies (Ribak *et al.* 1978, and personal communication 1977). Injection of colchicine to dam up the enzyme in the cell bodies greatly increased the number of GAD-positive perikarya in the hippocampal formation (Ribak *et al.* 1978). These perikarya were present in all the cortical cell layers and conformed to the short-axoned neurons described by Ramón y Cajal (1893) and Lorente de Nó (1934), including the basket cells. The granular and pyramidal cells were not stained. There was no evidence for the presence of GAD in glial cells.

The high affinity membrane transfer* of GABA in nerve ending particles in hippocampal homogenates was, like GAD, not affected by transection of ascending and commissural afferents (Storm-Mathisen 1975). In contrast, GABA uptake was heavily reduced along with GAD after axotomy in the striato-nigral system. In later autoradiographic studies (T. Taxt & J. Storm-Mathisen, in preparation) we have observed that the pattern of GABA uptake did not change after lesion of entorhinal afferents and intrahippocampal axons from CA3/CA4 pyramidal cells. The normal distribution of [^{3}H]GABA uptake, as seen in light microscope autoradiographs (Fig. 2), is very closely similar to that obtained by GAD immunohistochemistry (Barber & Saito 1976) with respect to the neuropil. In autoradiographs of sections from the interior of incubated slices a similar neuropil labelling, as well as uptake in perikarya of the type visualized by GAD immunohistochemistry, has previously been

---

*The shorter term 'uptake' is used hereafter, although there has been some dispute as to whether the highly selective accumulation of amino acids, as measured *in vitro*, represents net uptake or a homo-exchange phenomenon, perhaps related to release (Iversen 1975; Levi *et al.* 1978).

demonstrated in parts of the hippocampal formation (Hökfelt & Ljungdahl 1971; Iversen & Schon 1973). We found that the uptake was blocked in all layers by (—)-nipecotic acid and L-2,4-diaminobutyric acid, inhibitors of neuronal GABA uptake (Johnston et al. 1976; Lodge et al. 1976; Iversen & Kelly 1975), but was not affected by β-alanine, an inhibitor of glial uptake (Iversen & Kelly 1975). [^{3}H]-L-2,4-diaminobutyric acid showed the same distribution pattern as [^{3}H]GABA, but weaker. In contrast [^{3}H]-β-alanine, [^{3}H]glycine and [^{3}H]leucine at the same or higher radioactive and molar concentrations showed negligible uptake.

In conclusion, the capacity for GABA synthesis (GAD) is concentrated in the terminals of short-axoned neurons, including basket cells, in all layers of the hippocampal formation. The same structures are responsible for most of the high affinity membrane transfer of GABA.

GLUTAMATE AND ASPARTATE

Glutamate (Glu) and aspartate (Asp) are powerful excitants and occur in brain in high concentrations. The establishment of these substances as neurotransmitters has, however, been difficult due to the fact that both amino acids are important in general cell metabolism, and because it has been hard to provide evidence that markers associated with their alleged synaptic action are actually located in defined excitatory neurons. Significant reductions in Glu and Asp content in cortical regions have been demonstrated after lesions of pathways containing excitatory axons (Reiffenstein & Neal 1974; Harvey et al. 1975). However, reductions of 20–50% of such labile constituents as amino acids could be ascribed to functional changes, and one would like to see similar or larger changes in a more stable biochemical parameter. The work of Young et al. (1974) suggested that high affinity uptake could be a suitable parameter. They found that in cerebella of hamsters deficient in granular cells (which might use Glu or Asp as their excitatory transmitter) after neonatal virus infection the uptakes of Glu and Asp were reduced by up to 70%, the reduction being correlated to the extent of granular cell loss. The uptakes of other amino acids were not affected. Endogenous Glu was reduced by about 40%, the difference between the reductions in uptake and endogenous level possibly being a reflection of the amount of Glu situated outside granular cell elements. Other amino acids, including Asp, did not change. This suggests that the cerebellar granular cells do not contain a concentration of Asp that is higher than the average for the tissue, although they do have capacity for its uptake. This is in line with biochemical studies indicating that Glu and Asp are transported by very similar or identical membrane carriers (Balcar & Johnston 1972; Roberts & Watkins 1975), although there is evidence for some difference in the effect of lithium ions (Peterson & Raghupathy 1974; Davies & Johnston 1976). While Glu and Asp are probably released by different nerve endings, their uptake systems would therefore be poorly able to distinguish between the two amino acids. Davies & Johnston (1976) have found evidence that D-Asp is transported by the same membrane carrier as L-Asp. Theoretically, D-Asp is more useful in biochemical experiments than the naturally occurring amino acids, since it is not rapidly metabolized.

The hippocampal formation is very well suited for studying the possible localization of Asp and Glu to defined excitatory nerve endings, because this region contains a series of excitatory neurons, the axon terminals of which are localized in discrete laminae (Fig. 3; Andersen et al. 1966b; Blackstad et al. 1970; Hjorth-Simonsen 1972, 1973). In collaboration with L. L. Iversen of the University of Cambridge we have studied the laminar distribution of Glu uptake by incubating transverse hippocampal slices in Krebs phosphate solution with 1 μM [^{3}H]-L-Glu for 10 min at 25 °C (Iversen & Storm-Mathisen 1976; Taxt et al. 1977; Storm-Mathisen & Iversen 1978). The slices were rinsed, fixed in glutaraldehyde and osmium tetroxide, embedded in plastic and sectioned for light and electron microscopic autoradiography (Iversen & Bloom 1972). The light microscope preparations showed a very conspicuous laminar distribution with the highest uptake activity occurring in stratum radiatum and oriens of hippocampus and the inner part of the molecular layer of area dentata. There was a very sharp boundary towards subiculum, and the limit towards hilus fasciae dentatae was clearly evident. This distribution is closely similar to the distribution of nerve endings of axon branches from pyramidal cells in the ipsilateral and contralateral CA3 and CA4 (Blackstad 1956; Zimmer 1971; Gottlieb & Cowan 1972, 1973; Hjorth-Simonsen 1973; Swanson & Cowan 1977). Although considerably lower, there was also some uptake in the molecular layer of hippocampus and in the outer parts of the molecular layer of area dentata. The outer zone of the dentate molecular layer, containing the terminals of the lateral perforant path (Hjorth-Simonsen 1972), had a slightly lower density of silver grains in autoradiographs than the middle zone, which contains terminals of the medial perforant path. In all these layers the organization of the silver grains suggested their presence over small nerve terminals and fine axons.

In the mossy fibre layer (stratum lucidum) in CA3 and in hilus fasciae dentatae silver grains were organized in clusters with a size suggesting their presence over the giant mossy fibre terminals (Blackstad & Kjærheim 1961), which occur at these sites. In hilus the clusters could be seen lying along the

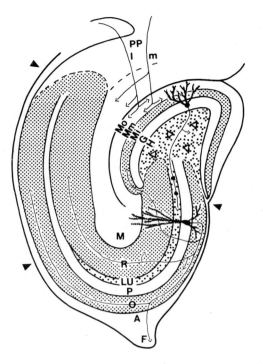

FIG. 3. Fine stippling indicates the areas with the highest concentrations of autoradiographic grains after uptake of [^{3}H]-L-glutamate, [^{3}H]-L-aspartate or [^{3}H]-D-aspartate (Iversen & Storm-Mathisen 1976; Storm-Mathisen 1977b; T. Taxt & J. Storm-Mathisen, in preparation). These areas (O, R, Mi) correspond to the target zones of ipsilateral and commissural axons from pyramidal cells in CA3 & 4 (see text for references). Those to R of CA1 are known as Schaffer collaterals. It is not known whether a single pyramidal cell can have axon ramifications to all the sites shown. The zones Mo, Mm and M have a moderate uptake activity, M and Mm somewhat higher than Mo. Mo and Mm are target areas of the lateral (l) and medial (m) entorhino-dentate fibres (perforant path, PP), respectively. Such fibres, as well as other fibres passing in the angular bundle, also terminate in M of hippocampus. Coarse spread-out dots in LU and H indicate labelled terminals of granular cell axons (mossy fibre terminals). H also contains labelled small terminals. See Fig. 1 for explanation of symbols.

dendrites and perikarya of the large multipolar cells. The electron microscopic autoradiographs confirmed that mossy fibre terminals were labelled. While most of these labelled terminals contained several grains, there were also many unlabelled mossy fibre terminals, some of which had a deteriorated structure. This, and the fact that mossy fibre endings were clearly less labelled in surface autoradiographs (see below) than in the autoradiographs on sections from the depth of embedded slices, would suggest that the lack of labelling in some of the mossy fibre terminals could be due to these large structures being more fragile and more susceptible to damage during the preparative procedures than small nerve terminals. However, the possibility that the capacity for membrane transport of Glu could be restricted to a subpopulation of mossy fibres should be considered.

Grain counts on electron micrographs taken systematically through the neuropil layers of area dentata from fissura hippocampi to hilus showed that the label was distributed as suggested by the light microscope autoradiographs, with the highest labelling intensity in hilus and the inner part of the molecular layer, and progressively lower labelling in the outer two parts of the molecular layer. These preparations were used to analyse the distribution of grains between the different cellular structures. The density of autoradiographic grains was highest over nerve endings in all the zones (2–4 times the average for all tissue elements). These structures had a grain density 3–6 times that in non-axonal tissue elements (dendrites, glia and space). Such a high ratio was found even for the mossy fibre terminals (4.6), although only a proportion of these appeared to be labelled. The axonal tissue elements (nerve endings and axons) had nearly 50% of all grains situated over them in hilus and the inner molecular layer, and 30–40% in the outer two zones of the molecular layer (unidentified 10–20%), but occupied only about 20% of the volume in this incubated material. Thus the grain density of axonal structures was 2–3 times the average for the tissue, and 3–5 times that in non-axonal structures. The values for terminals alone were slightly higher. The relative grain density over nerve terminals and axons was highest in the inner zone of the molecular layer. The grain density over glial processes was about one in all zones—that is, the same as the average for all tissue elements.

The figures given above need some qualification, because of the limited resolution of the autoradiographic technique. The half radius (the radius of a circle around a point of disintegration having 50% probability of containing the autoradiographic grain resulting from the β-particle emitted) is about 0.2 μm in the present type of material (Salpeter et al. 1969). This means that radioactive disintegrations in a small structure, such as a glial lamella and an axon branch, will have a relatively low probability of resulting in a silver grain situated over that structure. On the other hand, all structures will 'pick up' grains from neighbouring structures. To analyse these problems it is necessary to use a statistical approach, such as that developed by Parry & Blackett (1976). It may, however, be noted that the glial elements account for only a small proportion of the tissue volume (up to 9% in this material) and consequently would need a very high labelling intensity to account for a large proportion of the total uptake. The present material was not considered suitable for determining the proportion of labelled and unlabelled nerve endings, since only few nerve endings contained more than one grain, implying that many radioactive terminals had no grain.

The observation that the distribution of Glu uptake conformed to the distribution of excitatory nerve endings in the hippocampal formation prompted further investigations. In order to determine whether the laminar pattern could be due to variations in the penetration of labelled Glu in the tissue, a whole-mount autoradiographic method was adopted (Storm-Mathisen 1977b). Because of the short range of β-particles from ^{3}H this records mainly the radioactivity within 2 μm beneath the surface of the slice —that is, in the tissue that has been in immediate contact with the incubation mixture. This confirmed the laminar distribution pattern and showed that there were no apparent differences between the distributions of L-Glu, L-Asp and D-Asp uptake, whereas that of GABA was entirely different (Figs. 2 & 3) (T. Taxt & J. Storm-Mathisen, in preparation). The uptake in all layers could be essentially abolished by including 20–100 μM DL-*threo*-β-hydroxyaspartic acid (Balcar & Johnston 1972) or 100–200 μM L-aspartic acid β-hydroxamate (Roberts & Watkins 1975), inhibitors of Glu uptake in brain tissue.

When the afferents from area entorhinalis (the perforant path) and fibres passing in the angular bundle were interrupted, as previously described (Storm-Mathisen 1974), the autoradiographs showed a loss of uptake in the outer two zones of the molecular layer of area dentata (Storm-Mathisen 1977b)—that is, in the target zones of the perforant path (Hjorth-Simonsen 1972). When lesions were placed in CA3 (Storm-Mathisen 1977b) to sever fibres of pyramidal cells in CA3 and CA4 destined for CA1 and fascia dentata, there was a dramatic loss of uptake activity selectively in the stratum oriens and radiatum of CA1 and in the inner part of the molecular layer of area dentata—the target zones of the severed fibres (Hjorth-Simonsen 1973). The 'bleached' zones in CA1 stood out very sharply against the adjoining normally active subicular cortex and stratum lacunosum-moleculare of CA1. The distribution of GABA uptake was normal after these lesions.

To study the uptake activities quantitatively we did assays on homogenates, containing nerve ending particles, of samples dissected from various parts of the hippocampal formation. The methods have been described elsewhere (Storm-Mathisen 1977b; Divac *et al.* 1977). This biochemical approach confirmed the normal distribution visualized by autoradiography, in that stratum oriens and radiatum of CA1* had three times as high Glu uptake

---

*These samples also contain the pyramidal layer, but this constitutes a very small part of the volume and has a low uptake activity, to judge from autoradiographs.

activity as the outer part of the molecular layer of area dentata. The inner part of the molecular layer had 1.5 times the activity of the outer parts, but this probably represents an underestimate, since the inner zone is virtually impossible to dissect from fresh slices without contamination from the neighbouring zones.

After lesions of CA3 combined with commissurotomy on the other side the uptake activity for L-Glu or D-Asp in stratum oriens and radiatum of CA1 was reduced by 80–90% based on protein, compared to sham-operated or unoperated controls (Storm-Mathisen 1977b, 1978). The reduction was the same at six to 24 days. There was a slight increase in the uptake of GABA and in GAD activity, probably due to a concentration of the intrinsic nerve elements occurring when a large proportion of the nerve endings degenerate (compare below). Double reciprocal plots of uptake activity at different concentrations of D-Asp confirmed that the changes were due to reduction in the number of uptake sites rather than in the affinity of the membrane transport carrier, and that the measured uptake is high affinity uptake (K_m about 1.4 μM). The reductions in Glu and Asp uptake observed after these lesions are the largest reductions reported in any parameter associated with the proposed synaptic action of these amino acids. It is very difficult to explain these changes on the basis of trans-synaptic effects of the lesion or changes in glial elements, which are sparse normally (Nafstad & Blackstad 1966), but proliferate after denervation (Lynch et al. 1975). In different conditions Glu uptake has been found to occur in glia (Henn et al. 1974; Hökfelt & Ljungdahl 1972; McLennan 1976).

Injecting kainic acid locally to destroy perikarya with Glu receptors (Schwarcz & Coyle 1977) reduced the uptake of Glu by 45% in 1 mm slices of the hippocampal formation containing the injection site (Fonnum & Walaas 1978). The existence of far-reaching longitudinal ipsilateral fibres with Glu uptake and the use of the contralateral side for control (see below) may have contributed to making this reduction smaller than that in GAD (about 65%).

The contents of endogenous amino acids in stratum radiatum and oriens of CA1 at 5–10 days after such combined lesions was analysed by a double label dansylation procedure. The animals were killed by microwave irradiation or by decapitation into liquid nitrogen in order to avoid post-mortem changes in the amino acids. Serial 40 μm freeze-dried sections were prepared and dissected using alternate stained sections as reference (Storm-Mathisen 1972). There was a 40% reduction in Glu ($\alpha = 0.01$, Wilcoxon rank sum) and Asp ($\alpha = 0.02$), but no changes in the other amino acids (F. Fonnum & J. Storm-Mathisen, unpublished). This suggests that the nerve elements responsible for most of the uptake of Asp and Glu also contain a substantial proportion of the total tissue content of these amino acids.

The CA3/4-derived axons to hippocampus and area dentata are partly crossed (Blackstad 1956; Gottlieb & Cowan 1972, 1973; Swanson & Cowan 1977). To study the contribution of these fibres to the uptake of D-Asp in stratum oriens and radiatum of CA1, we dissected samples of these zones from slices of the dorsal quarter of the hippocampal formation on both sides in rats with parasagittal sections close to the midline on one side, severing commissural fibres and transecting the dorsal tip of the hippocampal formation unilaterally. Sham-operated animals received only incisions in the skull and meninges. The survival time was seven days. The first slice, containing the border of the lesion, was discarded. The samples were taken from the next 4–6 300 μm thick slices.

On the side opposite to the lesion there was only a 20% reduction in D-Asp uptake compared to sham-operated controls, while on the ipsilateral side, where longitudinal ipsilateral fibres would be transected in addition to the commissural ones, the reduction was about 40% in these experiments (Storm-Mathisen 1978). The ipsilateral fibres may course longitudinally as far as 3–4 mm (Hjorth-Simonsen 1973). In a previous study of Glu uptake (Storm-Mathisen 1977b) a similar reduction was obtained in two animals ipsilateral to parasagittal transections. This was fortuitously referred to as due to lesion of commissural afferents 'only'. Some degeneration of longitudinal GABAergic axons is to be expected close to a transection of the hippocampal formation. A very slight reduction (10%) in GAD was barely significant, and there was no detectable reduction in GABA uptake. Any depletion might, however, be obliterated by a simultaneous concentration by tissue shrinkage resulting from removal of a large proportion of the nerve endings (see above and below).

In two animals with lesions severing the CA1 from the subiculum samples were dissected from the part of the subiculum adjoining the presubiculum, but excluding the superficial half of the molecular layer. The dissected part receives fibres from CA1 and CA3 (Hjorth-Simonsen 1973; Swanson & Cowan 1977). The animals showed reductions of 42 and 63% in D-Asp uptake at three and 14 days, respectively. This preliminary result suggests that even the CA1/3-subiculum projection may contain acidic amino acid uptake.

In rats with unilateral lesions of the perforant path samples were dissected from the outer two zones of the molecular layer of area dentata on both sides, and compared to samples from the inner zone of the molecular layer and from stratum oriens and radiatum of CA1 (Storm-Mathisen 1977b).

In the perforant path zones there was a highly significant reduction of Glu uptake, by 40% based on protein. GAD, however, was increased by 30%, probably due to a large proportion of the tissue volume being lost on degeneration of the perforant path terminals (Matthews *et al.* 1976). Since GAD is likely to be situated in intrinsic neurons (see above), it is reasonable to take its activity as an index of the tissue elements remaining after the lesion. Glu uptake relative to GAD activity was reduced by 52% after the lesions. There were no changes in Glu uptake or GAD in the CA1 samples or in the samples from the inner part of the molecular layer. The terminals of the perforant path and pyramidal cell axons both constitute a large proportion of the terminals in their respective target areas, yet the uptake activity for acidic amino acids is much higher in the target area of the latter than in that of the former, and is also more reduced after lesions. Thus the uptake avidity of the nerve endings may differ considerably between different pathways.

In release experiments with slices from all layers of rat CA1 or area dentata Nadler *et al.* (1976) and Hamberger *et al.* (1978) have found that Ca^{2+}-evoked release of endogenous Glu and Asp from slices depolarized by 50 mM-KCl is reduced at nine days after lesions of excitatory nerve projections. Thus lesions of the contralateral hippocampus reduced the release of Glu from CA1, and release of Asp from CA1 as well as from area dentata. Bilateral entorhinal lesions reduced Glu release from area dentata, but increased the release of Asp. The latter finding may be related to the observation that commissural and ipsilateral CA3/4-derived fibres ending in the inner part of the molecular layer of area dentata proliferate after entorhinal lesions (Lynch *et al.* 1973) and seem to form active new synapses as soon as nine days after lesion of the perforant path afferents (West *et al.* 1975). These release experiments have been confirmed and extended by later studies in rabbit area dentata, also using exogenous amino acids synthesized *in situ* from precursors (Cotman & Hamberger 1978). Electrical stimulation of slices of fascia dentata evoked Ca^{2+}-dependent release of Glu (White *et al.* 1977). Release of Glu has also been observed *in vivo* from the alvear surface of cat hippocampus during stimulation of the perforant path, probably activating mossy fibres, and CA3/4 cells as well (Crawford & Connor 1972).

Electrophysiological results have shown that the dendrites of hippocampal pyramidal cells are sensitive to iontophoretically applied Glu (Dudar 1974; Schwartzkroin & Andersen 1975). At certain sensitive points, probably corresponding to dendritic branches of the cell recorded from, the sensitivity is high (1 nA for 100 ms) and the latency down to 10 ms. Asp is also active, but apparently less so than Glu, and the actions of both Glu and Asp are antagonized (better than that of acetylcholine) by L-glutamic acid diethyl ester (Spencer *et al.* 1976; Segal 1976). Segal (1976) found that glutamic acid diethyl ester antagonized the responses of CA1 cells to stimulation of commissural fibres, but not the responses to perforant path stimulation. However, the suitability of glutamic acid diethyl ester as an antagonist of Glu or Asp action (Clarke & Straughan 1977) is hardly good enough for us to place great emphasis on this observation. Another Glu analogue, DL-2-amino-4-phosphonobutyrate, antagonized the postsynaptic response to *in vitro* perforant path stimulation, but not the Glu release (White *et al.* 1977).

In conclusion, the terminals of several excitatory fibre systems in the hippocampal formation—entorhinal afferents, granular cell axons, and pyramidal cell axons—are responsible for a large proportion of the high affinity uptake of Glu and Asp in the tissue. However, the uptake avidity is considerably higher in the pyramidal than in the entorhinal axon terminals. The pyramidal cell terminals contain Asp and Glu at higher than average tissue concentrations. There is evidence that boutons of entorhinal fibres may release Glu, while those of commissural pyramidal cell axons may release Asp and possibly Glu. The electrophysiological data are compatible with a transmitter function of both amino acids in the hippocampal formation.

The results on the hippocampal formation are an important addition to the evidence that Glu and Asp are excitatory transmitters in the brain, and that high affinity uptake of Glu/Asp is useful as a marker for alleged glutamergic/aspartergic neurons. A picture is emerging of these amino acids as likely transmitters in several cortico-cortical and corticofugal pathways (Divac *et al.* 1977; McGeer *et al.* 1977; Kim *et al.* 1977; Lund Karlsen & Fonnum 1978; Storm-Mathisen & Woxen Opsahl 1978).

A NOTE ON AMINO ACIDS IN THE SEPTUM AND MAMILLARY BODIES

In view of the likelihood that excitatory neuron systems in the hippocampal formation use Asp or Glu as their transmitters it was of interest also to study the efferent projections. The hippocampal formation projects bilaterally in a topographically organized manner to the lateral parts of the septum (from CA1-3 and subiculum) and the mamillary bodies (from subiculum-parasubiculum) (Raisman *et al.* 1966; Swanson & Cowan 1977; Swanson, this symposium; Meibach & Siegel 1977). The projections appear to be excitatory (DeFrance *et al.* 1973; McLennan & Miller 1974*a, b*; MacLean 1975). There is also evidence that cells in the medial septum projecting to the hippocampus, as well as lateral septal neurons, are controlled by short-axoned inhibitory neurons, likely to use GABA as their transmitter, and that these interneurons are under the influence of the hippocampo-septal excitatory fibres.

The fimbria/fornix and fornix superior were completely transected on both sides in rats through a V-shaped incision in the skull about 2 mm behind the bregma. After nine days survival time the medial and lateral parts of the septum were dissected at 0–6 °C from frontal 350 μm slices of fresh brain essentially according to König & Klippel (1963). The medial samples contained the vertical limb of the nucleus of the diagonal band and the medial septal nucleus. The lateral samples contained the lateral nucleus, which includes the dorsal nucleus and parts of nucleus fimbriatus and medialis in the terminology of others (Białowas & Narkiewicz 1974; Harkmark et al. 1975). The mamillary bodies were 'scooped' out macroscopically. The uptake of D-Asp, L-Glu and GABA, and enzymes and protein were measured as before in homogenates containing nerve-ending particles (Divac et al. 1977).

The uptake of D-Asp was reduced to $34 \pm 4\%$ (S.E.M., $n = 7$, $P < 0.001$) in the lateral septum and to $48 \pm 6\%$ (S.E.M., $n = 7$, $P < 0.001$) in the mamillary bodies compared to unoperated controls (Storm-Mathisen & Woxen Opsahl 1978). The latter figure probably represents an underestimate, since the site of termination of the hippocampal fibres would occupy only a part of the macroscopically dissected mamillary body. There was no decrease in the uptake of D-Asp in the medial septum. Glutamic acid decarboxylase (GAD) and the uptake of GABA were not reduced in any of the locations. This, and the lack of change in septal GAD after hemisections behind the septum (Fonnum et al. 1977), is consistent with the presence of this enzyme and GABA uptake in interneurons in the septum, although the participation of axons from basofrontal regions, such as the medial forebrain bundle region, very rich in GAD (Fonnum et al. 1977), has not been excluded. Choline acetyltransferase showed minimal changes in the mamillary bodies and medial septum, but was increased to $151 \pm 11\%$ (S.E.M., $n = 7$, $P < 0.005$) in the lateral septum. This might be interpreted as due to a pile-up of enzyme in collaterals of the cut septo-hippocampal fibres. Such collaterals appear to terminate in a part of the dorsal septal nucleus (Harkmark et al. 1975), and might function to activate inhibitory interneurons. Another possibility is that cholinergic fibres proliferate after removal of the hippocampal input. The lack of reduction in GABA uptake, GAD, choline acetyltransferase and the mitochondrial enzyme carnitine acetyltransferase is good evidence that the losses observed in D-Asp uptake were not due to unspecific or transneuronal effects of the lesions.

When the normal activities in the septum and mamillary bodies were compared with those in stratum oriens and radiatum of CA1, an interesting distribution pattern was revealed (Table 2). The D-Asp uptake activity varied

TABLE 2

Normal distribution of transmitter markers in rat hippocampus CA1 (oriens and radiatum), septum and mamillary body

	n	[^{3}H]-D-Asp uptake	[^{14}C]GABA uptake	Glutamic acid decarbo-xylase	Choline acetyl-transferase	Carnitine acetyl-transferase
Hippocampus CA1, oriens + radiatum of the middle third	3	292 ± 19	36 ± 7	31 ± 3	109 ± 12	57 ± 3
Septum, lateral part	7	100 ± 7	100 ± 8	100 ± 10	100 ± 8	100 ± 4
Septum, medial part	4	34 ± 5	99 ± 14	90 ± 19	366 ± 18	68 ± 4
Mamillary body	7	54 ± 3	63 ± 5	69 ± 9	62 ± 3	72 ± 8

Relative units, based on protein, mean ± S.E.M. of n samples.

by a factor of nine between the medial septum and CA1, and that of the lateral septum was three times higher than that of the medial. The activity in the latter is similar to the activities remaining after lesions in CA1, lateral septum and mamillary bodies. GAD activity and GABA uptake, on the other hand, were highest in the septum and similar in the medial and lateral parts, in agreement with the notion that GABAergic interneurons are important in medial as well as lateral septum (McLennan & Miller 1974a, b). The ratio of GABA uptake to GAD was nearly constant between these regions, unlike the situation when hippocampus is compared to the basal ganglia and cerebellar nuclei (Storm-Mathisen 1975), which contain long-axoned, rather than short-axoned GABA neurons. Whereas the activity ratios between CA1 and lateral septum were inverse for D-Asp uptake and the GABA markers (3 and 0.3, respectively), choline acetyltransferase was equal at these two sites, suggesting similar densities of cholinergic innervation. The high activity of this enzyme in the medial septum is in agreement with its content of cholinergic perikarya.

In other experiments the uptake activities for [^{3}H]-D-Asp and [^{14}C]-L-Glu were compared in double labelling experiments in normal rats and rats with unilateral lesions of the septo-hippocampal formation (Storm-Mathisen & Woxen Opsahl 1978). The uptakes of D-Asp and L-Glu were equally reduced in septum and corpus mamillare on the operated side, somewhat less than after bilateral lesions. Moreover, the molar ratio of the two uptake activities was nearly constant in the various regions, with and without lesions, being 1.45 ± 0.03 (S.D., $n = 22$, $r = 0.997$, $P \ll 0.001$) in favour of D-Asp (1 μM of

D-Asp and L-Glu, 3 min incubation at 25 °C). These findings show that Glu and Asp uptake, in addition to being very similar biochemically, also have very similar localizations.

In summary, septal and mamillary terminals of fibres from the hippocampal formation are responsible for a large proportion of the Asp and Glu uptake activities in their target areas. In view of the findings suggesting that uptake of these amino acids may be useful as marker for putative 'glutamergic' or 'aspartergic' axon terminals (Young *et al.* 1974; Divac *et al.* 1977; and results on hippocampus dealt with above), it is reasonable to suggest that Glu or Asp could be transmitters in these projections. The results on 'GABAergic' markers are in line with the suggested importance of inhibitory GABAergic interneurons in the lateral and medial septum.

References

ANDERSEN, P., ECCLES, J. C. & LØYNING, Y. (1964) Pathway of post-synaptic inhibition in the hippocampus. *J. Neurophysiol.* 27, 608–619

ANDERSEN, P., HOLMQVIST, B. & VOORHOEVE, P. E. (1966*a*) Entorhinal activation of dentate granule cells. *Acta Physiol. Scand.* 66, 448–460

ANDERSEN, P., BLACKSTAD, T. W. & LØMO, T. (1966*b*) Location and identification of excitatory synapses on hippocampal pyramidal cells. *Exp. Brain Res. 1*, 236–248

ANDERSEN, P., BIE, B., GANES, T. & MOSFELDT LAURSEN, A. (1978) Two mechanisms for effects of GABA on hippocampal pyramidal cells, in *Iontophoresis and Transmitter Mechanisms in the Mammalian Central Nervous System* (Ryall, R. W. & Kelly, J. S. eds.), pp. 179–182, Elsevier/North-Holland, Amsterdam

AZMITIA, E. C. & SEGAL, M. (1978) An autoradiographic analysis of the differential ascending projections of the dorsal and median raphe nuclei in the rat. *J. Comp. Neurol.*, in press

BALCAR, V. J. & JOHNSTON, G. A. R. (1972) The structural specificity of the high affinity uptake of L-glutamate and L-aspartate by rat brain slices. *J. Neurochem.* 19, 2657–2666

BARBER, R. & SAITO, K. (1976) Light microscopic visualization of GAD and GABA-T in immunocytochemical preparations of rodent CNS, in *GABA in Nervous System Function* (*Kroc Foundation Series*, vol. 5) (Roberts, E., Chase, T. W. & Tower, D. B., eds.), pp. 113–132, Raven Press, New York

BARBIN, G., GARBARG, M., SCHWARTZ, J.-C. & STORM-MATHISEN, J. (1976) Histamine synthesizing afferents to the hippocampal region. *J. Neurochem.* 26, 259–263

BARBIN, G., GARBARG, M., LLORENS-CORTES, C., PALACIOS, J. M., POLLARD, H. & SCHWARTZ, J. C. (1977) Biochemical mapping of histaminergic pathways and cell-bodies in brain. *Proc. Int. Soc. Neurochem.* 6, 486

BIAŁOWĄS, J. & NARKIEWICZ, O. (1974) Acetylcholinesterase activity in the septum of the rat: a histochemical topography. *Acta Neurobiol. Exp.* 34, 573–582

BISCOE, T. J. & STRAUGHAN, D. W. (1966) Micro-electrophoretic studies on neurones in the cat hippocampus. *J. Physiol. (Lond.) 183*, 341–359

BJÖRKLUND, A., NOBIN, A. & STENEVI, U. (1973) The use of neurotoxic dihydroxytryptamines as tools for morphological studies and localized lesioning of central indolamine neurones. *Z. Zellforsch. Mikrosk. Anat.* 145, 579–501

BLACKSTAD, T. W. (1956) Commissural connections of the hippocampal region in the rat, with special reference to their mode of termination. *J. Comp. Neurol.* 205, 417–537

BLACKSTAD, T. W. & KJAERHEIM, Å. (1961) Special axo-dendritic synapses in the hippocampal cortex: electron and light microscopic studies on the layer of mossy fibers. *J. Exp. Neurol.* 117, 133–160

BLACKSTAD, T. W., FUXE, K. & HÖKFELT, T. (1967) Noradrenaline nerve terminals in the hippocampal region of the rat and the guinea pig. *Z. Zellforsch. Mikrosk. Anat.* 78, 463–473

BLACKSTAD, T. W., BRINK, K., HEM, J. & JEUNE, B. (1970) Distribution of hippocampal mossy fibers in the rat. An experimental study with silver impregnation methods. *J. Comp. Neurol.* 138, 433–450

CLARKE, G. & STRAUGHAN, D. W. (1977) Evaluation of the selectivity of antagonists of glutamate and acetylcholine applied microiontophoretically onto cortical neurones. *Neuropharmacology* 16, 391–398

COTMAN, C. W. & HAMBERGER, A. (1978) Glutamate as a CNS neurotransmitter: properties of release, inactivation and biosynthesis, in *Amino Acids as Chemical Transmitters* (Fonnum, F., ed.) *(NATO Advanced Study Institute Series)*, pp. 379–412, Plenum Press, New York

CRAWFORD, I. L. & CONNOR, J. D. (1972) Localization and release of glutamic acid in relation to the hippocampal mossy fibre pathway. *Nature (Lond.)* 244, 442–443

CURTIS, D. R., FELIX, D. & MCLENNAN, H. (1970) GABA and hippocampal inhibition. *Br. J. Pharmacol.* 40, 881–883

DAVIES, L. P. & JOHNSTON, G. A. R. (1976) Uptake and release of D- and L-aspartate by rat brain slices. *J. Neurochem.* 26, 1007–1014

DEFRANCE, J. F., KITAI, S. T. & SHIMONO, T. (1973) Electrophysiological analysis of the hippocampal-septal projections: II. Functional characteristics. *Exp. Brain Res.* 17, 463–476

DE WIED, D. (1977) Peptides and behavior. *Life Sci.* 20, 195–204

DIVAC, I., FONNUM, F. & STORM-MATHISEN, J. (1977) High affinity uptake of glutamate in terminals of corticostriatal axons. *Nature (Lond.)* 266, 377–378

DUDAR, J. D. (1974) *In vitro* excitation of hippocampal pyramidal cell dendrites by glutamic acid. *Neuropharmacology* 13, 1083–1089

DUDAR, J. D. (1975) The effect of septal nuclei stimulation on the release of acetylcholine from the rabbit hippocampus. *Brain Res.* 83, 123–133

FONNUM, F. (1970) Topographical and subcellular localization of choline acetyltransferase in rat hippocampal region. *J. Neurochem.* 17, 1029–1037

FONNUM, F. (1972) Application of microchemical analysis and sub-cellular fractionation techniques to the study of neurotransmitters in discrete areas of mammalian brain. *Adv. Biochem. Psychopharmacol*, 6, 75–88

FONNUM, F. & WALAAS, I. (1978) The effect of intrahippocampal kainic acid injections and surgical lesions on neurotransmitters in hippocampus and septum. *J. Neurochem.*, in press

FONNUM, F., STORM-MATHISEN, J. & WALBERG, F. (1970) Glutamate decarboxylase in inhibitory neurons. A study of the enzyme in Purkinje cell axons and boutons in the cat. *Brain Res.* 20, 259–275

FONNUM, F., GROFOVÁ, I., RINVIK, E., STORM-MATHISEN, J. & WALBERG, F. (1974) Origin and distribution of glutamate decarboxylase in the substantia nigra of the cat. *Brain Res.* 71, 77–92

FONNUM, F., WALAAS, I. & IVERSEN, E. (1977) Localization of GABAergic, cholinergic and aminergic structures in the mesolimbic system. *J. Neurochem.* 29, 221–230

FUXE, K. & JONSSON, G. (1974) Further mapping of central 5-hydroxytryptamine neurons: studies with the neurotoxic dihydroxytryptamines. *Adv. Biochem. Psychopharmacol.* 10, 1–12

GARBARG, M., BARBIN, G., FEGER, J. & SCHWARTZ, J.-C. (1974) Histaminergic pathway in rat brain evidenced by lesions of the medial forebrain bundle. *Science (Wash. D.C.)* 186, 833–834

GARBARG, M., BARBIN, G., BISCHOFF, S., POLLARD, H. & SCHWARTZ, J. C. (1976) Dual localization of histamine in an ascending neuronal pathway and in non-neuronal cells evidenced by lesions in the lateral hypothalamic area. *Brain Res.* 106, 333–348

GOTTLIEB, D. I. & COWAN, W. M. (1972) Evidence for a temporal factor in the occupation of available synaptic sites during the development of the dentate gyrus. *Brain Res. 41,* 452–456

GOTTLIEB, D. I. & COWAN, W. M. (1973) Autoradiographic studies of the commissural and ipsilateral association connections of the hippocampus and dentate gyrus of the rat. I. The commissural connections. *J. Comp. Neurol. 149,* 393–422

HAAS, H. L. & WOLF, P. (1977) Central actions of histamine: microiontophoretic studies. *Brain Res. 122,* 269–279

HABER, B., KURIYAMA, K. & ROBERTS, E. (1970) An anion stimulated L-glutamic acid decarboxylase in non-neural tissues: occurrence and subcellular localization in mouse kidney and developing chick embryo brain. *Biochem. Pharmacol. 19,* 1119–1136

HALARIS, A. E., JONES, B. E. & MOORE, R. Y. (1976) Axonal transport in serotonin neurons of the midbrain raphe. *Brain Res. 107,* 555–574

HAMBERGER, A., CHIANG, G., NYLÉN, E. S., SCHEFF, S. W. & COTMAN, C. W. (1978) Stimulus evoked increase in the biosynthesis of the putative neurotransmitter glutamate in the hippocampus. *Brain Res. 143,* 549–555

HARKMARK, W., MELLGREN, S. I. & SREBRO, B. (1975) Acetylcholinesterase histochemistry of the septal region in rat and human: distribution of enzyme activity. *Brain Res. 95,* 281–289

HARVEY, J. A., SCHOLFIELD, C. N., GRAHAM, L. T., JR & APRISON, M. H. (1975) Putative transmitters in denervated olfactory cortex. *J. Neurochem. 24,* 445–449

HENN, F. A., GOLDSTEIN, M. N. & HAMBERGER, A. (1974) Uptake of the neurotransmitter candidate glutamate by glia. *Nature (Lond.) 249,* 663–664

HJORTH-SIMONSEN, A. (1972) Projection of the lateral part of the entorhinal area to the hippocampus and fascia dentata. *J. Comp. Neurol. 146,* 219–231

HJORTH-SIMONSEN, A. (1973) Some intrinsic connections of the hippocampus in the rat: an experimental analysis. *J. Comp. Neurol. 147,* 145–162

HÖKFELT, T. & LJUNGDAHL, Å. (1971) Uptake of [^{3}H]noradrenaline and γ-[^{3}H]aminobutyric acid in isolated tissues of rat: an autoradiographic and fluorescence microscopic study. *Prog. Brain Res. 34,* 87–102

HÖKFELT, T. & LJUNGDAHL, Å. (1972) Application of cytochemical techniques to the study of suspected transmitter substances in the nervous system. *Adv. Biochem. Psychopharmacol. 6,* 1–36

HÖKFELT, T., ELDE, R., JOHANSSON, O., LJUNGDAHL, Å., SCHULTZBERG, M., FUXE, K., GOLDSTEIN, M., NILSSON, G., PERNOW, B., TERENIUS, L., GANTEN, D., JEFFCOATE, S. L., REHFELD, J. & SAID, S. (1978) Distribution of peptide-containing neurons in the nervous system, in *Psychopharmacology: A Generation of Progress* (Lipton, M. A., DiMascio, A. & Killam, K. F., eds.), pp. 39–66, Raven Press, New York

IVERSEN, L. (1975) High affinity uptake of neurotransmitter amino acids. *Nature (Lond.) 253,* 481

IVERSEN, L. L. & BLOOM, F. E. (1972) Studies on the uptake of [^{3}H]GABA and and [^{3}H]glycine in slices and homogenates of rat brain and spinal cord by electron microscopic autoradiography. *Brain Res. 41,* 131–143

IVERSEN, L. L. & SCHON, F. F. (1973) The use of autoradiographic techniques for the identification and mapping of transmitter specific neurones in CNS, in *New Concepts in Neurotransmitter Regulation* (Mandell, A. J., ed.), pp. 153–193, Plenum Press, New York

IVERSEN, L. L. & KELLY, J. S. (1975) Uptake and metabolism of γ-aminobutyric acid by neurones and glial cells. *Biochem. Pharmacol. 24,* 933–938

IVERSEN, L. L. & STORM-MATHISEN, J. (1976) Uptake of [^{3}H]glutamic acid in excitatory nerve endings in the hippocampal formation of the rat. *Acta Physiol. Scand. 96,* 22A–23A

JOHNSTON, G. A. R., KROGSGAARD-LARSEN, P., STEPHANSON, A. L. & TWITCHIN, B. (1976) Inhibition of the uptake of GABA and related amino acids in rat brain slices by the optical isomers of nipecotic acid. *J. Neurochem. 26,* 1029–1032

KÁSA, P. (1975) Histochemistry of choline acetyltransferase, in *Cholinergic Mechanisms* (Waser, P. G., ed.), pp. 271–281, Raven Press, New York

KATAOKA, K., BAK, I. J., HASSLER, R., KIM, J. S. & WAGNER, A. (1974) L-Glutamate decarboxylase and choline acetyltransferase in the substantia nigra and the striatum after surgical interruption of the strionigral fibres of the baboon. *Exp. Brain Res. 19,* 217–227

KIM, J.-S., HASSLER, R., HAUG, P. & PAIK, K.-S. (1977) Effect of frontal ablation on striatal glutamic acid level in rat. *Brain Res. 132*, 370–374

KÖNIG, J. F. R. & KLIPPEL, R. A. (1963) *The Rat Brain. A Stereotaxic Atlas of the Forebrain and Lower Parts of the Brain Stem*, 162 pp., Williams & Wilkins, Baltimore

KUHAR, M. J. & ROMMELSPACHER, H. (1974) Acetylcholinesterase-staining synaptosomes from rat hippocampus. Relative frequency and tentative estimation of internal concentration of free or 'labile bound' acetylcholine. *Brain Res. 77*, 85–96

KUHAR, M. J., AGHAJANIAN, G. K. & ROTH, R. H. (1972) Tryptophan hydroxylase activity and synaptosomal uptake of serotonin in discrete brain regions after midbrain raphe lesions: correlations with serotonin levels and histochemical fluorescence. *Brain Res. 44*, 165–176

LARSSON, L.-I. & REHFELD, J. F. (1977) Evidence for a common evolutionary origin of gastrin and cholecystokinin. *Nature (Lond.) 269*, 335–338

LEVI, G., BANAY-SCHWARTZ, M. & RAITERI, M. (1978) Uptake, exchange and release of GABA in isolated nerve endings, in *Amino Acids as Chemical Transmitters* (Fonnum. F., ed.) *(NATO Advanced study Institute Series)*, pp. 327–350, Plenum Press, New York

LEWIS, P. R., SHUTE, C. C. D. & SILVER, A. (1967) Confirmation from choline acetylase analyses of a massive cholinergic innervation to the rat hippocampus. *J. Physiol. (Lond.) 191*, 215–224

LINDVALL, O. & BJÖRKLUND, A. (1974) The organization of the ascending catecholamine neuron systems in the rat brain. *Acta Physiol. Scand. Suppl. 412*, 48 pp.

LODGE, D., JOHNSTON, G. A. R. & STEPHANSON, A. L. (1976) The uptake of GABA and β-alanine in slices of cat and rat CNS tissue: regional differences in susceptibility to inhibitors. *J. Neurochem. 27*, 1569–1570

LORENS, S. A. & GULDBERG, H. C. (1974) Regional 5-hydroxytryptamine following selective midbrain raphe lesions in the rat. *Brain Res. 78*, 45–56

LORENTE DE NÓ, R. (1934) Studies on the structure of the cerebral cortex. II. Continuation of the study of the ammonic system. *J. Psychol. Neurol. (Leipzig) 46*, 113–117

LUND KARLSEN, R. A. & FONNUM, F. (1978) Evidence for glutamate as a neurotransmitter in corticofugal fibres to the dorsal lateral geniculate body and the superior colliculus in rats. *Brain Res.*, in press

LYNCH, G., STANFIELD, B. & COTMAN, C. W. (1973) Developmental differences in post-lesion axonal growth in the hippocampus. *Brain Res. 59*, 155–168

LYNCH, G., ROSE, G., GALL, C. & COTMAN, C. W. (1975) The response of the dentate gyrus to partial deafferentation, in *Golgi Centennial Symposium: Perspectives in Neurobiology* (Santini, M., ed.), Raven Press, New York

MACLEAN, P. D. (1975) An ongoing analysis of hippocampal inputs and outputs: microelectrode and neuroanatomical findings in squirrel monkeys, in *The Hippocampus*, vol. 1: *Structure and Development* (Isaacson, R. L. & Pribram, K. H., eds.), pp. 177–211, Plenum Press, New York & London

MATTHEWS, D. A., COTMAN, C. & LYNCH, G. (1976) An electron microscopic study of lesion-induced synaptogenesis in the dentate gyrus of the adult rat. I. Magnitude and time course of degeneration. *Brain Res. 115*, 1–21

MCGEER, E. G., FIBIGER, H. C., MCGEER, P. L. & BROOKE, S. (1973) Temporal changes in amine synthesizing enzymes of rat extrapyramidal structure after hemitransections or 6-hydroxydopamine administration. *Brain Res. 52*, 289–300

MCGEER, P. L., MCGEER, E. G., SCHERER, U. & SINGH, K. (1977) A glutamergic corticostriatal path? *Brain Res. 128*, 369–373

MCLENNAN, H. (1976) The autoradiographic localization of L-[^{3}H]glutamate in rat brain tissue. *Brain Res. 115*, 139–144

MCLENNAN, H. & MILLER, J. J. (1974a) The hippocampal control of neuronal discharges in the septum of the rat. *J. Physiol. (Lond.) 237*, 607–624

MCLENNAN, H. & MILLER, J. J. (1974b) γ-Aminobutyric acid and inhibition in the septal nuclei of the rat. *J. Physiol. (Lond.) 237*, 625–633

MEIBACH, R. C. & SIEGEL, A. (1977) Efferent connections of the hippocampal formation in the rat. *Brain Res. 124*, 197–224

MILLER, L. P. & MARTIN, D. L. (1973) An artifact in the radiochemical assay of brain mitochondrial glutamate decarboxylase. *Life Sci. 13*, 1023–1032

MINCHIN, M. C. & IVERSEN, L. L. (1974) Release of [^{3}H]gamma-aminobutyric acid from glial cells in rat dorsal root ganglia. *J. Neurochem. 23*, 533–540

MOORE, R. Y. & HALARIS, A. E. (1975) Hippocampal innervation by serotonin neurons of the midbrain raphe in the rat. *J. Comp. Neurol. 152*, 163–174

NADLER, J. V., VACA, K. W., WHITE, W. F., LYNCH, G. S. & COTMAN, C. W. (1976) Aspartate and glutamate as possible transmitters of excitatory hippocampal afferents. *Nature (Lond.) 260*, 538–540

NADLER, J. V., WHITE, W. F., VACA, K. W. & COTMAN, C. W. (1977) Calcium-dependent γ-aminobutyric acid release by interneurons of rat hippocampal regions: lesion-induced plasticity. *Brain Res. 131*, 241–258

NAFSTAD, P. H. J. & BLACKSTAD, T. W. (1966) Distribution of mitochondria in pyramidal cells and boutons in hippocampal cortex. *Z. Zellforsch. Mikrosk. Anat. 73*, 234–245

OTSUKA, M., KONISHI, S., TAKAHASHI, T. & SAITO, K. (1976) Substance P and primary afferent transmission. *Adv. Biochem. Psychopharmacol. 15*, 187–191

PALKOVITS, M., SAAVEDRA, J. M., JACOBOWITZ, D. M., KIZER, J. S., ZÁBORSZKY, L. & BROWNSTEIN, M. J. (1977) Serotonergic innervation of the forebrain: effect of lesions on serotonin and tryptophan hydroxylase levels. *Brain Res. 130*, 121–134

PARRY, D. M. & BLACKETT, N. M. (1976) Analysis of electron microscope autoradiographs using the hypothetical grain analysis method. *J. Microsc. 106*, 117–124

PETERSON, N. A. & RAGHUPATHY, E. (1974) Selective effects of lithium on synaptosomal amino acid transport systems. *Biochem. Pharmacol. 23*, 2491–2494

PETRUSZ, P., SAR, M., GROSSMAN, G. H. & KIZER, J. S. (1977) Synaptic terminals with somatostatin-like immunoreactivity in the rat brain. *Brain Res. 137*, 181–187

PFAFF, D. & KEINER, M. (1974) Atlas of estradiol-containing cells in the central nervous system of the female rat. *J. Comp. Neurol. 151*, 121–158

PHILLIS, J. W. (1974) Evidence for cholinergic transmission in the cerebral cortex, in *Neurohumoral Coding of Brain Function* (Myers, R. D. & Drucker-Colin, R. R., eds.), pp. 57–77, Plenum Press, New York

PHILLIS, J. W. & LIMACHER, J. J. (1974) Excitation of cerebral cortical neurons by various polypeptides. *Exp. Neurol. 43*, 414–423

POTEMPSKA, A., SOSIŃSKA, H. & ODERFELD-NOWAK, B. (1977) On the mechanisms of increase of bound acetylcholine level in hippocampus after acute septal lesions in rats. *Proc. Int. Soc. Neurochem. 8*, 151

RAISMAN, G., COWAN, W. M. & POWELL, T. P. S. (1966) An experimental analysis of the efferent projection of the hippocampus. *Brain 89*, 83–108

RAMÓN Y CAJAL, S. (1893) *The Structure of Ammon's Horn*, translated by L. M. Kraft, 1968, 78 pp., Thomas, Springfield, Ill.

REIFFENSTEIN, R. J. & NEAL, M. Y. (1974) Uptake, storage and release of γ-aminobutyric acid in normal and chronically isolated cat cerebral cortex. *Can. J. Physiol. Pharmacol. 52*, 286–296

REIS, D. J., ROSS, R. A. & JOH, T. H. (1974) Some aspects of the reaction of central and peripheral noradrenergic neurons to injury, in *Dynamics of Degeneration and Growth in Neurons (Wenner-Gren Center International Symposium Series*. vol. 22) (Fuxe, K., Olson, L. & Zotterman, Y., eds.), pp. 109–125, Pergamon Press, Oxford

REZEK, M., HAVLICEK, V., HUGHES, K. R. & FRIESEN, H. (1976) Central site of action of somatostatin (SRIF): role of hippocampus. *Neuropharmacology 15*, 499–504

RIBAK, C. E., VAUGHN, J. E. & SAITO, K. (1978) Immunocytochemical localization of glutamic acid decarboxylase in neuronal somata following colchicine inhibition of axonal transport. *Brain Res. 140*, 315–332

ROBERTS, E., CHASE, T. N. & TOWER, D. B. (eds.) (1976) *GABA in Nervous System Function* (Kroc Foundation Series, vol. 5), Raven Press, New York

ROBERTS, P. J. & WATKINS, J. C. (1975) Structural requirements for the inhibition for L-glutamate uptake by glia and nerve endings. *Brain Res.* 85, 120–125

ROSS, R. A. & REIS, D. J. (1974) Effects of lesions of locus coeruleus on regional distribution of dopamine-β-hydroxylase activity in rat brain. *Brain Res.* 73, 161–166

ROSSIER, J. (1975) Immunohistochemical localization of choline acetyltransferase: real or artefact? *Brain Res.* 98, 619–622

SALPETER, M. M., BACHMANN, L. & SALPETER, E. E. (1969) Resolution in electron microscope radioautography. *J. Cell Biol.* 41, 1–32

SCHWARCZ, R. & COYLE, J. T. (1977) Striatal lesions with kainic acid: neurochemical characteristics. *Brain Res.* 127, 235–249

SCHWARTZ, J. C., BARBIN, G., GARBARG, M., POLLARD, H., ROSE, C. & VERDIRE, M. (1976) Neurochemical evidence for histamine acting as a transmitter in mammalian brain. *Adv. Biochem. Psychopharmacol.* 15, 111–126

SCHWARTZ, J. C., PALACIOS, J. M., BARBIN, G., QUACH, T. T., GARBARG, M., HAAS, H. L. & WOLF, P. (1978) Histamine receptors in mammalian brain: characters and modifications studied electrophysiologically and biochemically, in *Histamine Receptors* (Yellin, T.O., ed.), Spectrum, New York, in press

SCHWARTZKROIN, P. A. & ANDERSEN, P. (1975) Glutamic acid sensitivity of dendrites in hippocampal slices *in vitro*. *Adv. Neurol.* 12, 45–51

SEGAL, M. (1975) Physiological and pharmacological evidence for a serotonergic projection to the hippocampus. *Brain Res.* 94, 115–131

SEGAL, M. (1976) Glutamate antagonists in rat hippocampus. *Br. J. Pharmacol.* 58, 341–345

SEGAL, M. & BLOOM, F. E. (1974a) The action of norepinephrine in the rat hippocampus. I. Iontophoretic studies. *Brain Res.* 72, 79–97

SEGAL, M. & BLOOM, F. E. (1974b) The action of norepinephrine in the rat hippocampus. II. Activation of the input pathway. *Brain Res.* 72, 99–114

SEGAL, M. & BLOOM, F. E. (1976a) The action of norepinephrine in the rat hippocampus. III. Hippocampal cellular responses to locus coeruleus stimulation in the awake rat. *Brain Res.* 107, 499–511

SEGAL, M. & BLOOM, F. E. (1976b) The action of norepinephrine in the rat hippocampus. IV. The effects of locus coeruleus stimulation on evoked hippocampal unit activity. *Brain Res.* 107, 513–525

SEGAL, M., SIMS, K. & SMISSMAN, E. (1975) Characterization of an inhibitory receptor in rat hippocampus: a microiontophoretic study using conformationally restricted amino acid analogues. *Br. J. Pharmacol.* 54, 181–188

SHUTE, C. C. D. & LEWIS, P. R. (1961) The use of cholinesterase techniques combined with operative procedures to follow nervous pathways in the brain. *Bibl. Anat.* 2, 34–49

SMITH, C. M. (1974) Acetylcholine release from the cholinergic septo-hippocampal pathway. *Life Sci.* 14, 2159–2166

SNYDER, S. H., KUHAR, M. J., GREEN, A. I., COYLE, J. T. & SHASKAN, E. G. (1970) Uptake and subcellular localization of neurotransmitters in the brain. *Int. Rev. Neurobiol.* 13, 127–158

SPENCER, H. J., GRIBKOFF, V. K., COTMAN, C. W. & LYNCH, G. S. (1976) GDEE antagonism of iontophoretic amino acid excitations in the intact hippocampus and in the hippocampal slice preparation. *Brain Res.* 105, 471–481

STEVENS, R. & COWEY, A. (1973) Effects of dorsal and ventral hippocampal lesions on spontaneous alternation, learned alternation and probability learning in rats. *Brain Res.* 52, 203–224

STORM-MATHISEN, J. (1970) Quantitative histochemistry of acetylcholinesterase in rat hippocampal region correlated to histochemical staining. *J. Neurochem.* 17, 739–750

STORM-MATHISEN, J. (1972) Glutamate decarboxylase in the rat hippocampal region after lesions of the afferent fibre systems. Evidence that the enzyme is localized in intrinsic neurones. *Brain Res.* 40, 215–235

STORM-MATHISEN, J. (1974) Choline acetyltransferase and acetylcholinesterase in fascia dentata following lesion of the entorhinal afferents. *Brain Res.* 80, 181–197

STORM-MATHISEN, J. (1975) High affinity uptake of GABA in presumed GABA-ergic nerve endings in rat brain. *Brain Res. 84*, 409–427

STORM-MATHISEN, J. (1977a) Localization of transmitter candidates in the brain: the hippocampal formation as a model. *Prog. Neurobiol. 8*, 119–181

STORM-MATHISEN, J. (1977b) Glutamic acid and excitatory nerve endings: reduction of glutamic acid uptake after axotomy. *Brain Res. 120*, 379–386

STORM-MATHISEN, J. (1978) Localization of transmitter amino acids: application to hippocampus and septum, in *Amino Acids as Chemical Transmitters* (Fonnum, F., ed.) *(NATO Advanced Study Institute Series)*, pp. 155–173, Plenum Press, New York

STORM-MATHISEN, J. & BLACKSTAD, T. W. (1964) Cholinesterase in the hippocampal region. Distribution and relation to architectonics and afferent systems. *Acta Anat. 56*, 216–253

STORM-MATHISEN, J. & FONNUM, F. (1971) Quantitative histochemistry of glutamate decarboxylase in the rat hippocampal region. *J. Neurochem. 18*, 1105–1111

STORM-MATHISEN, J. & GULDBERG, H. C. (1974) 5-Hydroxytryptamine and noradrenaline in the hippocampal region: effect of transection of afferent pathways on endogenous levels, high affinity uptake and some transmitter-related enzymes. *J. Neurochem. 22*, 793–803

STORM-MATHISEN, J. & IVERSEN, L. L. (1978) Glutamic acid and excitatory nerve endings: selective uptake of [^{3}H]glutamic acid revealed by light- and electronmicroscopic autoradiography (in preparation)

STORM-MATHISEN, J. & WOXEN OPSAHL, M. (1978) Aspartate and/or glutamate may be transmitters in hippocampal efferents to septum and hypothalamus. *Neurosci. Lett.*, in press

SWANSON, L. W. & COWAN, W. M. (1977) An autoradiographic study of the organization of the efferent connections of the hippocampal formation in the rat. *J. Comp. Neurol. 172*, 49–84

SWANSON, L. W. & HARTMAN, B. K. (1975) The central adrenergic system. An immunofluorescence study of the location of cell bodies and their efferent connections in the rat utilizing dopamine-β-hydroxylase as a marker. *J. Comp. Neurol. 163*, 467–505

TAPIA, R., SANDOVAL, M.-E. & CONTRERAS, P. (1975) Evidence for a role of glutamate decarboxylase activity as a regulatory mechanism of cerebral excitability. *J. Neurochem. 24*, 1283–1285

TAXT, T., STORM-MATHISEN, J., FONNUM, F. & IVERSEN, L. L. (1977) Glutamate (GLU)/aspartate (ASP) in three defined systems of excitatory nerve endings in the hippocampal formation. *Proc. Int. Soc. Neurochem. 6*, 647

THIERRY, A. M., STINUS, I., BLANC, G. & GLOWINSKI, J. (1973) Some evidence for the existence of dopaminergic neurons in the rat cortex. *Brain Res. 50*, 230–234

UNGERSTEDT, U. (1971) Stereotaxic mapping of the monoamine pathways in the rat brain. *Acta Physiol. Scand. 83*, Suppl. 367, 1–48

WEINREICH, D. & HAMMERSCHLAG, R. (1975) Nerve impulse-enhanced release of amino acids from non-synaptic regions of peripheral and central nerve trunks of bullfrog. *Brain Res. 84*, 137–142

WEST, J. R., DEADWYLER, S. A., COTMAN, C. W. & LYNCH, G. S. (1975) Time-dependent changes in commissural field potentials in the dentate gyrus following lesions of the entorhinal cortex in adult rats. *Brain Res. 97*, 215–233

WHITE, W. F., NADLER, J. V., HAMBERGER, A., COTMAN, C. W. & CUMMINGS, J. T. (1977) Glutamate as transmitter of hippocampal perforant path. *Nature (Lond.) 270*, 356–357

WOOD, J. D. & PEESKER, S. J. (1974) Development of an expression which relates the excitable state of the brain to the level of GAD activity and GABA content, with particular reference to the action of hydrazine and its derivatives. *J. Neurochem. 23*, 703–712

WU, J.-Y. (1977) A comparative study of L-glutamate decarboxylase from mouse brain and bovine heart with purified preparations. *J. Neurochem. 28*, 1359–1367

YOUNG, A. B., OSTER-GRANITE, M. L., HERNDON, R. M. & SNYDER, S. H. (1974) Glutamic acid: selective depletion by viral induced granule cell loss in hamster cerebellum. *Brain Res. 73*, 1–13

ZIMMER, J. (1971) Ipsilateral afferents to the commissural zone of the fascia dentata, demonstration in decommissurated rats by silver impregnation. *J. Comp. Neurol. 142*, 393–416

Discussion

Weiskrantz: One question that arises is the correspondence between the fine details of the electrophysiology of excitation and inhibition and the assumed excitatory and inhibitory functions of these particular transmitters.

Andersen: There is a good correspondence. When glutamic acid is injected in the territories of excitatory synapses, you get short latency, remarkably effective activation. With regard to the distribution of GABA, there is also excellent agreement. Iontophoretic application of GABA around the soma of the pyramidal cells produces a fast and reasonably large hyperpolarization associated with an increase in conductance, as one would expect. What is lacking is a demonstration that these effects are associated with interneuronal activity. In addition, we have recently shown that GABA has another action. When delivered to the distal dendritic areas it produces a short latency depolarization, again associated with an increase in conductance (Langmoen et al. 1978). The distribution of this polarizing effect of GABA and the localization of glutamic acid decarboxylase shown by Dr Storm-Mathisen are in reasonably good agreement.

Azmitia: I have been asked to describe some of our recent findings on the serotonin innervation of the septo-hippocampal complex (Fig. 1). Our findings are based on the selective autoradiographic tracing of either the median raphe or dorsal raphe nuclei in combination with specific intracerebral microinjection of 5,7-dihydroxytryptamine (5,7-DHT) and 5,6-DHT. The dorsal raphe nucleus projects to both the septum and the hippocampus from the ventrolateral aspect of the medial forebrain bundle. The fibres to the septum extend dorsally in the septo-hypothalamic tract (TSHT) and appear to terminate in the accumbens nuclei and in the ventrorostral part of the lateral septal nuclei. The fibres to the hippocampus extend laterally from the medial forebrain bundle above the optic tract with the ansa lenticularis. The axons destined for the hippocampus sweep caudoventrolaterally through the amygdala to reach the entorhinal cortex. This projection innervates the most ventral part of the hippocampus and the dentate gyrus (molecular layer).

The projections from the median raphe account for the major part of the serotonin in the septo-hippocampal complex (80%) and ascend to the forebrain in the ventromedial aspect of the medial forebrain bundle. Two main branches are produced by these serotonin fibres in the forebrain: a supracallosal and an infracallosal tract. The supracallosal fibres extend above the corpus callosum in the diagonal bundle at the genu of the corpus callosum. The fibres travel caudally in the cingulum bundle and extend

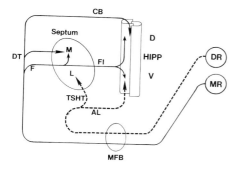

FIG. 1. (Azmitia). Diagrammatic representation of the main projections of the serotonin axons to the septo-hippocampal complex. The tracts were identified by the selective labelling of either the dorsal or median raphe nuclei with [^{3}H]proline and subsequent autoradiographic processing. The serotonin content of the tracts was established by localized microinjections of 5,7-dihydroxytryptamine, a neurotoxic drug. AL, ansa lenticularis; CB, cingulum bundle; D, dorsal and V, ventral hippocampus; DR, dorsal raphe; DT, diagonal tract; F, fornix column; FI, fimbria; L, lateral and M, medial septal nucleus; MFB, medial forebrain bundle; MR, median raphe; TSHT, septo-hypothalamic tract.

ventrally into the hippocampus at the splenium of the corpus callosum. These fibres *appear* to terminate in the most dorsal part of the hippocampus in the stratum moleculare-lacunosum of CA1, the stratum radiatum of CA2–CA3, and *do terminate* in the polymorphic region of the dentate gyrus. This latter termination site has been ultrastructurally verified and the serotonin axons make axo-dendritic connections (Gray type 1) in the polymorphic area.

The infracallosal fibres join the fornix fibres at the level of the anterior hypothalamus and extend dorsally through the medio-posterior septum where they branch into both the fimbria-fornix and the fornix dorsalis. The fimbrial fibres mainly innervate the intermediate regions of the hippocampus and have the same distribution pattern as the supracallosal fibres, but also project into the stratum oriens.

The median raphe nucleus innervation of the septum extends dorsally from the medial forebrain bundle in both the fornix and the diagonal band tracts. These fibres appear to terminate in the medial nucleus and the dorsal part of the lateral nucleus of the septum.

The importance of these projections is that they distribute to localized regions within the septo-hippocampal complex and may be selectively and individually removed by the serotonin neurotoxic drugs, 5,7-DHT and 5,6-DHT. In a series of studies made with A. Buchan and J. Williams at Cambridge University, it was found that interruption of either the supracallosal or infracallosal serotonin efferents to the hippocampus resulted in substantial reduction in the autoradiographically detectable fibres in specific regions of the hippocampus and in the hippocampal uptake of [^{3}H]serotonin but not of [^{3}H]noradrenaline. Since serotonin has been shown to modulate electrical activity in the hippocampus (Segal 1975; DeFrance et al. 1975), behavioural changes might be expected after lesions produced by these neurotoxic drugs. We have found changes in the orientation behaviour of rats in a novel situation and an increase in their locomotor activity measured in photocell cages after neurotoxic lesions confined to either the supracallosal or the infracallosal serotonin axons.

The injections (5 μg/4 μl) are made by a 15° lateral approach through a specially designed micropipette (tip ≈ 80 μm) so that non-specific damage to the brain is minimized. Behavioural changes found after these injections correlate very significantly ($P < 0.001$) only with depletion of dorsal hippocampal serotonin uptake. The injection appears to destroy only the distal part of the axon, since septal serotonin uptake levels usually remain unchanged after injection into the midline fimbria, despite substantial falls in hippocampal uptake.

This technique of intracerebral injection with serotonin neurotoxic drugs provided a means of removing a single inhibitory input from localized regions of the septo-hippocampal complex, thereby producing a subtle imbalance in a complex network of excitatory and inhibitory interconnections. Studies of this imbalance may provide insights into the electrophysiological and behavioural effects under the control of the septo-hippocampal system.

A final point concerns the plastic nature of these serotonin projections. We at Cambridge have preliminary evidence that destruction of the supracallosal serotonin axons with neurotoxic drugs results in collateral sprouting of the infracallosal serotonin axons after about one month. This collateral sprouting results in a restoration towards the normal pattern both in the morphological distribution in the hippocampus of the median raphe fibres and in the functional changes normally associated with supracallosal fibres. These studies suggest that the various serotonin inputs to the hippocampus do not convey unique information but acquire their specific functions within the hippocampus as a result of the neural elements with which they connect.

Zimmer: Dr Storm-Mathisen, you found that the uptake of glutamate in the perforant path zones was decreased by 40% down to 60% after lesion of the perforant path. The terminals of the perforant path make up at least 85% of the terminals in these zones (Matthews et al. 1976). Are you saying that 15% of the terminals are responsible for 60% of the glutamate uptake?

Storm-Mathisen: No. First, I think some caution is needed when interpreting the rather high figure of 85% (Matthews *et al.* 1976). Strictly, this was not based on the number of terminals, but on the number of terminal profiles with visible synaptic complexes. The size, shape, and also the ease of identification may affect the result when numbers of three-dimensional structures are estimated from sections (Weibel 1969). However, the proportion of the nerve endings supplied by the perforant path is probably very large. The reduction in [^{3}H]glutamate uptake observed after perforant path lesions was 40% based on protein, but since there is a concomitant reduction in tissue volume, the surviving structures will be concentrated. Allowing for this, the reduction is by 50% or so. It is difficult to tell what structures are responsible for the rest. In the electron microscope autoradiographs about 30 and 40% of the label was found over nerve terminals and axons in the outer and middle zones of the dentate molecular layer, respectively (Storm-Mathisen & Iversen 1978). These figures are probably somewhat underestimated because of the 'grain loss' from small structures. The biochemical data were obtained on homogenates, where one would expect terminals to be preserved to a greater extent than non-axonal structures, but it is not known how important the formation of 'dendrosomes' and 'gliosomes' may be. It should be noted that, as dealt with in my paper, the uptake activity of the perforant path terminals must be much lower than that of the terminals of CA3/4 pyramidal cells.

Zimmer: Would it be possible to combine the uptake with an acute lesion? Would a terminal in an initial state of degeneration, which could be identified in the electron microscope, still be able to take up glutamate?

Storm-Mathisen: Quite possibly; that is an experiment I would like to do.

Björklund: I would like to know more about how well the excitatory amino acids have been tied to a transmitter function. Having heard Dr Storm-Mathisen's paper one might believe that it is now completely settled that these amino acids are the transmitters of the perforant path and the commissural and ipsilateral connections, but how well are the rigid criteria for a transmitter function fulfilled? Secondly, is there any way to tell whether it is actually glutamate or aspartate which is the critical amino acid in each of these systems?

Storm-Mathisen: Uptake studies of this kind can't answer the latter question because the uptake systems are not specific, but the release studies done in Dr Cotman's and Dr Lynch's laboratory can distinguish them, because they examine release of *endogenous* transmitter.

Lynch: There is evidence that slices of tissue from the granule cell molecular layer and the stratum radiatum release aspartate and glutamate. There is a potassium-stimulated calcium-dependent release of these amino acids from these zones. By using selective lesions it has been shown that you can interfere with aspartate or glutamate release selectively (Nadler *et al.* 1976).

Storm-Mathisen: Strictly satisfying all criteria for identifying a transmitter is a very difficult problem in the central nervous system. At present we have to approach it indirectly. Uptake, content, and Ca^{2+}-dependent release of glutamate and aspartate appear to be selectively localized to excitatory neurons in the hippocampal formation. There are also some data on the excitatory action of these amino acids. We don't know what each terminal actually releases when it is active *in vivo*.

Segal: You seemed to allow a long survival time, up to 17 days, before you measured uptake. I wonder if the possible sprouting that takes place could affect your results. Secondly, and related to this, in one of your studies after contralateral cuts it looked as if in the far lateral (septal) pole of the hippocampus there is very little loss of uptake. Do you infer from this that the commissural input is heavier to the septal pole of the hippocampus than to the temporal pole?

Storm-Mathisen: Commissural fibres with glutamate/aspartate uptake were detected only in the septal portion of the hippocampal formation, but my figures for the temporal portion ($99 \pm 14\%$ of control uptake 4–8 days after contralateral cut) do not exclude a small connection to this site. The ipsilateral connections appear much more important than the commissural, so far as uptake of acidic amino acids is concerned. These findings are in agreement with anatomical data on fibre connections (see Swanson, this volume).

Post-lesion axonal growth (sprouting) might occur, but the uptake is reduced by the same amount (by about 85% based on protein content, compared to unoperated control rats) at 3–27 days after lesion of the axons of the ipsilateral CA3/4 pyramidal cells to CA1. Long and short survival times also gave similar reductions after parasagittal cuts (3–15 days) and perforant path lesions (6–17 days).

Swanson: Is there any evidence for postsynaptic amino acid receptors, and what sort of mechanisms are operating to inactivate the amino acids which are presumably released?

Storm-Mathisen: One imagines that the high affinity uptake is responsible for inactivation, but people who have studied this in homogenates or in synaptosomes (nerve terminal preparations) have had difficulties showing that there is a net uptake in the artificial conditions of *in vitro* experiments (Levi *et al.* 1974). I think this may be just a technical problem.

On the first point, there is evidence that glutamate and aspartate bind to receptors in brain tissue (Michaelis *et al.* 1974; Roberts 1974; De Robertis & de Plazas 1976). I don't think anybody has investigated receptors for these amino acids biochemically in the hippocampus, but Dr Andersen and others have shown that when you apply aspartate or glutamate to pyramidal or granular cell dendrites, they act. There are some data indicating that synaptic excitation is antagonized by putative glutamate antagonists, as I discussed in my paper.

Andersen: In the absence of calcium you still get fast and large depolarizations when you apply glutamic acid in reasonable doses to the places where anatomists have found synapses. What is lacking is the speed which people studying the muscle endplate find. The proposed transmitter must work within one millisecond, and the fastest recorded effect of glutamic acid is 12–15 ms. This is probably because we haven't got down to the actual point of the receptor yet. Otherwise, the case for glutamic acid as an excitatory transmitter is a good one.

Gray: Is the amino acid, once taken up, known to be destroyed or used again? What happens to it in these high affinity uptake systems?

Lynch: You can demonstrate in some circumstances that it is released.

Gray: In that case I find it puzzling that the nerve cell doesn't 'care' which amino acid it takes up, yet under physiological conditions only releases one of them, unless it has a selective destruction mechanism so that it only releases the right amino acid.

Storm-Mathisen: Part of this could be due to spatial segregation, for example by glutamate and aspartate nerve endings being situated in different places so that they would not see each other's transmitter. But there is a lot of work to be done on the compartmentation of these amino acids within the terminals.

Segal: In lobster it was demonstrated that excitatory fibres contain both aspartate and glutamate (Freeman 1976). It has also been suggested that aspartate enhances the efficacy of glutamate at postsynaptic sites, and that both amino acids may be released together (Freeman 1976). If so, you may be looking for a segregation that does not exist.

Weiskrantz: May I inject a different line of argument here, namely the question of the special affinity that the hippocampus seems to have to epilepsy —the fact that it has a low threshold for seizure activity. Is there any connection between the neurochemistry and possible mechanisms of epileptic spread in the hippocampus and, if so, does this suggest possible ways of controlling epilepsy, particularly psychomotor epilepsy?

Andersen: If you take individual nerve cells from the hippocampus and compare them with other nerve cells, their sensitivities to pentylene tetrazol, penicillin and heavy metal ions are much the same. So, any propensity of the hippocampal formation for producing epileptic discharge seems to lie more in its organization than in any specific property of the neuronal membrane itself.

References

DeFrance, J. F., McCrea, R. A. & Yoshihara, H. (1975) Effects of certain indoleamines in the hippocampal-septal circuit. *Exp. Neurol. 48*, 352–377
De Robertis, E. & de Plazas, S. F. (1976) Differentiation of L-aspartate and L-glutamate high affinity binding sites in a protein fraction isolated from cat cerebral cortex. *Nature (Lond.) 260*, 347–349
Freeman, A. R. (1976) Polyfunctional role of glutamic acid in excitatory synaptic transmission. *Prog. Neurobiol. 6*, 137–153
Langmoen, I. A., Andersen, P., Gjerstad, L., Mosfeldt Laursen, A. & Ganes, T. (1978) Two separate effects of GABA on hippocampal pyramidal cells *in vitro. Acta Physiol. Scand. 102*, 28–29A
Levi, G., Bertollini, A., Chen, J. & Raiteri, M. (1974) Regional differences in synaptosomal uptake of ^3H γ-aminobutyric acid and ^{14}C-glutamate and possible role of homoexchange processes. *J. Pharmacol. Exp. Ther. 188*, 429–438
Matthews, D. A., Cotman, C. & Lynch, G. (1976) An electron microscopic study of lesion-induced synaptogenesis in the dentate gyrus of the adult rat. I. Magnitude and time course of degeneration. *Brain Res. 115*, 1–21
Michaelis, E. K., Michaelis, M. L. & Boyarski, L. L. (1974) High affinity glutamic acid binding to brain synaptic membranes. *Biochim. Biophys. Acta 367*, 338–348
Nadler, J. V., Vaca, K. W., White, W. F., Lynch, G. S. & Cotman, C. W. (1976) Aspartate and glutamate as possible transmitters of excitatory hippocampal afferents. *Nature (Lond.) 260*, 538–540
Roberts, P. (1974) Glutamate receptors in the rat central nervous system. *Nature (Lond.) 252*, 399–401
Segal, M. (1975) Physiological and pharmacological evidence for a serotonergic projection to the hippocampus. *Brain Res. 94*, 115–131
Storm-Mathisen, J. & Iversen, L. L. (1978) Glutamic acid and excitatory nerve endings: selective uptake of [^3H]glutamic acid revealed by light- and electron microscopic autoradiography, in preparation
Weibel, E. R. (1969) Stereological principles for morphometry in electron microscopic cytology. *Int. Rev. Cytol. 26*, 235–302

Long-lasting facilitation of synaptic transmission

PER ANDERSEN*

The Institute of Neurophysiology, University of Oslo, Norway

Abstract After tetanization of several hippocampal pathways (10–50 Hz for 5–15 seconds) there is an increased synaptic transmission of long duration (long-lasting facilitation). The present investigation was undertaken on isolated hippocampal slices to study the mechanism of the effect.

The transverse hippocampal slice preparation *in vitro* allows the simultaneous testing of several afferent fibre systems on the same cell or population of cells. Tetanization of one group of afferent fibres to CA1 pyramids was followed by a long-lasting increase of synaptic transmission along the same fibres, whereas a control input line gave unchanged responses. Using the presynaptic volley as an indicator of the number of afferent impulses, the increased synaptic transmission appeared as an increased excitatory postsynaptic potential (EPSP), increased amplitude and reduced latency of the population spike, and an increased probability of firing of single units.

Intracellular recording showed increased EPSPs to afferents of the tetanized line, but no lasting change in membrane resistance or in the response to a depolarizing current pulse. Thus, the effect cannot be ascribed to a general postsynaptic excitability increase.

The specific changes in the synaptic transmission may be due either to an increased amount of liberated transmitter or to a local postsynaptic change near the tetanized synapses.

In the search for changes in synaptic transmission in the central nervous system following the use of a specific input line, frequency potentiation and subsequent long-lasting facilitation are of considerable interest. Frequency potentiation is a process by which the individual response to a synaptic volley increases while the system is stimulated at a particular rate. The phenomenon is found in many synapses, including junctions in the autonomic nervous

---

*Based on work done in collaboration with Drs T. Lømo, S. H. Sundberg, O. Sveen and H. Wigström.

system (Burnstock *et al.* 1964), spinal cord afferents like Ib afferents to ventral spino-cerebellar tract cells (Eccles *et al.* 1961), Ia afferents to thoracic motoneurons (Sears 1964) and descending volleys from the pyramidal tract to cervical motoneurons (Landgren *et al.* 1962). Also synapses terminating in various parts of the cerebral cortex share this property. Among such synapses are many within the hippocampal formation. In this structure, synaptic transmission is greatly enhanced if the stimulation rate is increased from about 0.2 Hz to higher values (Fig. 1). The potentiation is partly dependent on the rate of stimulation, 10–20 Hz being the optimal range, and partly on the duration of the stimulus train. The process progresses in a neuronal chain (Fig. 1) where the responses of the CA1 cells are potentiated much later than those of the CA3, which again lagged behind dentate granule cell recruitment. The optimal stimulus frequency also depends upon the anaesthetic level, being higher with reduced anaesthesia. However, above a certain stimulus rate, the responses start to decline, partly due to an exhaustion of the transmitter supply, and partly to the intense conduction change created by the converging synaptic inputs. In an early study (Andersen & Lømo 1967) we found that the underlying cause for the frequency potentiation in the hippocampal formation was an increased excitatory postsynaptic potential (EPSP) which grew in amplitude and duration to overwhelm the subsequent inhibitory postsynaptic potential (IPSP). The growth of the EPSP summated with the effect of the previous stimulation resulting in an overall depolarization. However, because of an intense conductance change, as judged by reduction of the spike size and eventually of the EPSP itself, the cell ceased to discharge

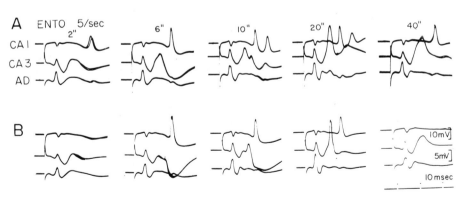

FIG. 1. Frequency potentiation in the hippocampal formation. Records from dentate area (AD), CA3 and CA1 during (A) and after (B) tetanic stimulation at 5 Hz to perforant path. Labels on top indicate time after onset and stop of the tetanus. Note sequence of recruitment, largest responses after 20 and smaller after 40 seconds of tetanus.

in spite of a considerable steady depolarization. After the stimulation, there was a gradual increase in the membrane potential but with a subsequent depression of activity. However, Lømo noticed (1968) that such periods of tetanic activation were followed by an increased granule cell response to a perforant path stimulus which was controlled to be the same as that applied before the tetanic stimulation period. The latter effect was called long-lasting facilitation. In further work, Bliss & Lømo (1973) showed that the increase in potentiation had an extremely long duration, sometimes several hours. Furthermore, the potentiated transmission appeared as a cumulative effect in that several tetanic stimulation periods added their effect to produce stepwise increases in the synaptic transmission. The same authors further found that the effect was dependent on the frequency of stimulation, but it was not directly related to the degree of frequency potentiation during the tetanus itself. Thus, high frequency stimulation usually gave an initial recruitment of new cells and, consequently, larger field responses. However, with continued tetanic stimulation, there was a reduction of the postsynaptic response. In spite of this behaviour, long-lasting potentiation to control stimuli was found for several hours after the high frequency tetanic stimulation. Similar results have also been found by Douglas & Goddard (1975) and by Lynch and his co-workers (Deadwyler et al. 1976).

The duration of the facilitation caused great interest, particularly since it followed tetanic stimulation within the physiological range of participating neurons and since the duration of the priming tetanic stimulus was relatively short (10–20 seconds). However, the facilitation was only studied for one input to a given population of neurons, and the process has only been studied with extracellular techniques. This approach is reasonable because of the statistical nature of the increased responses but cannot answer some of the questions about the mechanism of the induced changes.

Several important questions remained to be answered. Is the increased transfer specific to the tetanized pathway or is it a general change after any synaptic input to the cell? Is there a general increase of the excitability of the cell membrane? For an explanation of the increased population spike there are many possibilities: an increased release of transmitter, a changed membrane resistance, a lower firing threshold of the action potential, a greater synchrony of the afferent impulses, an increased input in spite of a constant stimulus, a specific change in the postsynaptic dendritic structures giving an increased synaptic depolarization, and, finally, a specific change in the synaptic structures mediating the input, facilitating the injected synaptic current which reaches the trigger-zone for the action potential.

In order to attack these problems several advances needed to be made in the experimental design. First, a proper study of the facilitation mechanism requires adequate input control, which means an ability to record the size and synchrony of the afferent volley. Furthermore, any question directed to the specificity of the action requires that the changes induced by the tetanized input line must be compared to responses either from other inputs to the same cell or population of cells, or to responses induced by direct stimulation of the cells themselves. Some of the problems require intracellular recording of long duration in order to see whether the changes are due to postsynaptic or presynaptic events.

With these requirements in mind, we tried to study the mechanism underlying the long-lasting facilitation using the transverse hippocampal slice (Skrede & Westgaard 1971). Schwartzkroin & Wester (1975) showed that tetanic stimulation in hippocampal slices was followed by increased responses of considerable duration. However, certain differences between the type of facilitation and that in intact animals were noted. For example, the potentiation in isolated slices occurs already after one tetanic period, and little facilitation was seen on subsequent tetani, in contrast to the commonly found step-wise increase after several tetani in the intact, anaesthetized animal. However, the similarities of the effect in slices and those obtained in intact preparations were sufficiently large to warrant a more detailed investigation.

In this paper it is described how it is possible to record the presynaptic afferent volley in slice preparations, and it is shown that this deflection can be used as an adequate monitor for the size of the input volley. Furthermore, our experiments with long-lasting intracellular recording did not show excitability changes of the cell membrane that can explain the long-lasting facilitated synaptic transmission. Moreover, it will be shown that the changed transmission is specific to the tetanized line, leaving a control line to the same cell unaltered. Finally, it appeared that the most likely cause for the increased transmission is a change in the EPSP, possibly due to an increased delivery of transmitter by the subsequent pulses in the tetanized pathway.

METHODS

Adult guinea pigs, weighing 250–400 g, were initially anaesthetized with ether. The brain was exposed and removed, divided into halves and the brainstem removed. By inserting a spatula into the lateral ventricle and freeing the septal and hippocampal ends of the hippocampus, it could be rolled out by gentle manipulation on to a wet filter-paper. After removal of surplus tissue, transverse slices, 350–400 µm thick, were cut nearly normal to the longitudinal axis of the hippocampus. The slices were removed from the knife with a soft brush and transferred to a dish with artificial cerebrospinal fluid and to the experimental chamber by a wide-bored pipette. In the experimental chamber the slices were placed on a net covered with lens paper, having artificial cerebrospinal fluid coming up from below and humidified and warm gas (95% O_2 and 5% CO_2) streaming from above. The bathing fluid was composed of (in mM): NaCl 124, KCl 5, KH_2PO_4 1.25, $CaCl_2$ 2, $MgSO_4$ 2, $NaHCO_3$ 26, glucose 10. The pH measured 7.4. Stimulating and recording electrodes were carried in manipulators and could be placed in identified structures under visual control. Recording electrodes were made of fibre-filled electrodes, backfilled with either 4 M-potassium acetate or 3 M-potassium chloride. The resistances measured from 10 to 20 megohm for extracellular work and from 80 to 250 megohm for intracellular use. Stimulation electrodes were monopolar, and consisted of glass- or lacquer-insulated electrolytically sharpened tungsten or platinum needles. The sizes were between 5 and 10 μm. The indifferent electrode for both recording and stimulation was the earth lead. Recording was made through a special-purpose amplifier with negative capacity control and possibilities for current injection in a bridge configuration.

For control of the experimental conditions the essential parts of the measurements were done on-line with the help of a digital computer. The computer delivered the results in real time to a graphic screen so that the experimental condition could be judged throughout the experiment. This proved to be particularly important for assessing the constancy of the presynaptic afferent volley.

Stimuli were delivered to two afferent lines to the cell, usually one system being fibres located in the middle of stratum oriens (layer of basal dendrites) and the other in the middle of stratum radiatum (layer of apical dendrites). In addition, stimuli were delivered by depolarizing pulses through the intracellular electrode. Extra- and intracellular recordings were made from cells in the pyramidal layer. In addition, recordings were made at the synaptic region of the two afferent lines. The afferent volley, and extracellular field potential associated with the EPSP (field EPSP), was measured at both input stages. In addition, the size of the intracellular EPSP, the firing level of the spike and the occurrence of action potentials from the tested cells were continuously monitored. Further, we measured continuously the membrane potential, membrane resistance and the latency to a directly evoked action potential in response to a constant depolarizing pulse and the firing level of this spike. The latter three responses were used to assess the general excitability of the cell.

RESULTS

Recording of the afferent volley

The major afferent fibre system in the hippocampal formation runs parallel to the ventricular surface and the pyramidal layer (Fig. 2A, B). When the recording electrode was located at the same level relative to the pyramidal layer as that of the stimulating cathode, a typical extracellular potential was recorded (Fig. 2C, D, lower traces). This consisted of an initial diphasic wave with a latency of 0.5–1 ms and a total duration of the same order (Fig. 2C, arrow). After this initial deflection there was a negative wave (asterisk) which could carry a population spike on its peak (Fig. 2G, star). Because of the recording situation this population spike was positive, the opposite polarity of the signal recorded from the pyramidal layer.

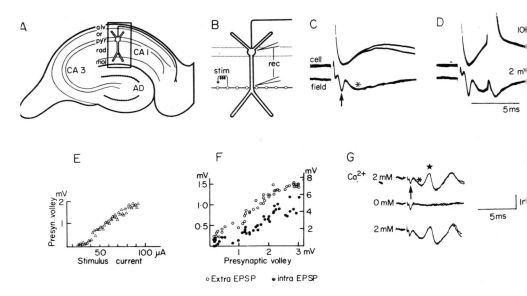

FIG. 2. Features of hippocampal field potentials. A. Diagram of a hippocampal slice. Boxed-in area is shown enlarged in B; stim. and rec. indicate stimulating and recording electrodes. C. Upper trace shows intracellular record from CA1 pyramidal cell in response to stimulation of radiatum fibres. Lower trace is simultaneous recording from the synaptic layer. Arrow points to the presynaptic volley and asterisk indicates the field EPSP. D. As C but stronger stimulus. E. Graph with the size of the presynaptic volley plotted against stimulus current. F. Field EPSP (open circles) and intracellular EPSP (filled circles) plotted against the amplitude of the corresponding presynaptic volley. G. Effect on field potentials of removal of extracellular calcium (0.0 M) and the recovery following re-administration of Ca^{2+} (2 M).

When the recording electrode was moved towards the ventricular surface or towards the hippocampal fissure, the responses changed. The initial deflection was reduced in amplitude, and the EPSP was reduced and eventually reversed when the electrode was placed on the other side of the cell body. When a recording was made intracellularly from one of the activated cells, the extracellular negative wave occurred simultaneously with an intracellular EPSP (Fig. 2C, D, upper traces). However, the initial diphasic deflection had no counterpart in the intracellular record.

Since the amplitude of the initial deflection followed the distribution of afferent fibres (Ramon y Cajal 1893; Lorente de Nó 1934) and preceded the onset of the EPSP, it was assumed that it represented the impulse conduction in the afferent fibres. This hypothesis was verified by removing calcium from the artificial cerebrospinal fluid. This procedure abolished both the field EPSP and the superimposed population spike, and the intracellular EPSP with the action potential (Fig. 2G). However, the initial diphasic deflection was left intact for a long period of time.

As expected for a fibre volley, the latency increased linearly with increasing recording distance, whereas the amplitude was reduced and the duration increased. Obviously, this behaviour is due to the composition of the compound potential by individual action potentials of different fibres, conducting with different velocity because of their varying diameter. Thus, with conductance distances of more than 1.2 mm, there was little left of the presynaptic volley. This is a main limitation of the technique, representing the largest allowable distance between the stimulating cathode and the

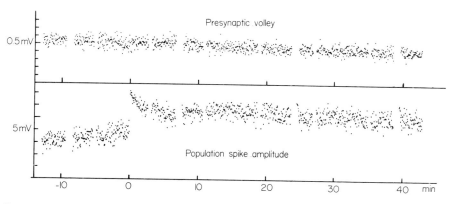

FIG. 3. Amplitude of presynaptic volley *(upper half)* and of CA1 population spike *(lower half)* to a constant current stimulus (90 μA) delivered to radiatum fibres. At 0, a 50 Hz tetanus was delivered for 5 s, same strength.

recording electrode. Both radiatum and oriens fibres were shown to conduct with a velocity of 0.3 m s^{-1} (range 0.24–0.35). The size of the presynaptic volley showed a linear relationship to the stimulating current (Fig. 2E). Likewise, there was a linear relationship between the size of the presynaptic volley and that of both the extra- and intracellularly recorded EPSP (Fig. 2F). These data support the idea that the size of the presynaptic volley can be taken as a reliable measure of the number of afferent fibres impinging upon the recording area.

Although a short-term experiment showed a near linear relationship between the stimulating current and the size of the presynaptic volley, stimulation over longer periods showed a moderate but definite variation in the presynaptic volley amplitude in spite of constant current stimulation. This feature makes the stimulating current an inadequate measure of a constant input (Fig. 3).

Pattern of change after tetanic stimulation

The possibility that tetanic stimulation could give rise to an increased synaptic transmission was tested by a number of techniques.

A simple method was to keep a test stimulus current constant, and measure the size of the presynaptic volley (Fig. 3, upper half) and of the population spike of the pyramidal layer (lower half) before and after a tetanus which was delivered at zero time. The variability and the slow decline of the presynaptic volley in spite of a constant stimulus current illustrate the necessity of referring any change to the input volley actually recorded. In this case the long-lasting increase occurred in spite of a slow decline in the presynaptic volley. A better method was to construct input/output curves which relate the size and latency of the presynaptic volley to the amplitude of the field EPSP, and similarly to the size and latency of the population spike recorded from the pyramidal layer. This potential is an index of the number of cells brought to discharge (Andersen *et al.* 1971). Furthermore, the relation between the presynaptic volley and the probability of isolated unit discharges was measured. Finally, input/output curves with the presynaptic volley relative to the intracellular EPSP, firing level and latency and occurrence of the action potential were measured. The same types of input/output curves were repeated after the tetanized period for a total time of 25–75 min.

Short periods of tetanic stimulation (10 Hz for 10 seconds or 50 Hz for 5 seconds) gave rise to consistently changed input/output curves for a number of functions. Thus, the curve relating the size of the field EPSP to the presynaptic volley was steeper than the control curve. Similarly, the amplitude

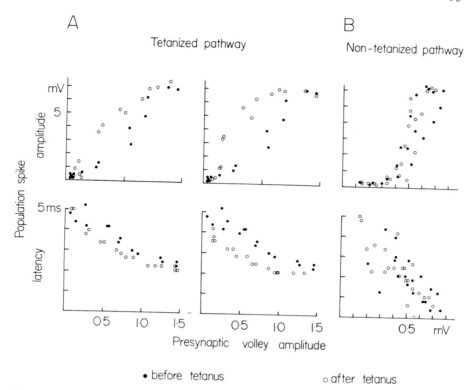

Fig. 4. Specific changes in synaptic transmission of tetanized input. A. Upper panels give relation between the amplitudes of the presynaptic volley in oriens fibres and the corresponding CA1 population spike before (filled circles) and after (open circles) a tetanus of 50 Hz, 5s delivered to oriens fibres. Left panel is taken 5 minutes and right panel 30 minutes after the tetanus. Upper panel in B gives corresponding values for stimulation of radiatum fibres which were not tetanized. Lower panels give the corresponding data for the relation between the prevolley size and the latency of the population spike. Filled circles give values before and open circles measurements after the tetanus. Automatic measurements.

of the population spike was increased and its latency decreased (Fig. 4). With unitary discharges, both the latency and the probability of discharge were increased. Finally, the EPSPs showed an increase corresponding to the increased probability and reduced latency of discharge.

Specificity of the induced changes

This point was tested by alternating stimulation of oriens and radiatum fibres and recording the responses of receiving cells as a function of the size of the appropriate presynaptic volley (Andersen *et al.* 1977). The input/

output curves for the population spike relative to the size of the presynaptic volley changed for the tetanized line (Fig. 4A), whereas no consistent change occurred in the non-tetanized line (Fig. 4B). Filled circles give the results before and open circles the values after the tetanus.

The measured changes applied both to the amplitude of the population spike, indicating the number of participating cells (Fig. 4A, upper row), and to the peak latency, indicating the average latency of discharge (lower row).

The changes were of long duration, commencing a few seconds after the tetanic stimulation and lasting for 30-75 minutes. Usually, however, the effect had an immediate peak after the tetanic period. This initial peak of heightened responses lasted about 1-2 minutes and was possibly related to the ordinary post-tetanic potentiation (Larrabee & Bronk 1947; Lloyd 1949; Eccles & Krnjević 1959; Gage & Hubbard 1966a, b). However, after this initial period, there was a long-lasting period with increased responses over those that a corresponding presynaptic volley created before the tetanus. These heightened responses with reduced latency showed a small decline, but responses well above control values were noted from 30 to 75 minutes. We have not attempted to measure the total duration of the effect.

Effect of the probability of discharge

When units were isolated and recorded from extracellularly, their probability of discharge was plotted as a function of the input volley before and for various times after a tetanus of 10-50 Hz for 5-10 seconds. Individual units showed an increased probability of discharge, as indicated by a smaller volley being necessary to drive the cell. The increased probability of firing was also associated with reduced latency as related to the size of the presynaptic volley. Again, the changes of long duration were associated with the tetanized line only; no changes were seen in the non-tetanized control line. Only just after the stimulus was it possible to see any changes in the control line, usually a reduced response. However, this situation was most intense during the post-tetanic potentiation period, and disappeared within two minutes, occasionally as late as five minutes. This decline was probably associated with an increased conductance, as indicated by other tests (see below).

Changes in the EPSP amplitude

With intracellular recording from cells in which it was possible to keep the membrane potential and resistance constant for sufficiently long control and experimental periods, a number of tests were made. By alternating

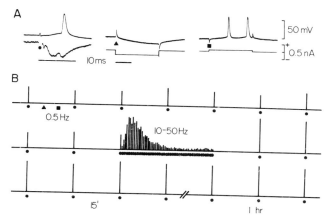

FIG. 5. Procedure used to test for long-lasting facilitation. A. Upper traces are, from left, orthodromic response of CA1 pyramid to stimulation of radiatum fibres (dot), hyperpolarizing current pulse (triangle) and depolarizing current pulse (square). Lower trace monitors field potential or current pulses. B. Diagram of the experiment which started with a control period in which the three stimuli were delivered in sequence. After the tetanus, the tests were repeated, although only the orthodromic stimuli are indicated. Larger responses for prolonged time.

orthodromic stimulation (Fig. 5, circles), hyperpolarizing pulses (triangles) and depolarizing pulses (squares) the effect on synaptic transmission and general excitability characteristics were measured before and after a tetanic period (Fig. 5B). At given intervals, input/output curves were made, relating the size of the EPSP to the presynaptic volley. This relation was changed after the tetanic stimulation (Fig. 6A, B). The change was less obvious than the change in the field EPSP. This was partly related to the greater variability of the EPSPs when they were plotted against the presynaptic volley (Fig. 6C, D). The nature of this variability is not known. However, the overall relationship of the EPSP amplitude to the presynaptic volley showed consistent and definite changes after the tetanic period. Likewise, the latency to the action potential was reduced in relation to the size of the presynaptic volley. Furthermore, the probability of an action potential emerging at a given size of the presynaptic volley was increased. All these changes were of long duration.

Lack of changes in the general excitability of the cell

A possible explanation of the increased responses to a given presynaptic volley is that the excitability of the membrane has been changed. However, this was found not to be the case as tested by depolarizing constant pulses

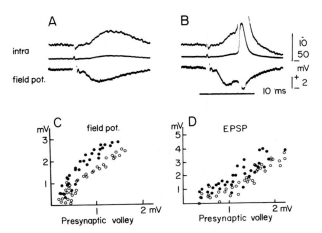

FIG. 6. Increased EPSP during long-lasting facilitation. A. Upper two traces give intracellular EPSP in CA1 pyramid in response to stimulation of radiatum fibres at low and high gain. Low trace gives response at synaptic layer. B. Response to same stimulus strength 15 minutes after tetanizing test line (50 Hz, 5s). C. Relation between the amplitudes of the presynaptic volley (abscissa) and the size of the field EPSP (ordinate) in the same experiment. Measurements made before and after the tetanization are given by open and filled circles, respectively. D. As C, but ordinate now gives intracellular EPSP.

across the membrane. The resulting depolarization of the membrane did not change (Fig. 7). Neither did the latency to the first directly evoked action potential change significantly, in spite of a clear change in the latency and probability of synaptic activation of the tetanized line. Thus, we could not find any evidence for a general increase in postsynaptic excitability to explain the observed results.

Another measure of the excitability is the membrane resistance. This was tested by delivering constant hyperpolarizing current pulses across the membrane. The membrane resistance did not show any long-lasting changes which could be associated with the increased synaptic transmission (Fig. 7, MR). However, during the tetanic stimulation and for 1–3 minutes immediately after, there was a considerable reduction in membrane resistance. However, in all cells this vanished, and after 1–3 minutes the input resistance was back to the pre-tetanic value. Thus, the increased postsynaptic responses after tetanic stimulation could not be explained by a changed membrane resistance. The latter observation of unchanged membrane excitability and membrane resistance is in contrast with the observations of Lynch et al. (1977) who found a reduced size of the population spikes in a non-tetanized line after tetanic stimulation and also a reduced reactivity to iontophoretically applied glutamic acid (Lynch et al. 1976).

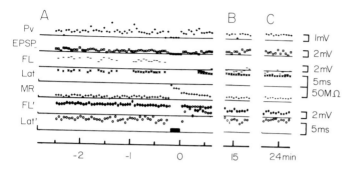

FIG. 7. Unchanged membrane excitability during long-lasting facilitation. A. The different lines illustrate the size of various parameters of a CA1 pyramid before, during and immediately after tetanization of radiatum fibres (50 Hz, 10s, heavy bar). Pv, presynaptic volley amplitude; EPSP, EPSP amplitude; FL, firing level of synaptically driven action potentials; Lat, latency of the same action potentials; MR, membrane resistance (distance of crosses from overlying line); FL', firing level of directly elicited action potentials; Lat', latency of such action potentials. B and C. As A, but taken 15 and 24 minutes after the tetanus.

DISCUSSION

Specificity of long-lasting facilitatory results

An important point is the finding that the increased synaptic transmission only applies to that line which previously has been tetanized. This is in accord with similar findings by Lynch et al. (1977), who found increased dentate population spikes with reduced latency in a tetanized line whereas the non-tetanized line showed not an improved, but a reduced transmission. The latter point could not be verified in the present experiment. Also in our study there was a short period immediately after the tetanized period where the responses were reduced, most likely due to the concomitant reduction of the input resistance as measured by hyperpolarizing pulses (Fig. 7). This reduced excitability was, however, of short duration (usually less than one minute). It showed itself not only by reduced membrane resistance pulses but also by a flattening of both the EPSP and IPSP, and also by a reduction and even blockade of the action potential. This result no doubt followed from the increased synaptic conductance imposed by the large number of impulses creating intense transmitter release during the tetanic stimulation.

Nature of the increased synaptic responses

Since the responses cannot be ascribed to a general change in the excitability of the membrane as indicated by the effects of constant depolarizing pulses, or to a changed membrane resistance, shown by the responses to hyperpolarizing current pulses, the increased EPSP with the augmented probability of firing and the reduced latency can be due either to an increased amount of transmitter released by the presynaptic volley or to a specific change in the intracellular current pathway. A possible explanation would be that a part of the current path is subject to a reduced resistance so that the same synaptic current injected peripherally could make itself better felt in the soma, where the spike most probably is initiated (Andersen *et al.* 1976).

Possible physiological role of the increased synaptic transmission

An increased efficiency of synaptic transmission after a short but intense series of afferent volleys would be a very useful way of modulating transmission in a cortical network. Since the duration and intensity of the increased synaptic transmission are a function of the number and frequency of the afferent stimuli, there seems to be a well-controlled mechanism by which the system could be made to work appropriately. The best effects were seen by stimulation at 50 Hz. However, it may well be that other stimulating ranges are better, since no systematic attempt was made to assess the best rates. Even 50 Hz seems to be in the high range if the effect is induced in normal circumstances. However, lower frequencies also were effective. Thus, clear effects were seen after stimulation at 10 Hz for 10 seconds. By recording from awake, behaving animals (O'Keefe & Dostrovsky 1971; O'Keefe 1976; Ranck 1973) it has been found that certain cells may fire in the region of 10–30 Hz for periods of more than 10 seconds. Thus, for individual axons it is possible that the stimulation parameters used may apply in physiological conditions. However, it is also possible that the reported results need the synchronous activation of a group of neighbouring fibres. Because this is a more demanding requirement, it is less likely to occur in physiological conditions.

One indication that the reported system is probably not used as rigorously as in the present experimental conditions in physiological situations is the fact that a short period of tetanic stimulation gave increased synaptic transmission for periods of several days (Bliss & Gardner-Medwin 1973). This applied to tetanic stimulation in awake rabbits where a reduced variability, increased population spike and reduced latency were dominant features. The extremely long duration after a short tetanic stimulation makes it difficult to couple the phenomenon to a particular physiological role. However, in physiological conditions, probably a smaller number of cells would be taking part in the afferent barrage, making the results probably less dramatic and shorter-lasting than the one in the Bliss and Gardner-Medwin experiments. In conclusion, the present experiments and those of several other groups have shown that short-lasting tetanic stimulation creates a situation with increased synaptic transmission for a long period of time. The effect may well occur in physiological situations. Because of its duration it might be a useful method for modulating the discharge pattern of hippocampal neurons.

References

ANDERSEN, P., BLISS, T. V. P. & SKREDE, K. K. (1971) Unit analysis of hippocampal population spikes. *Exp. Brain Res. 13*, 208–221

ANDERSEN, P. & LØMO, T. (1967) Control of hippocampal output by afferent volley frequency. *Prog. Brain Res. 27*, 400–412

ANDERSEN, P., SILFVENIUS, H., SUNDBERG, S. H. & SVEEN, O. (1976) Effects of remote dendritic synapses on hippocampal pyramids. *J. Physiol. (Lond.) 266*, 100P

ANDERSEN, P., SUNDBERG, S. H., SVEEN, O. & WIGSTRÖM, H. (1977) Specific long-lasting potentiation of synaptic transmission in hippocampal slices. *Nature (Lond.) 266*, 736–737

BLISS, T. V. P. & GARDNER-MEDWIN, A. R. (1973) Long-lasting potentiation of synaptic transmission in the dentate area of the unanaesthetized rabbit following stimulation of the perforant path. *J. Physiol. (Lond.) 232*, 357–374

BLISS, T. V. P. & LØMO, T. (1973) Long-lasting potentiation of synaptic transmission in the dentate area of the anaesthetized rabbit following stimulation of the perforant path. *J. Physiol. (Lond.) 232*, 331–356

BURNSTOCK, G., HOLMAN, M. E. & KURIYAMA, H. (1964) Facilitation of transmission from autonomic nerve to smooth muscle of guinea-pig vas deferens. *J. Physiol. (Lond.) 172*, 31–49

DEADWYLER, S. A., GRIBKOFF, V., COTMAN, C. W. & LYNCH, G. (1976) Long-lasting chances in spontaneous activity of hippocampal neurons following stimulation of the entorhinal cortex. *Brain Res. Bull. 1*, 1–7

DOUGLAS, R. M. & GODDARD, G. V. (1975) Long-term potentiation of the perforant path-granule cell synapse in the rat hippocampus. *Brain Res. 86*, 205–215

ECCLES, J. C. & KRNJEVIĆ, K. (1959) Presynaptic changes associated with post-tetanic potentiation in the spinal cord. *J. Physiol. (Lond.) 149*, 274–287

ECCLES, J. C., HUBBARD, J. I. & OSCARSSON, O. (1961) Intracellular recording from cells of the ventral spinocerebellar tract. *J. Physiol. (Lond.) 158*, 486–516

GAGE, P. W. & HUBBARD, J. I. (1966a) The origin of the post-tetanic hyperpolarization of mammalian motor nerve terminals. *J. Physiol. (Lond.) 184*, 335–352

GAGE, P. W. & HUBBARD, J. I. (1966b) An investigation of the post-tetanic potentiation of end-plate potentials at a mammalian neuromuscular junction. *J. Physiol. (Lond.) 184*, 353–375

LANDGREN, S., PHILLIPS, C. G. & PORTER, R. (1962) Minimal synaptic actions of pyramidal impulses on some alpha motoneurones of the baboon's hand and forearm. *J. Physiol. (Lond.) 161*, 91–111

LARRABEE, M. G. & BRONK, D. W. (1947) Prolonged facilitation of synaptic excitation in sympathetic ganglia. *J. Neurophysiol. 10*, 139–154

LLOYD, D. P. C. (1949) Post-tetanic potentiation of response in monosynaptic reflex pathways of the spinal cord. *J. Gen. Physiol. 33*, 147–190

LORENTE DE NÓ, R. (1934) Studies on the structure of the cerebral cortex. II. Continuation of the study of the Ammonic system. *J. Psychol. Neurol. (Lpz.) 46*, 113–177

LØMO, T. (1968) Nature and distribution of inhibition in a simple cortex (dentate area). *Acta Physiol. Scand. 74*, 8–9A

LYNCH, G. S., DUNWIDDIE, T. & GRIBKOFF, V. (1977) Heterosynaptic depression: a postsynaptic correlate of long-term potentiation. *Nature (Lond.) 266*, 737–739

LYNCH, G. S., GRIBKOFF, V. K. & DEADWYLER, S. A. (1976) Long term potentiation by a reduction in dendritic responsiveness to glutamic acid. *Nature (Lond.) 263*, 151–153

O'KEEFE, J. (1976) Place units in the hippocampus of the freely moving rat. *Exp. Neurol. 51*, 78–109

O'KEEFE, J. & DOSTROVSKY, J. (1971) The hippocampus as a spatial map. Preliminary evidence from unit activity in the freely-moving rat. *Brain Res. 34*, 171–175

RAMON Y CAJAL, S. (1893) Über die feinere Struktur des Ammonshornes. *Z. Wiss. Zool. 56*, 615–663

RANCK, J. B., JR (1973) Studies on single neurons in dorsal hippocampal formation and septum in unrestrained rats. Part I. Behavioral correlates and firing repertoires. *Exp. Neurol. 41*, 461–531

SCHWARTZKROIN, P. & WESTER, K. (1975) Long-lasting facilitation of a synaptic potential following tetanization in the *in vitro* hippocampal slice. *Brain Res. 89*, 107–119

SEARS, T. (1964) Investigations on respiratory motoneurones of the thoracic spinal cord. *Prog. Brain Res. 12*, 259–273

SKREDE, K. K. & WESTGAARD, R. H. (1971) The transverse hippocampal slice: a well-defined cortical structure maintained *in vitro. Brain Res. 35*, 589–593

Discussion

Weiskrantz: Is potentiation a general property of synapses stimulated in this way in the nervous system? Would a slice through the optic radiations to layer 4 of the striate cortex show such changes, or is there something special about the hippocampus? The assumption behind much of the work presented at this symposium, which is a perfectly reasonable assumption, is that the hippocampus is an organ with special proclivities for learning.

Andersen: It has been found in many systems—pyramidal tract synapses onto motoneurons; Ib fibres to spinocerebellar tract cells; Ia afferents onto respiratory motoneurons; and there are a lot of examples of frequency potentiation in the autonomic nervous system.

Segal: Can all pathways in the hippocampus be potentiated?

Andersen: The radiatum/CA1 synapse is the only one in which that has been established, but it is likely that the same applies to the perforant path/granule cell synapse, and the mossy fibres onto the CA3 cells; it certainly applies to both the oriens and radiatum fibres in CA1.

Gray: Has this been examined for the septal projections to the hippocampus, and not found?

Andersen: It has not been looked at for this connection with sufficient stringency.

Grossman: You are talking about long-lasting changes: what are the limits? What is long-lasting?

Andersen: It depends! With anaesthetized preparations this effect lasts 15–30 minutes. The less anaesthesia used, the longer it lasts. And if it has something to do with what Bliss & Gardner-Medwin (1973) found in the rabbit, it can last up to three days.

Koella: May we speculate a little bit more about the mechanism? I am sure you are aware of the newly discovered evidence for a presynaptic receptor system modulating transmitter release in synapses. Is it possible that after tetanization there is a kind of subsensitivity of the presynaptic receptors and in this way you would increase transmitter release in response to afferent impulses? As you don't see a change in the amplitude of the presynaptic volley, you probably can exclude a hyperpolarization of the presynaptic terminals as the cause of increased transmitter release.

Andersen: It is possible. We are trying to see whether glutamic acid, or increased potassium, do change the presynaptic volley. With a given input volley we see a change in EPSP, and the simplest explanation is that for unknown reasons there is an increased transmitter release. Another possible mechanism may be the facilitating calcium current of Heyer & Lux (1976) which may be important in a small-volume fibre such as those we have here. The extremely thin fibres (0.1-0.2 μm) could have difficulties with their mitochondria not being able to take up the calcium fast enough.

Another explanation relates to the synaptic current that is injected into the cell. It must pass through the spine and has to find its way up to the spine trigger zone, which probably is at the initial axon. If this current path decreases its resistivity it will give the observed effect. In a number of these spines there is a spinal apparatus, a set of bags with amorphous substance in between. If an osmotic change occurs here, water may be leaving the bags, which would shrink and reduce the resistance of the neck. The EPSPs would increase in amplitude.

Molnár: We made some observations on cats which may provide a partial answer to Dr Weiskrantz's question. When we investigated the modality of the hippocampal neurons, if we stimulated with single pulses they appeared to be uni-, or bimodal neurons; if, however, we stimulated with short bursts of impulses we observed multimodality in practically all of the recorded hippocampal units (Molnár & Arutyunov 1969).

Secondly, the activation of hippocampal neurons is long-lasting (Arutyunov & Molnár 1969; Molnár 1973), and this is similar to what Dr Vinogradova found in rabbits. This long-lasting activation was restricted to the hippocampal neurons, in our hands. I wonder if you can see a mechanism similar to your data on post-tetanic potentiation, Dr Andersen, behind this observation, made on a very different level?

Thirdly, I was excited to hear your data, because in another series of experiments we found that with certain lateral hypothalamic stimulations the animal (cat) seemed to 'remember' the direction of his stimulation-elicited movement for 45–60 seconds (which was the time lag by which the hippocampal units were activated) if tested by another stimulating electrode, normally eliciting locomotion of opposite direction (Grastyán *et al.* 1968; Molnár & Grastyán 1972). Is it possible that post-tetanic potentiation is responsible for the long-lasting activation of hippocampal neurons by short series of impulses of peripheral and central origin?

Andersen: It is not easy to take the concept of long-lasting potentiation and use it to explain the occurrence of a change in the whole neuronal network. It can only explain increased transmission in a single synapse. To explain the changes in a longer chain you have to study each element to see where the alterations come from.

Molnár: From the septum, probably?

Andersen: I'm afraid I cannot answer that.

Storm-Mathisen: In CA3/4, and probably also in CA1, there are axon collaterals going back into and mingling with the same population of cells that they originate from.

Andersen: In the response we see, a change occurs too fast to be due to recurrent collaterals. It seems to be a monosynaptic effect.

Storm-Mathisen: But you are talking about a long-lasting effect.

Andersen: Yes. Bliss & Lømo (1973) discussed whether the effect could be due to a self-reexciting process through interneurons. We tried to answer that by feeding in antidromic activation as a way of discharging cells, but it had no effect on the synaptic transmission. So it is unlikely to be a self-reexciting or a population effect. It looks as if it is coupled to events going on at the synaptic site.

Björklund: Have you tested the sensitivity of the pyramidal cells to different transmitters?

Andersen: No, but Dr Lynch has done that. He feels he can do it, while I think you can't. If you have a glutamate electrode within a sensitive area, which seems to be the synapses on the spines, it only needs to be moved 12–15 μm to give a 50% reduction of the effect. To me, it seems difficult to keep the distance to the receptor sufficiently constant over the necessary length of time. Neither is it possible, I think, to compare the effect of an iontophoretic injection over different cells.

Lynch: If you look immediately before or after the tetanus, which in our case can be less than 1 s, you can see a certain change in the excitation produced by iontophoretically applied glutamic acid. One can ask if this effect could last 10 minutes, as we did, but no longer.

I find this effect that you have with the synapses of equal strength fascinating,

Dr Andersen. In this system are you confident that you are not seeing dendritic spikes?

Andersen: We do see them, but we have to unmask them by hyperpolarizing the soma. Perhaps long-lasting potentiation is due to a better synchronization of dendritic spikes, but I don't think so. If you see the response to depolarizing pulses, you can find out how much you have to depolarize the soma before a spike occurs and compare that to how much you have to depolarize with a synapse to get the same spike. It appears to be exactly the same. You and Dr Björklund are asking whether there is a synaptic change. When we test the postsynaptic excitability it does not change, except for a minute or so after the tetanus. The latter observation may explain your finding, Dr Lynch. There is a tremendous conductance increase which takes a minute or two to wear off.

Lynch: The frequencies that produce that effect (the reduction in the responsiveness of the target cells to glutamate) are not the best ones for producing potentiation. But there may be several forms of plasticity inherent in this system, and potentiation may be only one of them.

Ranck: In all these frequency potentiation studies, not only is one firing cells at a particular frequency, but one is firing a group of cells synchronously whose axons run close together. Consequently, in translating these results into how it works in a normal animal, perhaps it requires not only that a single cell fires at a particular frequency but also that cells with synaptic endings adjacent to it are firing at the same frequency—if not synchronously, at least at about the same time. Indeed, many of the suggested mechanisms of potentiation which involve the accumulation or depletion of certain substances invoke that kind of event. This is an even more exciting feature of potentiation because it implies the interaction of a group of cells.

Andersen: That is very important, and David Marr (1969) has made a main point of that. Does the cell improve by delivering its discharge, or by the influence of its neighbours? I don't think the first possibility applies in this case because if we deliver the tetanus in the form of antidromic activation we do not see long-lasting potentiation. With a weak input which does not discharge the cell the EPSP still shows a long-lasting change. I think we can say that long-lasting potentiation does not depend on the discharge of the cell, but it does depend on the synaptic activity.

Gray: Has anyone looked to see whether long-lasting potentiation depends on protein synthesis? Many of the mechanisms that you have been talking about for the short-term effects seem implausible when it comes to long-lasting effects.

Lynch: Potentiation does not depend on protein synthesis, because we can block it with cycloheximide without blocking potentiation. We can maintain the slice for an hour in the drug, which is when we stop the experiment, because by then the slice has begun to be affected by the cycloheximide.

Gray: How then do Per Andersen's comments about mopping up the calcium apply for periods of an hour?

Andersen: The delayed calcium current is not as long as that, but it may take a long time to remove all intracellular calcium into the mitochondria. It has not been measured in the hippocampal neurons.

Storm-Mathisen: I don't think protein or peptide synthesis can be disregarded; nerve endings (synaptosomes) can synthesize proteins and this is increased on stimulation (Wedege *et al.* 1977; Hernández *et al.* 1976). Synaptic activity could also induce transmitter-specific peptide synthesis (Reichelt & Edminson 1977).

Ursin: Dr Andersen has been teasing the behaviourists by saying that memory is just a bag of water in a dendritic spine! We should return at some point to discuss the prerequisites we require from a model of memory. An increased response to one stimulus due to repetition of that same stimulus is by definition sensitization. You can hook it up to be part of a model of memory, but it requires some other conditions.

Weiskrantz: It is necessary but not sufficient for a model.

Andersen: It is not memory, of course. You can *use* it to construct models for memory, however.

Weiskrantz: One question that one would like to ask is whether you had frequency-specific changes.

Andersen: In intact animals there is an optimal frequency around 10–20 Hz.

Ursin: Since this is a system subjective to sensitization, does it matter in what order the stimulation is done? Do you have to stimulate in a random order, or can you use the same programme throughout?

Andersen: I don't know if it matters whether you do it randomly; I don't think anyone has asked that, because it is difficult enough as it is!

Rawlins: I wonder whether anyone has demonstrated frequency potentiation in free-moving animals without at the same time inducing seizure activity; is it well established that the elicitation of frequency potentiation is possible in unanaesthetized subjects without at the same time producing changes, such as after-discharges, which might be regarded as being pathological?

Gray: If you *can't* do this in a free-moving animal without producing seizures, it is not a good model of memory. It may be a good model of the way epilepsy develops.

Vinogradova: We work with anaesthetized rabbits and they show beautiful frequency potentiation and also long-lasting potentiation in certain systems of hippocampal synapses without any epileptic discharge in cellular and EEG activity or change in behaviour. Thus it is not a pathological phenomenon.

To return to the possible mechanism of long-lasting potentiation, which seems to be one of the most important for us, do you, Dr Andersen, think that some postsynaptic mechanism may also be involved in this phenomenon? What do you think of the data of Van Harreveld & Fifkova (1975)? The same results were obtained in our department by Dr Petrovskaya for postsynaptic spines on which the mossy fibre synapses terminate. The work was done with physiological control on potentiated and control hippocampal slices incubated *in vitro*, with the usual osmium tetroxide/glutaraldehyde fixation. There was a 40% increase in the width of the spines' stalks in potentiated slices. (The electron microscopic analysis was done blind: Bragin 1978). Might the change in the longitudinal resistance of the spines be important in long-lasting changes in the effectiveness of synapses?

Andersen: Absolutely; that is what I tried to say. However, I think one can object to these experiments, because they have not shown that the systems they looked at were potentiated. If you have a soma, dendrite and spine, what you see is the effect at the soma to an input at the dendrites. We can say that there is no great change in the postsynaptic excitability at the soma but it is definite that there is more synaptic current flowing at the dendritic spines. What we cannot say is whether there are any postsynaptic changes in the vicinity of the spines, because we have not the technique to look at that yet.

References

ARUTYUNOV, V. S. & MOLNÁR, P. (1969) Hippocampal neuronal activity in narcotized cats. *Bull. Acad. Sci. Georgian SSR 56*, 425–428

BLISS, T. V. P. & GARDNER-MEDWIN, A. R. (1973) Long-lasting potentiation of synaptic transmission in the dentate area of the unanaesthetized rabbit following stimulation of the perforant path. *J. Physiol. (Lond.) 232*, 357–374

BLISS, T. V. P. & LØMO, T. (1973) Long-lasting potentiation of synaptic transmission in the dentate area of the anaesthetized rabbit following stimulation of the perforant path. *J. Physiol. (Lond.) 232*, 331–356

BRAGIN, A. G. (1978) Posttetanic changes in ultrastructure of gigantic synapses in the hippocampal field CA_3. *Doklady Academii Nauk SSSR*, in press

GRASTYÁN, E., SZABÓ, I., MOLNÁR, P. & KOLTA, P. (1968) Rebound, reinforcement and self-stimulation. *Commun. Behav. Biol. Part A2*, 235–266

HEYER, C. B. & LUX, H. D. L. (1976). Properties of a facilitating calcium current in pace-maker neurones of the snail, *Helix pomatia*. *J. Physiol. (Lond.) 262*, 319–348

HERNÁNDEZ, A. G., LANGFORD, G. M., MARTINEZ, J. L. & DOWDALL, M. J. (1976) Protein synthesis by synaptosomes from the head ganglion of the squid, *Loligo pealli*. *Acta Cient. Venez. 27*, 120–123

MARR, D. (1969) A theory of cerebellar cortex. *J. Physiol. (Lond.) 202*, 436-470

MOLNÁR, P. (1973) The hippocampus and the neural organization of mesodiencephalic motivational functions, in *Recent Developments of Neurobiology in Hungary*, vol. 4 (Lissák, K., ed.), pp. 93-173, Akadémiai Kiadó, Budapest

MOLNÁR, P. & ARUTYUNOV, V. S. (1969) Hippocampal response to peripheral stimuli in anaesthetized cats. *Bull. Acad. Sci. Georgian SSR 56*, 661-664

MOLNÁR, P. & GRASTYÁN, E. (1972) The role of inhibition in motivation and reinforcement, in *Inhibition and Learning* (Boakes, R. A. & Halliday, M. S., eds.), pp. 401-430, Academic Press, London & New York

REICHELT, K. L. & EDMINSON, P. D. (1977) Peptides containing probable transmitter candidates in the central nervous system, in *Peptides in Neurobiology* (Gainer, H., ed.), pp. 171-181, Plenum Press, New York

VAN HARREVELD, A. & FIFKOVA, E. (1975) Swelling of dendritic spines in the fascia dentata after stimulation at the perforant fibers as a mechanism of post-tetanic potentiation. *Exp. Neurol. 49*, 736-749

WEDEGE, E., LUQMANI, Y. & BRADFORD, H. F. (1977) Stimulated incorporation of amino acids into proteins of synaptosomal fractions induced by depolarizing treatments. *J. Neurochem. 29*, 527-537

Cholinergic mechanisms and short-term potentiation

J. F. DEFRANCE, J. C. STANLEY, J. E. MARCHAND and R. B. CHRONISTER*

*Department of Neurobiology and Anatomy, The University of Texas Medical School at Houston and Department of Anatomy, University of South Alabama**

Abstract Acutely prepared rabbits were used to study, electrophysiologically, tetanic and post-tetanic potentiation of the pathway from the medial septal region to hippocampal field CA1. It was found that tetanic potentiation, evoked by short stimulus trains, was maximal at 6–8 Hz. Responses recovered from post-tetanic potentiation in 5–35 seconds.

Acetylcholine, physostigmine, and cyclic GMP each had an excitatory effect on pyramidal cell responses when applied in stratum radiatum. The time course studies showed that these effects outlasted the duration of the injection current by many minutes.

Phosphodiesterase inhibitors (e.g., isobutyl methyl xanthine) prolonged the time course of recovery with test responses which were post-tetanically potentiated. K^+, on the other hand, selectively enhanced tetanic potentiation.

It is suggested, with respect to the potentiation phenomena, that K^+ acted primarily presynaptically to facilitate transmitter release, whereas cyclic GMP acted primarily postsynaptically for the enhancement of pyramidal cell excitability.

One of the most interesting features of hippocampal physiology is the potentiation of evoked responses under certain conditions of stimulation. The potentiation can become long-lasting, to extend from a few seconds to many hours (Alger & Teyler 1976; Andersen & Lømo 1967; Andersen et al. 1961, 1977; Bliss & Lømo 1973; Izquierdo & Vasquez 1968; Lømo 1971). These observations have led to speculation that the changes that produce such potentiation might also be related to those involved in 'learning' (Andersen & Lømo 1967).

We report here certain aspects of hippocampal potentiation observed after medial septal stimulation, and discuss possible underlying mechanisms for the potentiation.

METHODS

Rabbits were acutely prepared under urethane or urethane-chloralose anaesthesia. The dorsal aspects of the hippocampal formation and septal region were exposed by removing the cortex and corpus callosum. This allowed for precise positioning of the stimulating and recording electrodes. Fig. 1A shows the stimulating and recording arrangement.

Monosynaptic responses were recorded in hippocampal field CA1 with microelectrodes after activation of: (1) the septal-hippocampal pathway which takes its origin primarily from the medial septal region (i.e., the medial septal nucleus and the nucleus of the diagonal band of Broca), and (2) the contralateral hippocampal field CA3 (CCA3). Stimulation of the medial septal region and CCA3 was done with microelectrodes (1–50 megohm).

To study the potentiation phenomena, single-barrel microelectrodes were used to record population excitatory postsynaptic potentials (EPSPs) and population spikes in CA1. To study the influence of various drugs on the normal and potentiated responses, multibarrel electrodes were used. This array included a recording barrel and a current-summing barrel in addition to the barrels containing the drugs. The compounds studied were: acetylcholine (0.1–0.5 M, pH 6.7), guanosine 3′:5′-monophosphate (cyclic GMP, 0.1–0.5 M, pH 6.0), atropine sulphate (0.1 M, pH 6.2), physostigmine (0.5 M, pH 6.5), isobutyl methyl xanthine (0.1 M, pH 6.5) and K^+Cl^- (0.1 M, pH 6.8). Theophylline was administered peripherally (2.0–6.0 mg/kg).

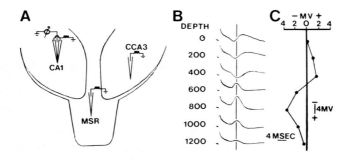

FIG. 1. A. Schema of experimental arrangement. Abbreviations: CA1, ipsilateral hippocampal field CA1; CCA3, contralateral hippocampal field CA3; MSR, medial septal region.
B. Field profile through CA1 after microstimulation of the medial septal region. In this and in subsequent figures depths are given in μm from the alvear surface.
C. Plot of response amplitude at 7.3 ms (line) after stimulation.

RESULTS AND DISCUSSION

Single stimuli (0.4–0.8 Hz) presented to the medial septal region characteristically generate field potentials in hippocampal field CA1 similar to those shown in Fig. 1B. At the alvear surface the dominant response is a positivity followed by a negativity. The positivity shows an apparent reversal between 400 and 600 μm to a negativity. This negativity is interpreted to be a sign of extracellular EPSPs. The negativity typically reached a maximum between 600 and 800 μm (stratum radiatum). The maximal response for the profile in Fig. 1C occurred at 800 μm. The negativity then diminishes in amplitude.

Low rates (0.4–0.8 Hz) of septal stimulation typically did not generate population spikes. When population spikes did occur, they were maximal, as might be predicted, in stratum pyramidale (400–500 μm).

Tetanic potentiation (response enhancement during a stimulus train) and post-tetanic potentiation (response enhancement after a stimulus train) of hippocampal responses have been demonstrated by a number of investigators (Andersen et al. 1961, 1977; Bliss & Lømo 1973). In our work, potentiation was produced and manipulated with a train of 16 stimuli. This number of stimuli was adequate to develop post-tetanic potentiation while reducing the time duration of the potentiation to the point where it could be more easily manipulated. Figs. 2 and 3 illustrate both tetanic and post-tetanic potentiation.

Fig. 2 illustrates the effects of various stimulus frequencies on the responses in stratum radiatum of CA1. In each case, 16 stimuli were employed. The record specimens are shown in Fig. 2A and maximum amplitudes plotted in Fig. 2B. Each record specimen is the average of the 16 responses. Also plotted are the control responses at 0.4 Hz taken just before the train of potentiating stimuli. The 16 stimuli at 0.4 Hz and the 16 stimuli at a given test frequency constituted a trial. The inter-trial interval was 120 seconds. This interval turned out to be especially critical to the development of post-tetanic potentiation. As can be seen in Fig. 2B, the control responses (0.4 Hz, solid lines) show little effect of the preceding trial. But, there was marked tetanic potentiation (interrupted lines). The U-shaped curve in Fig. 2B was typical for tetanic potentiation of CA1 responses to stimulation of the medial septal region. As the frequency of stimulation increases from control (0.4 Hz) to 6 or 8 Hz, the amplitude of the averaged response monotonically increases. In the instance shown in Fig. 2, the population EPSP increased uncontaminated by a population spike up to 2 Hz. At 4 Hz a population spike appeared and continued to develop through 8 Hz. After 8 Hz, there is a gradual decline in the population response. In the vast majority of the cases

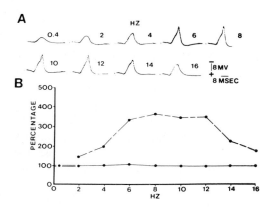

FIG. 2. Effects of frequency of stimulation of medial septal region on the CA1 response: 120 s inter-trial interval.
A. Series showing responses (average of 16) at the indicated stimulus frequency (0.4–16 Hz). In each case, 16 stimuli were delivered. The response at 0.4 Hz was taken to be the control (100%).
B. Plot of peak amplitudes in terms of percentage of control of the responses during tetanic stimulation (interrupted lines). Also plotted are the control values (solid lines) taken just before a tetanic stimulus trial. Depth was 609 μm.

examined, the maximal tetanic potentiated response occurred at either 6 Hz or 8 Hz. These observations are similar to those of Alger & Teyler (1976), using a hippocampal slice preparation. It is important to keep in mind that these observations relate to the use of relatively short stimulus trains, and involved a relatively small locus of excitation within the hippocampus.

As the inter-trial interval was decreased, the probability of the appearance of post-tetanic potentiation increased. Fig. 3 shows a series taken with an inter-trial interval of 30 seconds. The responses of Fig. 3 are predominantly population EPSPs. There was no indication of spike activity. Here, only frequencies up to 6 Hz were considered. Again, the maximum tetanic potentiation occurred at 6 Hz. But now, in contrast to the series with a 120 second interval (Fig. 2), there was a marked enhancement of the control test responses. After the 6 Hz train, the test response was 200% of control. Thus, we can operationally speak of two types of response enhancements: *tetanic* and *post-tetanic*. Whether or not the two forms of potentiation share common mechanisms is now the question of interest. The following experiments were addressed to the problem of mechanisms.

It has been proposed that potentiation phenomena reflect primarily presynaptic changes (Andersen *et al.* 1977). It has also been proposed that both presynaptic and postsynaptic changes are involved (Bliss & Lømo 1973).

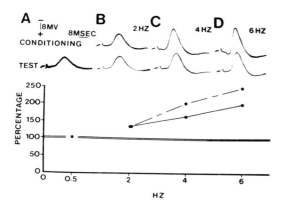

FIG. 3. Effects of frequency of stimulation of the medial septal region on the CA1 response: 30 s inter-trial interval.
A. Test control response at 0.5 Hz. B, C and D. Upper traces are the averaged responses at 2, 4 and 6 Hz, respectively. The lower traces in B, C and D are the averaged responses at 0.5 Hz taken 30 s after tetanic stimulation. Both tetanic (interrupted lines) and post-tetanic (solid lines) are plotted as percentage of control. All responses are an average of 16. Depth was 616 μm.

We sought clues as to the site of change by investigating the action of acetylcholine in stratum radiatum.

It has been well established by a number of diverse techniques that the transmitter in the septal-hippocampal junction is acetylcholine (Dudar 1975; Lewis & Shute 1967; Mathisen & Blackstad 1964; Mosko et al. 1973; Smith 1974; Szerb et al. 1977; Yamamura & Snyder 1974). It has been reported that the iontophoretic administration of acetylcholine can both enhance and depress spontaneously active pyramidal cell firing (Bird & Aghajanian 1976; Biscoe & Straughan 1966). Since the direction of effect is contingent upon, to some extent, the baseline rate of firing (Szabadi et al. 1977) as well as on complicated presynaptic effects of acetylcholine (Szerb et al. 1977), we chose to test the action of acetylcholine against the non-cholinergic input from CCA3. Both septal afferents and CCA3 afferents terminate in stratum radiatum of CA1. However, the available evidence indicates that a portion of the septal projection terminates immediately below the pyramidal cell layer, while the CCA3 input comes in more distally on the apical dendritic tree (Andersen et al. 1961; Mosko et al. 1973). Therefore, injections of acetylcholine were made in the proximal dendrite field.

The strategy used was to challenge a submaximal CCA3 response: one in which there was only a relatively small population spike. This approach seems best in allowing the direction of the response to follow the actual direction of the administered drug effect (DeFrance et al., in preparation).

Excitatory (or facilitatory) effects were easily demonstrable with acetylcholine application (Fig. 4). However, of real interest is the time course of the effect. As has been found in the cerebral cortex (Krnjević & Phillis 1963) and in the hippocampus (Biscoe & Straughan 1966), the excitation induced by acetylcholine is slow to develop and slow to decline, as opposed for example, to glutamate-induced excitation (Biscoe & Straughan 1966; Krnjević & Phillis 1963). This experimental series, illustrated in Fig. 4, was divided up into a number of 30 second epochs. At the start of each epoch, an average of eight responses was taken at a frequency of 0.5 Hz. The first three epochs (1.5 min) of the series are the controls before administration of acetylcholine. These baseline controls (established as the 100% level) were relatively stable. Acetylcholine was then injected with 26 nA for the next four epochs (2.0 min). As the injection proceeded, the negativity of the CCA3 response increased in amplitude. The response even continued to increase beyond the termination of the injection current. The CCA3 response finally achieved its maximum amplitude 2.0 min after the cessation of the injection current. The response remained at an elevated level for an additional 5 min, whereupon it fell to a point below control levels. This post-excitatory depression was usually present.

Hence, acetylcholine has a prolonged excitatory effect on pyramidal firing when applied in the proximal part of stratum radiatum. This excitatory effect could be antagonized by the administration of atropine, so there is every reason to believe that cholinergic mechanisms were being affected.

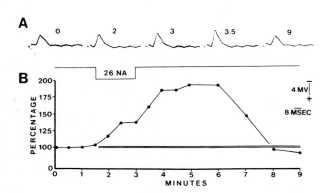

FIG. 4. Effects of acetylcholine in CA1.
A. Record specimens (average of eight responses) of responses to CCA3 stimulation: shown for epochs of time (t) = 0, 2, 3, 4.5 and 9 min of the series.
B. Plot of peak amplitudes as percentage of control (100%) at $t = 0$. The first eight responses of each epoch were the responses averaged. Depth was 600 μm.

The long time course for the excitatory effect of acetylcholine hints that cholinergic mechanisms might be involved in the potentiation process.

While acetylcholine facilitated the CCA3 response, so, somewhat surprisingly, did physostigmine. In a paradigm similar to that for acetylcholine, physostigmine applied to the proximal stratum radiatum induced a facilitation that was slow to develop and slow to decline (Fig. 5). Fig. 5A shows the CCA3 response at supramaximal stimulus intensity. Submaximal responses before, during and after the injection of physostigmine are shown in Fig. 5B. The peak amplitudes for the whole series are plotted in Fig. 5C. Responses, averaged eight times, were taken at times coincident with the start of each 30 second epoch. With the control responses showing stability in the 1.0 and 1.5 minute epochs, administration of physostigmine began in the 2 minute epoch. The CCA3 response increased monotonically throughout the injection, peaking at the end of the 2 minute injection period. The responses then began a gradual decline and eventually fell below control levels (Fig. 5C, 5 min).

The administration of physostigmine (a cholinesterase inhibitor) should be expected to increase local levels of acetylcholine, assuming that there is some minimal spontaneous input coming in via the septal-hippocampal pathway. The facilitation by acetylcholine of the CCA3 response in CA1 predicts that there should be heterosynaptic potentiation of the CCA3 response with septal conditioning responses. Heterosynaptic facilitation (or potentiation) has been demonstrated in other hippocampal pathways by Izquierdo & Vaśquez (1968). Furthermore, this facilitation is likely to

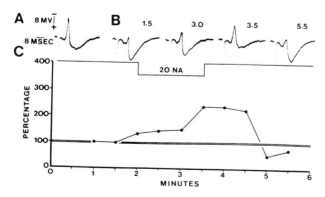

FIG. 5. Effects of physostigmine in CA1.
A. CA1 response to supramaximal CCA3 stimulation.
B. Record specimens of physostigmine series: shown are those for $t = 1.5, 3.0, 3.5$ and 5.5 min epochs.
C. Peak amplitudes plotted as percentage of control (100%, $t = 1$ min). Depth was 602 μm.

be of relatively long duration. The occurrence of heterosynaptic potentiation could mean that the potentiation by acetylcholine reflects postsynaptic changes. Acetylcholine, then, by increasing the excitability of the pyramidal cell, would enhance the probability of pyramidal discharge for a given CCA3 input.

Lee *et al.* (1972) found that cyclic GMP levels in cerebral cortex increase with the addition of acetylcholine. Recently, interest has developed in the possibility that cyclic GMP might function as a 'second messenger' for acetylcholine (Stone *et al.* 1975) in much the same manner as cyclic AMP functions for the catecholamines (Siggins *et al.* 1971; Segal & Bloom 1974).

Therefore, a logical next step was to evaluate the effects of cyclic GMP on CA1 pyramidal cells. The paradigm employed was the same as that for the studies of acetylcholine and physostigmine (Fig. 6). Fig. 6A shows the CCA3 evoked response at supramaximal intensity. The stimulus intensity was then reduced for testing (see Fig. 6B, $t=0$). Record specimens (each representing an average of eight responses) are shown for $t=0$ (before cyclic GMP), $t=1.5$, 2.5 (during cyclic GMP injection), and $t=7.0$ and 11.0 (after cyclic GMP injection). The peak amplitudes are plotted in Fig. 6C. In this series, the test response showed a rapidly developing facilitation. The test response was enhanced to 135% of the control response during the first epoch of injection. This rapid time-to-effect cannot be taken as typical, however, since there was considerable variation between different experiments, and between different injections with the same electrode. Nevertheless, facilitation of the CCA3 response was commonly observed. Suppression

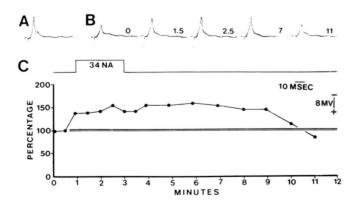

FIG. 6. Effects of cyclic GMP in CA1.
A. CA1 response to supramaximal CCA3 stimulation.
B. Record specimens of cyclic GMP series: shown are those at $t=0$, 1.5, 2.5, 7 and 11 min epochs.
C. Peak amplitudes plotted as percentage of control (100%, $t=0$). Depth was 600 μm.

of the CCA3 evoked response was not observed, except as a post-excitatory depression (see Fig. 6c, $t=11$). The duration of the cyclic GMP-induced excitatory effect is noteworthy. The test response was still elevated seven minutes after the termination of the injection.

In contrast to the effects of acetylcholine, simultaneous treatment or pretreatment with atropine failed to block the excitatory response.

The fact that the effects of cyclic GMP parallel those of acetylcholine fits well with the findings of Stone *et al.* (1975), where cyclic GMP paralleled the excitatory effects of acetylcholine at muscarinic synapses.

The above results indicate that facilitation of the CCA3 response by the administration of acetylcholine, physostigmine and cyclic GMP has a relatively prolonged time course. What possible implications might these observations have for mechanisms of the potentiation of the CA1 responses to trains of septal stimuli? Stimulation of CCA3 was used as the test to side-step the problems of interaction of acetylcholine and physostigmine with presynaptic cholinergic receptors (Szerb *et al.* 1977). In the course of this, the data obtained indicate that acetylcholine, physostigmine and cyclic GMP enhance pyramidal cell excitability. This is not to say that all, or even most, of the septal potentiation reflects postsynaptic hyperexcitability. Manipulation of cyclic GMP levels may, however, be a most sensitive test of the hypothesis that an increase in postsynaptic excitability is a factor in the observed potentiation. Theophylline, as a phosphodiesterase inhibitor, should act to elevate intracellular cyclic GMP levels.

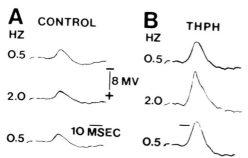

FIG. 7. Effects of theophylline (THPH) on responses to stimulation of the medial septal region.
A. Control series before theophylline. Shown are record specimens for the control response (0.5 Hz), the tetanic response (2.0 Hz), and the post-tetanic response (0.5 Hz, lower trace) at an inter-trial interval of 30 s.
B. Test series taken approximately 18 min after injection of 6 mg of theophylline. Shown are the control response (0.5 Hz), the tetanic response (0.2 Hz), and the post-tetanic response (0.5 Hz, lower trace) at a 30 s inter-trial interval. Line near the peak of the post-tetanic response indicates peak amplitude of control response to theophylline. All responses are an average of 16. Depth was 517 μm.

Fig. 7 shows the effects of theophylline on the septal responses in stratum radiatum of CA1. Fig. 7A is the control series taken before theophylline administration. The upper trace (0.5 Hz) is the control response, composed of extracellular EPSPs. The middle trace is the average of 16 responses at 2.0 Hz. The lower trace is the post-control, taken 30 seconds after the train. The series of responses in Fig. 7B was taken ~18 min after the injection of 6 mg theophylline. The pre-train control is clearly enhanced after administration of the inhibitor. At 2.0 Hz, the responses are facilitated to the point that a population spike appears in the average. Significantly, the post-train control shows marked post-tetanic potentiation. Systemic injection of theophylline, therefore, enhanced both the responses and potentiation.

When theophylline was injected systemically it was not easy to separate tetanic from post-tetanic effects. Some of the problems of systemic administration could be obviated by the iontophoretic administration of the phosphodiesterase inhibitor, isobutyl methyl xanthine (Fig. 8). Record

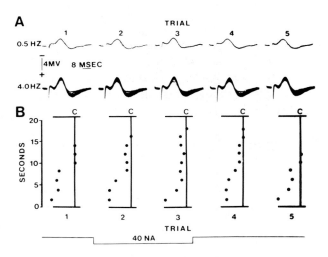

FIG. 8. Effects of isobutyl methyl xanthine on CA1 responses to stimulation of the medial septal region. The experiment was conducted in five trials. Each trial consisted of a control series (0.5 Hz), a tetanic series (16 stimuli at 4.0 Hz), and a post-tetanic recovery series (0.5 Hz).
A. Non-averaged record specimens of a control response (0.5 Hz) taken before tetanic stimulation for trials 1–5. The tetanic responses (4.0 Hz) are shown as the superposition of the 16 individual responses. Administration of isobutyl methyl xanthine was initiated 30 s before trial 2 and continued throughout trial 3. There was a 120 s inter-trial interval.
B. Plot of the time course to recovery (pre-tetany peak amplitude, C) for each trial. Along the vertical axis is the time (s) subsequent to the termination of tetanic stimulation. Along the horizontal axis is the post-tetanic response amplitude. The distance from C (control, 100%) to the end of the horizontal axis for each trial represents 200%. Depth was 740 μm.

specimens for the series are shown in Fig. 8A. Records at 0.5 Hz are the control responses before delivery of 16 stimuli at 4.0 Hz. The 16 responses are superimposed, not averaged. Trial 1 is pre-drug (control series); trials 2 and 3 are during the injection, and trials 4 and 5 are post-drug. The time course for the return of the response at 0.5 Hz to the control level is plotted in Fig. 8B. The typical time course for recovery varies from 5 to 35 seconds with these stimulation paradigms. This is similar to the findings of Alger & Teyler (1976) with short stimulus trains.

An examination of the control (0.5 Hz) responses and the potentiated (4.0 Hz) responses before, during, and after the application of isobutyl methyl xanthine reveals only little response enhancement due to this inhibitor. On the other hand, its application significantly protracted the time course for recovery of single stimuli at 0.5 Hz, as can be seen from a comparison of trials 2 and 3 with trials 1, 4 and 5. These data indicate that the time course to recovery is preferentially affected by isobutyl methyl xanthine after tetanic stimulation. Tetanic potentiation itself is less affected.

Taken together, these data indicate that acetylcholine, acting through an agent such as cyclic GMP, may have induced significant long-term facilitatory effects via postsynaptic action. The question of possible presynaptic components of tetanic and post-tetanic potentiation remains, however. There is substantial evidence that an enhancement in the efficacy of transmitter release contributes heavily to potentiation (Andersen et $al.$ 1977; Bliss & Lømo 1973). Moreover, it has been suggested by Izquierdo & Vaśquez (1968) that K^+ ions are important in potentiation. The following data support the notion that K^+ facilitates transmission presynaptically and this in turn assists in tetanic potentiation. It should be emphasized that in these experiments K^+ was applied iontophoretically with low ejection currents in stratum radiatum in conjunction with stimulation of the medial septal region. Fig. 9A shows the record specimens for the control (0.5 Hz) responses and tetanically potentiated (4.0 Hz) responses. In contrast to the effects of isobutyl methyl xanthine, K^+ usually enhanced both the amplitude of the population EPSP (Fig. 9A, Trial 2, 0.5 Hz) and the amount of tetanic potentiation (Fig. 9A, Trials 2 and 3, 4.0 Hz). Trial 3 is interesting in that the control response showed little enhancement of amplitude due to K^+ application, while the tetanically potentiated response was the largest of all the trials. In further contrast to the action of isobutyl methyl xanthine, K^+ does not prolong the duration of post-tetanic potentiation.

These results suggest that the accumulation of K^+ might be important in the development of tetanic potentiation. K^+ may enhance tetanic potentiation by increasing the probability of transmitter release (Gage & Quastel 1965;

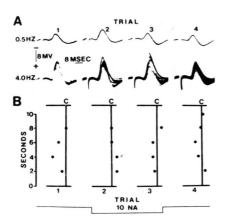

FIG. 9. Effects of K$^+$ on CA1 responses to stimulation of the medial septal region. The experiment was conducted in four trials. Each trial consisted of a control series (0.5 Hz), a tetanic series (16 stimuli at 4.0 Hz), and a post-tetanic recovery series.
A. Non-averaged record specimens of a control response (0.5 Hz) taken before tetanic stimulation. The tetanic responses are shown as the superposition of the 16 individual responses. K$^+$ administration commenced at the start of trial 2 and continued throughout trial 3. There was an interval of 120 s between trials 1 and 2 and between trials 3 and 4. Trials 2 and 3 were taken without interruption.
B. Plot of the time course to recovery (pre-tetany peak amplitude, C) for each trial. The distance from C (control, 100%) to the end of the horizontal axis for each trial represents 200%. Depth was 601 μm.

Parsons et al. 1965). The accumulation of intracellular cyclic GMP, on the other hand, may be more important in the development of post-tetanic potentiation and, consequently, in the long-term effects of repetitive stimulation of the septal input. This is not to claim that the two mechanisms are independent. Indeed, we have some reason to believe that the duration of post-tetanic recovery varies with the amount of facilitation. It is not unreasonable to expect this. But the fact that tetanic and post-tetanic potentiation may be operationally separated indicates to us that a single mechanism is not sufficient to account for the totality of observation.

ACKNOWLEDGEMENTS

The authors wish to thank Drs Y. Clement-Cormier and S. Strada for their advice, and for the isobutyl methyl xanthine. The assistance of Katherine Taber and Mrs Lillian Puccio is also cheerfully acknowledged.
This study was supported by NSF GB 35532.

References

ALGER, B. E. & TEYLER, T. J. (1976) Long-term and short-term plasticity in the CA1, CA3, and dentate regions of the rat hippocampal slice. *Brain Res. 110*, 463–480
ANDERSEN, P. & LØMO, T. (1967) Control of hippocampal output by afferent volley frequency. *Prog. Brain Res. 27*, 400–412
ANDERSEN, P., BRULAND, H. & KAADA, B. R. (1961) Activation of field CA1 of the hippocampus by septal stimulation. *Acta Physiol. Scand. 51*, 29–40
ANDERSEN, P., SUNDBERG, S. H., SVEEN, O. & WIGSTRÖM, H. (1977) Specific long-lasting potentiation of synaptic transmission in hippocampal slices. *Nature (Lond.) 266*, 736–737
BIRD, S. J. & AGHAJANIAN, G. K. (1976) The cholinergic pharmacology of hippocampal pyramidal cells: a microelectrophoretic study. *Neuropharmacology 15*, 273–282
BISCOE, T. J. & STRAUGHAN, D. W. (1966) Micro-electrophoretic studies of neurones in the cat hippocampus. *J. Physiol. (Lond.) 183*, 341–359
BLISS, T. V. P. & LØMO, T. (1973) Long-lasting potentiation of synaptic transmission in the dentate area of the anaesthetized rabbit following stimulation of the perforant path. *J. Physiol. (Lond.) 232*, 331–356
DUDAR, J. D. (1975) The effect of septal nuclei stimulation on the release of acetylcholine from rabbit hippocampus. *Brain Res. 83*, 123–133
GAGE, P. W. & QUASTEL, D. M. J. (1965) Dual effects of potassium on transmitter release. *Nature (Lond.) 206*, 625–626
IZQUIERDO, I. & VÁSQUEZ, B. (1968) Field potentials in rat hippocampus: monosynaptic nature and heterosynaptic post-tetanic potentiation. *Exp. Neurol. 21*, 133–146
KRNJEVIĆ, K. & PHILLIS, J. W. (1963) Iontophoretic studies of neurones in the mammalian cerebral cortex. *J. Physiol. (Lond.) 165*, 274–304
LEE, T. P., KUO, J. F. & GREENGARD, P. (1972) Role of muscarinic cholinergic receptors in regulation of guanosine 3′, 5′-monophosphate content in mammalian brain, heart muscle, and intestinal smooth muscle. *Proc. Natl. Acad. Sci. U.S.A 69*, 3287–3298
LEWIS, P. R. & SHUTE, C. C. D. (1967) The cholinergic limbic system: projections to the hippocampal formation, medial cortex, nuclei of the ascending cholinergic reticular system, and the subfornical organ and supraoptic crest. *Brain 90*, 521–540
LØMO, T. (1971) Potentiation of monosynaptic EPSPs in the perforant path-dentate granule cell synapse. *Exp. Brain Res. 12*, 46–63
MATHISEN, J. S. & BLACKSTAD, T. W. (1964) Cholinesterase in the hippocampal region. *Acta Anat. 54*, 216–253
MOSKO, S., LYNCH, G. & COTMAN, C. W. (1973) The distribution of septal projections to the hippocampus of the rat. *J. Comp. Neurol. 152*, 163–174
PARSONS, R. L., HOFMANN, W. W. & FIEGEN, G. A. (1965) Presynaptic effects of potassium ion on the mammalian neuromuscular junction. *Nature (Lond.) 208*, 590–591
SEGAL, M. & BLOOM, F. E. (1974) The action of norepinephrine in the rat hippocampus. I. Iontophoretic studies. *Brain Res. 72*, 79–97
SIGGINS, G. R., HOFFER, B. J. & BLOOM, F. E. (1971) Studies on norepinephrine-containing afferents to Purkinje cells of rat cerebellum. III. Evidence for mediation of norepinephrine effects by cyclic 3′,5′-adenosine monophosphate. *Brain Res. 25*, 535–553
SMITH, C. M. (1974) Acetylcholine release from the cholinergic septo-hippocampal pathway. *Life Sci. 14*, 2159–2166
STONE, T. W., TAYLOR, D. A. & BLOOM, F. E. (1975) Cyclic AMP and cyclic GMP may mediate opposite neuronal responses in the rat cerebral cortex. *Science (Wash. D.C.) 187*, 845–847
SZABADI, E., BRADSHAW, C. M. & BEVEN, P. (1977) Excitatory and depressant neuronal responses to noradrenaline, 5-hydroxytryptamine and mescaline: the role of baseline firing rate. *Brain Res. 126*, 580–583

Szerb, J. C., Hadhazy, P. & Dubar, J. D. (1977) Release of ^{3}H acetylcholine from rat hippocampal slices: effects of septal lesions and of graded concentrations of muscarinic agonists and antagonists. *Brain Res. 128*, 285–291

Yamamura, H. I. & Snyder, S. H. (1974) Post-synaptic localization of muscarinic cholinergic receptor binding in rat hippocampus. *Brain Res. 78*, 320–326

Discussion

Gray: On a point of technique: if you stimulate the septal area, you can produce evoked potentials in the hippocampus and driven theta waves by the same stimulus (James *et al.* 1977). How do you pick your part of the response out of the total response?

DeFrance: We see very little theta. We are using microstimulating techniques where there is, probably, a sphere of activation of 20–100 μm of tissue, and that is not sufficient to start theta. Earlier, before we used these microstimulating techniques, we were able to drive theta, but that produces an uninterpretable situation, when studying potentiation. But here the baselines (or controls) are stable, because of the absence of slow waves.

Gray: Do you always use urethane anaesthesia?

DeFrance: We use various anaesthetics: urethane, Nembutal and urethane-chloralose, and Surital–chloralose combinations, to protect ourselves against a systematic effect of the anaesthetic.

Segal: You imply that the pathway from the septum terminates in a laminated fashion in the radiatum. That doesn't correspond to the anatomy, from what we have heard. There is also a cholinergic innervation of stratum oriens.

DeFrance: My understanding is that on the pyramidal cell in CA1 there is a cholinergic input coming in on the proximal apical dendrites and a non-cholinergic input more distally.

Segal: I don't think one would see such a big difference in the innervation. In any case, I couldn't find such nice evoked potentials to stimulation of the septum in the rat (M. Segal, unpublished work 1976).

DeFrance: We use the rabbit, but I think there is very little species difference here. At least there is some cholinergic input in stratum radiatum, just below the pyramidal cells.

Srebro: Dr DeFrance recorded septal input in the CA1 field. As far as histochemical preparations for acetylcholinesterase are concerned, it appears that septal input in this field is relatively smaller and closer to the pyramidal layer than in the CA3 field.

Swanson: In the rat, most of the septal input to Ammon's horn is to field CA3, instead of field CA1.

Lynch: Yes; most of it is to the stratum oriens and most of it is to CA3. In the stratum radiatum of CA1 in rat the input from septum is quite diffuse and not the kind of projection expected to generate powerful responses.

DeFrance: The response of the medial septal region to microstimulation is atropine-sensitive, so we think we are working with a cholinergic input. We didn't show the effects of atropine, but we were able to block the effects of septal stimulation with atropine. The atropine effect itself is complicated because it can serve both as agonist and antagonist (because of the presence of presynaptic muscarinic receptors).

Storm-Mathisen: On the question of lamination, I think we may be deceived by the beautiful pattern we see in sections stained for acetylcholinesterase. For instance, in stratum radiatum of CA1, which has the lowest staining intensity, the activities of acetylcholinesterase and of the enzyme synthesizing acetylcholine (choline acetyltransferase) are no less than 30% of the highest value in the hippocampal formation (supragranular zone in area dentata). While our subjective impression of a stained section tends to be that of a black-and-white picture, microdensitometry agrees well with the graded distribution found on biochemical assay of microdissected samples (Storm-Mathisen 1970). On subcellular fractionation, the major part of the two enzymes is found in nerve ending particles (Fonnum 1970). We may therefore expect all parts of the pyramidal and granular cell dendrites to be influenced by cholinergic synapses.

Andersen: When acetylcholine is delivered onto the dendrites, there is excitation of a large proportion of the radiatum (Dingledine *et al.* 1978).

Segal: I also wonder about the pharmacology. I know there is the hypothesis of cyclic GMP as the second messenger of acetylcholine but I don't know how far it has been shown in the brain. I have tried myself to inject cyclic GMP iontophoretically, with little effect (M. Segal, unpublished work 1975). It has to cross the postsynaptic membrane, which is a problem with cyclic nucleotides. Theophylline and isobutyl methyl xanthine also have very potent effects towards calcium, and that might be why you got the effects you have. Another problem is how you would influence responsiveness to the pathway that is not actually recorded using iontophoresis, which should be a very local effect.

DeFrance: It's not so unreasonable. If we consider for argument's sake that the septal input that we are concerned with comes close to the soma in the CA1 pyramidal cell, and the contralateral CA3 input goes also to the apical dendrites (although some goes to oriens), we are saying that if we apply a small amount of acetylcholine high up in stratum radiatum, the effect of acetylcholine plus the effect of the contralateral CA3 input will equal the response that we see. This use of an indifferent pathway is for a proper interpretation. If we were to use the septal input with the acetylcholine application to challenge the input, what we might see, because of the presence of presynaptic receptors (which have recently been well characterized), is a suppression of the response. This is why we used an 'indifferent pathway' analysis, in conjunction with Dr Andersen's type of population EPSP and spike analysis. Furthermore, what you can do with such a pair of pathways (a tandem input) is to look at the interaction between the pathways and demonstrate a weak type of heterosynaptic facilitation, even using as few as 16 stimuli. This makes sense in a way, because the effects must be spatially summated. There is a spatial problem, so you have to deal with a space constant, so you would not expect a simple summation of the potentiations. You get a grading of potentiations.

Andersen: It is perhaps dangerous to draw too many inferences about synaptic events, because we don't know much about this. Kelly's group (Dingledine *et al.* 1978) has shown that when you drop acetylcholine on different parts of CA1 pyramidal cells, there is a clear depolarization. However, in contrast to most synapses, there was no conductance increase. So you would not expect this sort of competition you describe *during* the application. What they saw, strangely enough, was a resistance increase, which lags after the depolarization—something which could explain some of your increased field potential.

Lynch: We too have been interested in the idea that potentiation involves cyclic nucleotides and membrane phosphoproteins. We have found that repetitive stimulation of the type that produces potentiation causes changes in specific synaptic membrane proteins. However, of the particular phosphoproteins that are changed, the one that we are most interested in, which has a molecular weight of about 40 000, doesn't seem to be phosphorylated by cyclic nucleotides. This leads me to suspect that it is involved with calcium-sensitive kinases.

Gray: If we take Dr DeFrance's paper to indicate that the septal projection to the hippocampus is able to produce frequency potentiation as opposed to tetanic potentiation, and if we take it that the same pathway is being used as that which produces theta (and it would be difficult to suppose that it is *not*, or at least a closely related pathway), then (since we know that throughout the entire waking iife of the animal, whenever it is doing anything interesting it is busy broadcasting theta), this implies that the hippocampus is normally working under frequency potentiation and/or post-tetanic potentiation. I find that set of concepts difficult to grasp.

DeFrance: It seems as if the pyramidal cells like 6–8 Hz, irrespective of the pathway. We see effects in other pathways—the contralateral CA3 pathway, the subicular pathway—along with the septal pathway. We hoped we would see substantial differences, but we have not. It seems as if the pyramidal cell at CA1 likes the 6–8 Hz frequency.

Weiskrantz: We have been concentrating on the interpretation of the local effects and forgetting your claim that there is something unique about potentiation in this system, as against the nucleus accumbens.

DeFrance: I should add that we are using a similar potentiation paradigm in the nucleus accumbens using an 'indifferent input'. We record no potentiation and no long-term effects in response to the application of acetylcholine to neurons in the nucleus accumbens. So there seems to be something unique about the pyramidal cell, or about the synapses upon it, which gives it the ability to produce long-lasting changes in excitability, whether this is a translation of membrane resistance changes or some other changes in the membrane which make synaptic transmission more efficacious.

Weiskrantz: The general point that Dr Gray raised is the relationship between findings at this level and more general functional and behavioural questions.

Gray: I was suggesting that people have thought of frequency potentiation, coming in from the perforant path, as something that might happen occasionally in order to do something special, handle information in a particular way at a critical point in a behavioural or some other functional transaction. Whereas if the hippocampal response to septal inputs in the theta range of frequencies is also to be regarded as the same sort of phenomenon, it is going on all the time.

DeFrance: We are only talking about a single synapse and mechanisms surrounding the single synapse. There is a whole neural circuit to contend with and the U-shaped curve that you find for tetanic potentiation is not a curve generated only by that single synaptic mechanism, but is a circuit effect also. If one looks at only the population EPSPs and not population spikes, one doesn't see such a dramatic U-shaped curve. If we look only at population EPSPs at these frequencies, there is a build up between 6 and 8 Hz but almost no fall-off.

Andersen: A point we should remember is that when we do these kinds of experiments we excite many cells in parallel. In order to really believe that long-lasting potentiation has a physiological role, the behaviourists should demand that we can demonstrate the phenomenon with a single-cell input.

References

DINGLEDINE, R., DODD, J. & KELLY, J. S. (1978) Acetylcholine-evoked excitation of cortical neurones. *J. Physiol. (Lond.)*, in press

FONNUM, F. (1970) Topographical and subcellular localization of choline acetyltransferase in rat hippocampal region. *J. Neurochem. 17*, 1029–1037

JAMES, D. T. D., MCNAUGHTON, N., RAWLINS, J. N. P., FELDON, J. & GRAY, J. A. (1977) Septal driving of hippocampal theta rhythm as a function of frequency in the free-moving male rat. *Neuroscience 2*, 1007–1017

STORM-MATHISEN, J. (1970) Quantitative histochemistry of acetylcholinesterase in rat hippocampal region correlated to histochemical staining. *J. Neurochem. 17*, 739–750

General discussion I

MONOAMINERGIC INPUTS TO THE HIPPOCAMPUS

Björklund: The septo-hippocampal system receives its catecholamine inputs from three different sources: the noradrenergic neurons of the locus coeruleus; the noradrenergic neurons of the medullary A1 and A2 cell groups; and the dopaminergic neurons of the ventral tegmental area (group A10).

The locus coeruleus provides probably the sole catecholamine input to the hippocampal formation, and in addition it projects to both the medial and lateral septal nuclei (Ungerstedt 1971; Segal & Landis 1974; Jones & Moore 1977; Lindvall & Stenevi 1978). The axons reach the hippocampus via three routes: along the hippocampal fimbria and the dorsal fornix; around the corpus callosum via the cingulum bundle; and finally along a ventral route passing through the ventral amygdaloid bundle and the amygdalapiriform region (Lindvall & Björklund 1974; Swanson & Hartman 1975). According to Storm-Mathisen & Guldberg (1974) this latter ventral route is responsible for about 60% of hippocampal noradrenaline, and the dorsal route for about 40%. The densest termination of the locus coeruleus occurs in the hilar zone of the dentate gyrus and in stratum radiatum of CA3 (Blackstad *et al.* 1967).

A second noradrenergic input to the septo-hippocampal system originating in medulla oblongata has recently been identified by Lindvall & Stenevi (1978). These afferents terminate in both the medial and lateral septal nuclei. The dopaminergic innervation is confined to the lateral septal nucleus (Lindvall 1975; Lindvall *et al.* 1977; Lindvall & Stenevi 1978). These axons originate in the medial part of the mesencephalic A10 cell group and give rise to two types of innervation in the lateral septum: first, a dense aggregation of fine-varicose terminals outlining the fornix in the medial part of the lateral septal nucleus; secondly, terminals forming basket-like pericellular arrangements that are located mainly in the medial and ventral parts of the nucleus. Both the noradrenergic and the dopaminergic afferents reach the septum along the medial forebrain bundle.

Segal: Using autoradiography and/or horseradish peroxidase we found a heavy locus coeruleus innervation of the medial septum and less of the lateral septum (Segal & Landis 1974; Pickel *et al.* 1974; Segal 1976).

Weiskrantz: There is a question of the interpretation of density of staining, Dr Björklund; what does it tell us about the functional or quantitative strength of a projection?

Björklund: I don't think that quantitative relations are equal to importance; but the absence of a projection is a different matter. From an anatomical point of view, dopaminergic mechanisms occur in the septo-hippocampal system solely in the lateral septum. There is so far no anatomical evidence for such projections to the hippocampus or medial septum.

Robinson: Dr Björklund, concerning the noradrenergic input to the hippocampus, there appears to be some controversy about the route taken by noradrenergic fibres innervating the hippocampus. Using the glyoxylic acid fluorescence method you (Lindvall & Björklund 1974; also see Ungerstedt 1971) have reported that noradrenergic fibres reach the hippocampus both via the fornix system and caudally via fibres which course above the corpus callosum. However, using an immunofluorescence technique with dopamine-β-hydroxylase as a marker, Dr Swanson found no dopamine-β-hydroxylase-containing fibres in the fimbria (Swanson & Hartman 1975). He suggested that the dopamine-β-hydroxylase input to the hippocampus arises entirely from caudally directed fibres. Could you comment?

Swanson: Quantitatively, an adrenergic input through the fimbria, if present, is the least significant.

Björklund: One can reconcile our data by saying that the locus coeruleus axons running through the fimbria constitute the minority population. We are convinced that locus coeruleus axons pass through the fimbria and the dorsal fornix to reach the hippocampus.

Gray: Does that mean that the supracallosal projection is the strongest of all, or just that the fornix-fimbria projection is less powerful than the supracallosal or the ventral amygdalar ones?

Storm-Mathisen: If we cut all the fimbria and dorsal fornix input by a parasagittal cut (some of the fibres passing via the retrosplenial cortex might escape this lesion) we get about 50% loss of noradrenaline uptake and if we cut everything by a hemisection, we get over 90% loss (Storm-Mathisen & Guldberg 1974). I think that Swanson & Hartman (1975) suggested that there might be some dopamine terminals in the hippocampus. This could account for the apparent disagreement between our results and theirs, since such terminals would take up noradrenaline. Is there any more evidence for dopamine terminals in the hippocampus?

Björklund: Not in the hippocampus proper. The suggestion made by Dr Swanson about dopamine was based on the failure of dopamine-β-hydroxylase staining to reveal axons running in the fimbria. I think it is possible to explain this negative finding in other ways.

Swanson: There is some biochemical evidence that levels of dopamine may be 10-20% of the levels of noradrenaline in the hippocampus. The interpretation problem there is whether these are precursor levels of dopamine. It is not certain that there is no dopamine in the hippocampal formation, especially since cells in the ventral tegmental area can be retrogradely labelled after injections of horseradish peroxidase into Ammon's horn and the dentate gyrus (Wyss 1977).

Björklund: This is not entirely conclusive, I think. As you suggest, dopamine levels 10-20% of those of noradrenaline are in many cases found when dopamine is the precursor of noradrenaline.

Swanson: Many adrenergic fibres pass through the medial septal complex on their way over the genu of the corpus callosum, so it is difficult to know whether the staining in this region represents terminals or fibres of passage.

Segal: The fibres passing through are more ventral than those terminating in the medial septum itself. From our autoradiography data, it also looks as if the major pathways are in the supracallosal-cingulum area and not in the fornix.

Björklund: We have also made morphological studies of neuronal plasticity, using transplants of rat embryonic neural tissue to the hippocampus, from the locus coeruleus and substantia nigra (Björklund *et al.* 1976). The lesion removes entorhinal input to the dorsal part of the hippocampus, whereas other inputs, like the septal and commissural inputs, are intact. We remove the normal adrenergic innervation to the hippocampus at the time of transplantation by injecting 6-hydroxydopamine (250 μg intraventricularly). A locus coeruleus transplant gives a certain picture: the cells grow into the hippocampal formation, particularly dorsally, probably along routes normally used by perforant path fibres. The fibres end up in the hippocampal formation with a pattern reminiscent of the normal adrenergic innervation: the hilar zone gets the densest supply; the molecular layer gets the most scattered supply. The locus coeruleus axons thus seem to produce a relatively normal pattern.

If we take a dopamine transplant, the axons end up predominantly in the outer part of the dentate molecular layer, fitting quite nicely to the zone of the entorhinal perforant path fibres. There is also some extension into the hilar zone and a partial projection to the inner molecular layer. CA3 and CA1 receive only few fibres. Normally the hippocampus does not receive a dopaminergic projection, so the transplant creates an anomalous input.

In an attempt to see whether the hippocampus accepts the new fibres and whether the new fibres create functional contacts, we have, in collaboration with Dr Piers Emson in Cambridge (Stenevi *et al.* 1977) studied biochemical parameters of dopaminergic receptors. The dopamine-sensitive adenylate cyclase is in most cases related to the dopaminergic postsynaptic receptor. This cyclase is not normally demonstrable in the hippocampus and does not appear after lesion of the occipital cortex. However, after the grafting of the embryonic ventral mesencephalon to the occipital cortex, a dopamine-sensitive cyclase appears in the hippocampus in parallel to the ingrowth of dopaminergic fibres. This cyclase is probably postsynaptic, because in control rats, where we have removed the transplant after $1\frac{1}{2}$ months, the cyclase persists in the hippocampus.

From these experiments we suggest that postsynaptic elements in the hippocampus have a plasticity in the degree that they can adapt their receptive machinery to a new input. Perhaps this has some bearing on general plasticity problems in the hippocampus.

NICOTINIC TRANSMISSION IN THE HIPPOCAMPUS

Segal: The excitatory cholinergic septo-hippocampal path is commonly assumed to terminate on muscarinic receptors since antimuscarinic agents, such as atropine, readily antagonize excitatory responses to acetylcholine applied iontophoretically and also antagonize some forms of hippocampal theta rhythm. We have used the radiolabelled peripheral nicotinic receptor ligand, ^{125}I-labelled α-bungarotoxin (^{125}Iα-Btx), to characterize the distribution of nicotinic receptors in the rat brain (Segal *et al.* 1978). We first found that ^{125}Iα-Btx binds in the hippocampus preferentially to nicotinic receptors since its binding can be prevented by pretreating the tissue with nicotinic agonists (nicotine) or antagonists (*d*-tubocurarine) but not muscarinic agonists (oxotremorine) or antagonists (atropine) (Fig. 1). Nicotine is about 10 times more potent than acetylcholine as a nicotinic agonist. We then studied the distribution of α-Btx binding sites in the brain. Next to the hypothalamus (26 fmol/mg protein) the hippocampus is about the richest in the brain in α-Btx binding sites (25 fmol/mg protein). For comparison, the cerebellum contains only minute amounts of binding sites (<1 fmol/mg protein). Using autoradiography, we found that the nicotinic binding sites are concentrated in the dentate hilus and in the outer third of stratum oriens (Fig. 2).

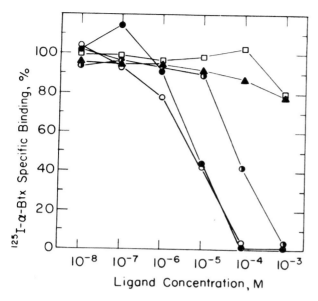

FIG. 1. (Segal). The effect of cholinergic ligands on specific $^{125}I\alpha$-Btx binding to hippocampal homogenate. Aliquots containing 0.2 mg protein were preincubated for 25 min at 25°C in buffer containing the appropriate ligand concentration. Reaction was started by addition of $^{125}I\alpha$-Btx (10 nM) and terminated 60 min later. ○---○, d-tubocurarine; ●---●, nicotine; ◐---◐, acetylcholine; □---□, atropine; △---△, oxotremorine. (From Segal et al. 1978.)

The question now is what the nicotinic receptors do in the hippocampus. Using microiontophoresis we find that acetylcholine excites bursting (pyramidal?) cells in the hippocampus and that atropine readily reverses this excitation (Fig. 3). Nicotinic antagonists are ineffective in this respect. I could detect a small population of cells, the non-bursting ('theta cells', interneurons?) type of cells which do not respond to acetylcholine (Fig. 3). When applied iontophoretically, nicotine reduces spontaneous activity of these neurons, without affecting the activity of the bursting cells. The suppressing action of nicotine is antagonized by d-tubocurarine (Fig. 4) or by gallamine. These results indicate the existence of excitatory muscarinic receptors on presumably pyramidal neurons and inhibitory nicotinic receptors on presumably interneurons. I then tested the possible contribution of the nicotinic receptors to the theta rhythm. Like carbachol, the intraventricular injection of nicotine produces theta rhythm (Fig. 5) in the awake rat. The intraventricular injection of d-tubocurarine reduces septally driven and spontaneous theta rhythm but not the rhythm associated with walking (Fig. 6).

In summary, there are nicotinic receptors in the rat hippocampus and they

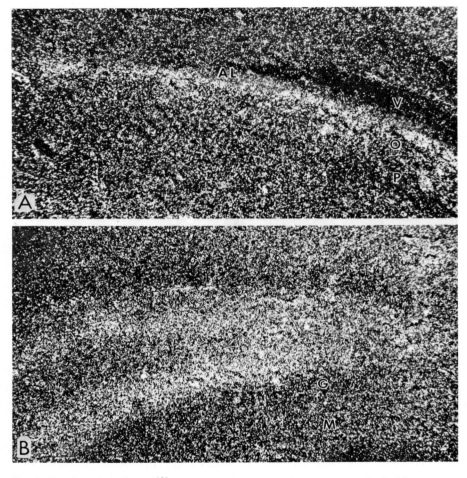

FIG. 2. (Segal). A. Labelling of ^{125}Iα-Btx in the outer third of stratum oriens in the hippocampus region CA1. Dark field photography. 63×.
B. Labelling of ^{125}Iα-Btx in the dentate hilus. AL, alveus; V, ventricle; O, stratum oriens; P, stratum pyramidale; G, stratum granulare; M, stratum moleculare. (From Segal et al. 1978.)

may have a different physiological role from that of the muscarinic receptor already known to exist in the hippocampus.

Grossman: Ross and I (Ross & Grossman 1974) have shown that intrahippocampal injections of nicotine have effects on behaviour in a number of test situations that provide measures of inhibitory control (such as DRL or Sidman avoidance) that are directly opposite to those of muscarinic compounds such as methylcholine. Nicotinic and muscarinic receptor blockers

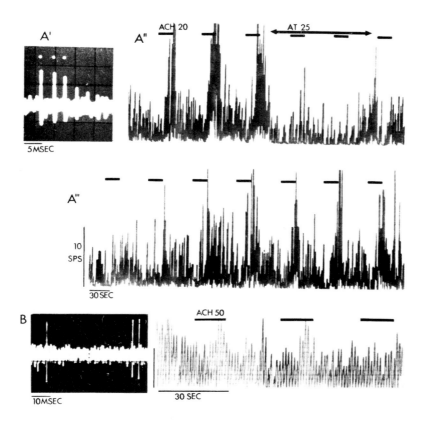

FIG. 3. (Segal). A. Excitatory responses of a bursting cell (A′) to the iontophoretic application of 20 nA pulses of acetylcholine (ACh, short bars) and the antagonizing of these responses by atropine (A″ and A‴ are a continuous, 1 second-integrated record of cellular activity).
B. The relative lack of effects of acetylcholine towards activity of a non-bursting cell. SPS, spikes per second.

(mecamylamine and scopolamine respectively) produced appropriate, opposite effects, suggesting that we were not merely recording non-specific reactions to one or the other of these compounds. (Whenever nicotine facilitated responding, the nicotinic receptor blocker mecamylamine, as well as the muscarinic compound methylcholine, produced inhibitory effects. Whenever methylcholine facilitated responding, the muscarinic receptor blocker scopolamine, as well as nicotine, resulted in a decrease in responding.) These results suggest that the hippocampal mechanisms that exert inhibitory influences over non-reinforced behaviour may have nicotinic as well as muscarinic

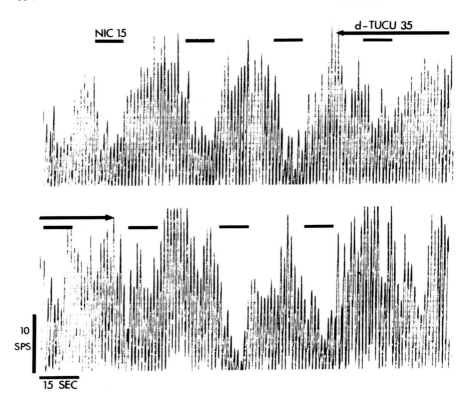

FIG. 4. (Segal). The antagonism of the inhibitory effects of nicotine (NIC, applied with a current of 15 nA) towards spontaneous activity of a non-bursting cell by d-tubocurarine (d-TUCU).

inputs that appear to have opposite influences on the output of the system.

Segal: If we compare the nicotinic with the muscarinic receptors (and work has been done on muscarinic receptors in the hippocampus: Kuhar & Yamahura 1976), it appears that the nicotinic receptor distribution fits the septo-hippocampal termination fields much better than the muscarinic receptor distribution.

Grossman: Blocking the nicotinic receptors in some parts of the hippocampus produces effects on non-reinforced behaviour that are quite similar to the changes seen after septal lesions (Ross & Grossman 1974).

Vanderwolf: I am not sure what one can conclude from an intraventricular injection, because atropine will abolish one kind of theta if you inject it systemically (intraperitoneally) but not if you inject it into the ventricle (Whishaw *et al.* 1976). With nicotine there also might be a problem of it

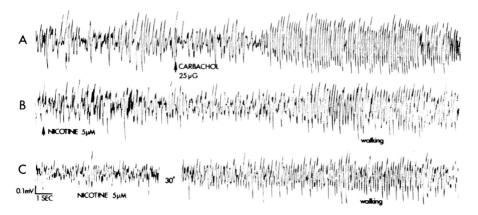

FIG. 5. (Segal). Encephalographic record from the dorsal hippocampus of a freely moving rat. The injection of carbachol (A) or nicotine (B and C) intraventricularly produces theta rhythm.

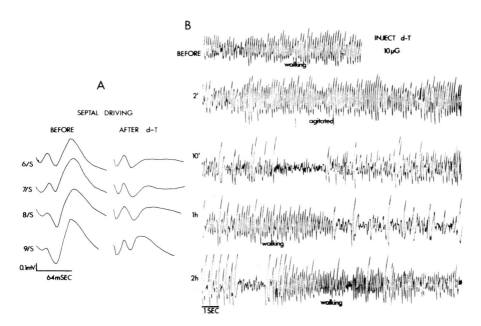

FIG. 6. (Segal). A. Averaged evoked hippocampal responses to medial septal stimulation applied at various rates before and after the intraventricular injection of d-tubocurarine (10 μg). Stimulation parameters, 0.5 ms pulses of 50 nA each applied at the specified rate. Each record is an average of 32 repetitions.

B. An encephalographic record of rat hippocampus before and at several time points after the intraventricular injection of d-tubocurarine.

entering the hippocampus. Whishaw et al. (1977) have injected nicotine and mecamylamine intravenously and see no blockade of the theta.

Segal: I wouldn't expect nicotine to do that, because it is an agonist. If you use nicotinic antagonists, you would expect some effects, but you would also block the neuromuscular junction and unless you used artificial respiration you would block the theta rhythms altogether.

Black: You said that *d*-tubocurarine and gallamine should block theta; but when they are injected systemically, they don't block it (Black *et al.* 1970; Black & Young 1972).

Segal: That is puzzling, indeed. There might be a concentration difference in that only a small amount of the drug gets into the hippocampus. It is still a problem; why do the nicotinic antagonists have such a small effect, in view of the wide distribution of the receptors?

Gray: Can anyone present a good case that the septal control of theta activity in the hippocampus is in fact cholinergic? I don't believe that such a case can be made at present. We know that the largest septal projection to the hippocampus is cholinergic; we know that the pacemaker cells controlling theta are in the medial septum. But, in my view, we don't know very much about the route followed by the fibres from the septum, or what transmitter they use in controlling theta. So we may be arguing whether the control is nicotinic or muscarinic before we know that it's cholinergic at all!

Weiskrantz: How would you be helped if you did know?

Gray: We should be helped in knowing whether to relate the results of certain drug studies to changes in hippocampal theta activity, in the way Professor Grossman mentioned, and in the way that Dr Carlton (Carlton & Markiewicz 1971), for example, has attempted to relate the effects on behaviour of anticholinergic drugs to changes in hippocampal function. It would be nice to know if that could be related to changes in hippocampal electrical activity.

Weiskrantz: Those drug effects can act over a complex network without necessarily implying anything about a specific transmitter between septum and hippocampus.

Vanderwolf: In my paper I shall be presenting some data relevant to this (see pp. 199–221).

Molnár: I am unable to answer Dr Gray's question, but let me mention some related information. One would like to know whether the inhibitory interneurons (the basket cells) in the hippocampus have direct or indirect afferentation from the septum, which is part of Dr Gray's problem. To state this in a somewhat different way: one would like to know, whether the pacemaker (or, rather, periodic filter: see my later comment, p. 424)

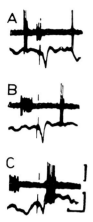

FIG. 1. (Molnár). Effect of spontaneous interneuron (probably basket cell) burst on the reactivity of hippocampal pyramidal unit to light flash. Cat anaesthetized with Nembutal (45 mg/kg. i.p.). *Upper trace:* CA3 units; *lower trace:* visual cortex. Calibration: 1mV and 80 ms.

function of the septum is realized through inhibitory synapses on the soma of the pyramidal cells or through excitatory synapses ending on inhibitory interneurons. In curarized cats we find cells which appear to be inhibitory interneurons (Molnár et al. 1971; Molnár 1973). In the cat dorsal hippocampus where—working on rats—Dr Ranck found theta cells, we found that quite frequently two kinds of spike activity were picked up simultaneously by the microelectrode, one bigger than the other. In the case documented here (Fig. 1) we were frustrated in our attempt to elicit evoked unit firing by light flashes. If we measured the latency of the larger spike activity from the end of the burst or smaller spike, however, we detected a remarkably stable latency. Our idea that the origin of the smaller spike might be the basket cells of dorsal hippocampus was substantiated pharmacologically. It is known that inhibitory interneurons of the spinal cord, the Renshaw cells, can be inactivated by chloralose (Haase & van der Meulen 1961). Using this analogy, when we gave the cats chloralose, the smaller spike activity was completely abolished and the inhibitory effect of septal impulses on pyramidal cell activity was also wiped out. This suggests that the inhibitory interneurons, or at least the presynaptic endings from the septum on these interneurons, might be sensitive to chloralose. If the Dale principle holds water, the appropriate synapse, on the spinal cord analogy, should be cholinergic.

SPECIFIC PATHWAYS BETWEEN SEPTUM AND HIPPOCAMPUS

Weiskrantz: Can we address ourselves to the problem of which specific pathways make which specific interconnections between septum and hippocampus, particularly fornix versus fimbria and their relations with medial versus lateral septum? There is also the question as to whether the CA1 output goes forward into fimbria or not.

Gray: On the last point, could Dr Andersen restate the dilemma posed by the physiological data as against the anatomical data, which Dr Vinogradova also mentioned earlier?

Andersen: The question is this: if we consider the two major groups of pyramidal cells, CA1 and CA3, we should like to know whether the output of CA1 is exclusively backwards towards the subiculum, or whether some CA1 pyramids project into the fimbria. Similarly, one would like to know whether all the CA3 cells project into the fimbria to the lateral septal nucleus and whether they all have a Schaffer collateral which projects back into CA1. Physiologically you can test this by stimulating the respective output fibres. You get very good antidromic invasion of CA1 cells after stimulation of the alveus in a strip running from the recording spot and back into the subiculum. The response stops abruptly at the CA1/CA3 border. Alveus stimulation does not give antidromic invasion of CA3 cells unless you happen to stimulate the Schaffer collaterals. Conversely, if you do stimulate the fimbria it is easy to activate all CA3 cells antidromically. Again, the response vanishes abruptly when the electrode is moved into CA1.

Because of the dramatic changes in antidromic field potentials when crossing the CA1/CA3 border, it is disturbing to hear that if you inject proline in CA1 you see projections through the fimbria. Can it be that the fibres pick up the amino acid at their nodes? I think I saw labelled CA3 cells in Dr Swanson's pictures.

Swanson: You could question the results of these autoradiographic studies. However, in this context, we have recently confirmed the pathway from field CA1 to the septum with horseradish peroxidase histochemistry.

Andersen: That doesn't answer the question, though, because the axons could come in other ways. Some of the CA1 axons which go to subiculum might go on to the septal region.

Swanson: That is conceivable, but such a pathway is not shown autoradiographically; a pathway through the fimbria *is* shown.

Andersen: Can we say anything about the proportion of the connections? It would be helpful for further investigations if we could say that CA1 cells *preponderantly* go to the subiculum and CA3 cells preponderantly go to fimbria.

Zimmer: You said that *all* CA3 cells could be driven by stimulating the ipsilateral fimbria; this implies that every CA3 cell sends an axon or an axon collateral through the fimbria.

Swanson: Yes, but they could be going through the commissure.

Zimmer: Just how many collaterals does a CA3 cell have? There is certainly one to the fimbria, and possibly one to the longitudinal association paths within CA3; another one goes to CA1 as a Schaffer collateral; then still another collateral goes further posteriorly to layer 4 of the entorhinal area. The collateral in the fimbria could go to the contralateral hippocampus as a commissural fibre or go to the septum, or both.

Andersen: The axon going into the fimbria goes on to the other side, and branches off to the lateral septum where it ends in an onion-shaped shell. The axonal branch that bends back, called the Schaffer collateral, ends abruptly (anatomically and physiologically) just at the border towards the pro-subiculum.

Zimmer: Does one CA3 cell have so many branches? Dr Swanson mentioned the possibility, by comparing the numbers of pyramidal cells in the hippocampus and the number in the fornix, that not all of them send their axons through the fornix. Are there different subpopulations of pyramidal cells in CA3?

Andersen: In rabbits we recorded from individual CA3 cells; 167/167 pyramidal cells were activated from the fimbria! Of those, 162 were activated from the properly placed electrode in the lamella, at the Schaffer collateral level. So, the vast majority of CA3 cells have branched axons. I am unable to establish the longitudinal association path, and maybe it does not exist?

Rawlins: The longitudinal association pathway may not have been demonstrated in rabbits, but I think it may be there in rats. Conversely the Schaffer system is perhaps not so well demonstrated in rats as it has been in rabbits (Rawlins & Green 1977); it might be that the relative importance of the two pathways is reversed between the two species.

Incidentally, if one is considering the number of axon branches from CA3 cells, Lugaro has been quoted (Hjorth-Simonsen 1973) as saying that they may give rise to up to six collaterals.

Andersen: Those data were derived from Golgi material, which is difficult to interpret, so we have to be cautious.

Srebro: I can only add to what Dr Swanson said that it is very difficult to confirm a projection from the CA3 field to the septum using horseradish peroxidase technique. In a large number of rat brains with intraseptal injections of horseradish peroxidase we could demonstrate many retrogradely labelled cells in the subiculum as well as the neighbouring pyramids of the

CA1 field but we were unable to detect any transport to the CA3 pyramids.

As far as the distribution of the cholinergic neurons in the septum is concerned, it might be of interest to comment here that a distribution pattern of acetylcholinesterase (AChE) in the human septum is very similar to that described in the rat brain (Harkmark et al. 1975). All major AChE-positive elements present in the rat septum can be found in the septal region of human brain. Similarly, there is a very close correspondence of the distribution pattern of AChE-positive elements in the hippocampal regions of the rat and human brain (Mellgren et al. 1977).

Zimmer: We have forgotten the question of whether there are cholinergic cells intrinsic to the hippocampus. At least at certain stages of development, and in transplants of hippocampal tissue to the cerebellum or the brainstem, there are cells in the hilus (CA4) that stain for acetylcholinesterase.

Srebro: After medial septal lesions in the rat it is possible to see in AChE preparations many raisin-like stained cells in the hilus of fascia dentata and along the pyramids in the regio inferior.

Segal: After lesions of the fornix, do they disappear?

Srebro: No, in fact these cells can be seen *more* clearly, because of disappearance of the neuropil stain. How much of this intracellular stain represents 'true' cholinergic neurons has yet to be established, bearing in mind Dr Storm-Mathisen's caution. Still, we are quite positive that this stain represents an intraneuronal acetylcholinesterase since we use a very short incubation time and, subsequently, counterstain our preparations with thionin.

Storm-Mathisen: What I said in my paper on cholinesterase representing cholinergic neurons in the hippocampus needs a qualification. If you cut the septo-hippocampal pathway you lose over 90% of the choline acetyltransferase, but only 80–85% of the AChE. The remaining AChE could be accounted for by the perikarya which show up in stained preparations. These are probably mostly non-cholinergic. There is only one place where some choline acetyltransferase and AChE appears to be left in the hippocampus after septal lesion, namely in part of the molecular layer of subiculum and CA1 ('site 31') (Storm-Mathisen 1972, 1974). That could be due to terminals of intrinsic cholinergic neurons, but these would only account for a very small proportion of the total choline acetyltransferase in the hippocampal formation.

Vinogradova: Basket cells are extremely AChE-positive, and their staining completely disappears after connections from the septum have been interrupted.

Srebro: No; as I said before, the AChE-positive cells show up more clearly after the lesions on a pale background around the pyramidal cell layer and in the hilus.

Vinogradova: That is strange, because we have studied the de-septalized hippocampus with AChE staining. In the basket cells AChE-positive staining disappears after de-septalization in the rabbit (Kultas *et al.* 1974).

Srebro: Unfortunately, we have no histochemical preparations after the septal lesions in the rabbit, so I don't know the answer.

Gray: I would like to have some idea of the quantitative importance of the projection from the hippocampus to the medial septum, as distinct from the lateral septum, and the quantitative significance of the projection that travels through the fornix, as distinct from through the fimbria, to the hippocampus from the septum. I would also like an idea of the quantitative significance of the non-cholinergic projection from the septum to the hippocampus, so that we can know, in considering behavioural studies, which of these inputs and outputs between the septum and hippocampus are really important.

DeFrance: On the first point, and in defence of an anterior projection from CA1 to septal nuclei, recently, for other reasons, we placed a microstimulating electrode in the stratum oriens of CA1 and also a microstimulating electrode in stratum radiatum. Then we stimulated and recorded through the medial septal region. Actually we were looking for antidromic responses, but what we found is that stimulation of stratum oriens and perhaps the alveus evoked responses which suggest that there is an input into the medial septal region, whereas stimulation of stratum radiatum did not necessarily do so. So our finding is not merely a backfiring of CA3 neurons projecting into the dorsoseptal area.

Andersen: Are you sure this is not antidromic activation, or activity by collaterals after antidromic activation?

DeFrance: They are spatially distinct. In the dorsal septal region we find primarily monosynaptic responses. If they are activated by recurrent collaterals, what must be happening is that we are backfiring neurons in the diagonal band which are giving off ascending collaterals. That is possible but not probable. A better solution is to say that there is a small number of fibres projecting anteriorly from CA1 and they may not be the same cells, but we consistently find monosynaptic responses both in the medial part and in the lateral part of the dorsal septal region.

References

BJÖRKLUND, A., STENEVI, U. & SVENGAARD, N.-A. (1976) Growth of transplanted monoaminergic neurones into the adult hippocampus along the perforant path. *Nature (Lond.) 262*, 787–790

BLACK, A. H. & YOUNG, G. A. (1972) Electrical activity of the hippocampus and cortex in dogs operantly trained to move and to hold still. *J. Comp. Physiol. Psychol. 79*, 128–141

BLACK, A. H., YOUNG, G. A. & BATENCHUK, C. (1970) Avoidance training of hippocampal theta waves in flaxedilized dogs and its relation to skeletal movement. *J. Comp. Physiol. Psychol. 70*, 15–24

BLACKSTAD, T. W., FUXE, K. & HÖKFELT, T. (1967) Noradrenaline nerve terminals in the hippocampal region of the rat and the guinea pig. *Z. Zellforsch. Mikrosk. Anat. 78*, 463–473

CARLTON, P. L. & MARKIEWICZ, B. (1971) Behavioral effects of atropine and scopolamine, in *Pharmacological and Biophysical Agents and Behavior* (Furchtgott, E., ed.), pp. 346–374, Academic Press, New York

HAASE, J. & VAN DER MEULEN, J. P. (1961) Die spezifische Wirkung der Chloralose auf die recurrente Inhibition tonischer Motoneurone. *Pflügers Arch. Ges. Physiol. 274*, 272–280

HARKMARK, W., MELLGREN, S. I. & SREBRO, B. (1975) Acetylcholinesterase histochemistry of the septal region in rat and human: distribution of enzyme activity. *Brain Res. 95*, 281–289

HJORTH-SIMONSEN, A. (1973) Some intrinsic connections of hippocampus in the rat: an experimental analysis. *J. Comp. Neurol. 147*, 145–161

JONES B. E. & MOORE, R. Y. (1977) Ascending projections of the locus coeruleus in the rat. II. Autoradiographic study. *Brain Res. 127*, 23

KUHAR, M. J. & YAMAMURA, H. I. (1976) Localization of cholinergic muscarinic receptors in rat brain by light microscopic radioautography. *Brain Res. 110*, 229–243

KULTAS, K. N., SMOLIKHINA, T. I., BRAZHNIK, E. S. & VINOGRADOVA, O. S. (1974) The effect of septal afferent lesion upon the acetylcholinesterase activity in the short-axon neurons of the hippocampus. *Doklady Academii Nauk. 216*, 462–463 (in Russian)

LINDVALL, O. (1975) Mesencephalic dopaminergic afferents to the lateral septal nucleus of the rat. *Brain Res. 87*, 89

LINDVALL, O. & BJÖRKLUND, A. (1974) The organization of the ascending catecholamine neuron systems in the rat brain, as revealed by the glyoxylic acid fluorescence method. *Acta Physiol. Scand.*, Suppl. 412, 1–48

LINDVALL, O. & STENEVI, U. (1978) Dopamine and noradrenaline neurons projecting to the septal area in the rat. *Cell Tissue Res.*, submitted

LINDVALL, O., BJÖRKLUND, A. & DIVAC, I. (1977) Organization of mesencephalic dopamine neurons projecting to neocortex and septum. *Adv. Biochem. Psychopharmacol. 16*, 39–46

MELLGREN, S. I., HARKMARK, W. & SREBRO, B. (1977) Some enzyme histochemical characteristics of the human hippocampus. *Cell Tissue Res. 181*, 459–471

MOLNÁR, P. (1973) The hippocampus and the neural organization of mesodiencephalic motivational functions, in *Recent Developments of Neurobiology in Hungary*, vol. 4 (Lissák, K., ed.), pp. 93–173, Akadémiai Kiadó, Budapest

MOLNÁR, P., ARUTYUNOV, V. S. & NARIKASHVILI, S. P. (1971) Extracellular unit analysis of pyramidal and a kind of interneuronal (probably basket cell) activity in the dorsal hippocampus of the cat. *Acta Physiol. Acad. Sci. Hung. 40*, 387–391

PICKEL, V., SEGAL, M. & BLOOM, F. E. (1974) A radioautographic study of the efferent pathways of the nucleus locus coeruleus. *J. Comp. Neurol. 155*, 15–41

RAWLINS, J. N. P. & GREEN, K. F. (1977) Lamellar organisation in the rat hippocampus. *Exp. Brain Res. 28*, 335–344

ROSS, J. F. & GROSSMAN, S. P. (1974) Intrahippocampal application of cholinergic agents and blockers: effects on rats in DRL and Sidman avoidance paradigms. *J. Comp. Physiol. Psychol. 86*, 590–600

SEGAL, M. (1976) Brainstem afferents to the rat medial septum. *J. Physiol. (Lond.) 261*, 617–631

SEGAL, M. & LANDIS, S. C. (1974) Afferents to the septal area of the rat studied with the method of retrograde axonal transport of horseradish peroxidase. *Brain Res. 82*, 263–268

SEGAL, M., DUDAI, Y. & AMSTERDAM, A. (1978) Distribution of an α-bungarotoxin-binding cholinergic nicotinic receptor in rat brain. *Brain Res.*, in press

STENEVI, U., EMSON, P. & BJÖRKLUND, A. (1977) Development of dopamine sensitive adenylate cyclase in hippocampus reinnervated by transplanted dopamine neurons: evidence for new functional contacts. *Acta Physiol. Scand.*, Suppl. 452

STORM-MATHISEN, J. (1972) Glutamate decarboxylase in the rat hippocampal region after lesions of the afferent fibre systems. Evidence that the enzyme is localized in intrinsic neurones. *Brain Res. 40*, 215–235

STORM-MATHISEN, J. (1974) Choline acetyltransferase and acetylcholinesterase in fascia dentata following lesion of the entorhinal afferents. *Brain Res. 80*, 181–197

STORM-MATHISEN, J. & GULDBERG, H. C. (1974) 5-Hydroxytryptamine and noradrenaline in the hippocampal region: effect of transection of afferent pathways on endogenous levels, high affinity uptake and some transmitter-related enzymes. *J. Neurochem. 22*, 793–803

SWANSON, L. W. & HARTMAN, B. K. (1975) The central adrenergic system. An immunofluorescence study of the location of cell bodies and their efferent connections in the rat utilizing dopamine-β-hydroxylase as a marker. *J. Comp. Neurol. 163*, 467–505

UNGERSTEDT, U. (1971) Stereotaxic mapping of the monoamine pathways in the rat brain. *Acta Physiol. Scand.*, Suppl. 367

WHISHAW, I. Q., ROBINSON, T. E. & SCHALLERT, T. (1976) Intraventricular anti-cholinergics do not block cholinergic hippocampal RSA or neocortical desynchronization in the rabbit or rat. *Pharmacol. Biochem. Behav. 5*, 275–283

WHISHAW, I. Q., BLAND, B. H. & BAYER, S. A. (1977) Postnatal hippocampal granule cell agenesis in the rat: effects on two types of rhythmical slow activity (RSA) in two hippocampal generators. *Brain Res.*, in press

WYSS, J. M. (1977) Hypothalamic and brainstem afferents to the hippocampal formation in the rat. *Abstr. Soc. Neurosci. 3*, 209

Neuronal aspects of septo-hippocampal relations

O. S. VINOGRADOVA and E. S. BRAZHNIK

Department of Memory Problems, Institute of Biophysics, Puschino-on-Oka, USSR

Abstract In unanaesthetized, conscious rabbits, in unstressful conditions, the neurons of the hippocampus and septum were investigated extracellularly during the presentation of a series of varied sensory stimuli. In the *normal hippocampus* these stimuli evoke habituating reactions of tonic (more usually, inhibitory) type in field CA3, with the addition of 'specific' patterned, and phasic reactions in field CA1. After complete *septo-hippocampal disconnection* the proportion of tonic (especially, of inhibitory) reactions in the hippocampus decreases. Theta bursts in the neuronal activity are absent; reactions to repeated sensory stimuli do not habituate. After *lesion of the cortical perforant path* to the hippocampus the majority of reactions in both fields are of tonic type. The proportion of neurons with regular theta bursts increases. Habituation is completely absent. A high correlation appears between the sensory reactions and the effects of midbrain reticular formation stimulation in the same neurons. *The combination of both lesions* does not significantly change the spontaneous activity of hippocampal neurons (except for the absence of the theta bursts). An increase in the level of activity of hippocampal neurons (by physostigmine), or rhythmic stimulation of the remaining synaptic systems, does not restore their rhythmic theta activity.

In *the septum deprived of hippocampal input* the normal level of reactivity to sensory stimuli and the normal types of reaction are preserved. The proportion of neurons with theta bursts increases. The typical linear and rapid habituation of reactions disappears and is replaced by an unlimited increment in effects during repeated presentations of sensory stimuli.

Discussion concerns the synchronizing and inhibitory influences of the septum on the hippocampus, and the role of the hippocampus in the organization of decremental processes (habituation) in the septum and brainstem structures.

An understanding of septo-hippocampal relations is necessary for approaching many problems of modern neuropsychology and neurophysiology. The

© USSR Copyright Agency.

complex interconnections between these two structures, and the position of the septum as a cross-roads for many important forebrain pathways, make it difficult to approach their general functional significance with the usual methods of lesion and stimulation.

Though some differences exist between the behavioural effects of septal and hippocampal lesions, many authors also recognize common consequences of these treatments (Beatty & Schwartzbaum 1968; Clody & Carlton 1969; Duncan & Duncan 1971; Glick et al. 1974). At least in part this depends on unavoidable damage to important afferent and efferent paths to one structure during lesion of the other. Some effects of presumed septal lesions may result also from concomitant damage of other topographically close, but functionally separate systems (Turner 1970; Paxinos 1975; Ursin 1976). The same problem affects experiments on electrical and chemical stimulation of the septum, where current spread or diffusion may involve various structures not related to the septum itself. Thus, further analytical approaches are necessary in investigating the septum and hippocampus, if one is to appreciate their specific functions and interactions.

One of the most intriguing aspects of septo-hippocampal relations is the nature of the theta rhythm and its so-called 'behavioural correlates'.

Though the main ideas about the origin of the hippocampal theta rhythm were developed in the sixties, discussion of this problem still continues. Some authors have suggested that the hippocampal theta is an intrinsic phenomenon, based on recurrent inhibition by the basket interneurons (Andersen et al. 1964; Spencer & Kandel 1961), while others have shown that hippocampal EEG theta activity depends on the integrity of septo-hippocampal connections (Petsche & Stumpf 1960; Stumpf et al. 1962; Parmeggiani 1967; Poletaeva 1968; Thomson & Laget 1966; Nikitina & Boravova 1972). Several more recent studies have supported this idea by demonstrating a high correlation between septal unit discharges and the hippocampal EEG and only weak and unstable correlation between hippocampal unit discharges and EEG activity (Gogolák et al. 1968; Macadar et al. 1970; Morales et al. 1971). Nevertheless even the authors of the concept of the septal origin of hippocampal theta activity have accepted that recurrent inhibition participates in the organization of the theta waves (Petsche et al. 1965). Recently, McLennan & Miller (1974, 1976) have suggested that the theta bursts of the septal neurons depend on hippocampal influences. Thus, even on this basic point of septo-hippocampal relations we are still far from a clear understanding.

The same is true for the state of the hippocampal neurons during EEG theta activity. Some authors, mainly working at the behavioural level, suggest that concomitantly with the theta state, hippocampal neurons are invariably activated (see, for example, Vanderwolf 1971), while others, more often electrophysiologists, think that hippocampal neuronal activity may be inhibited in this state (Shaban 1970; Artemenko 1972; Franzini *et al.* 1975).

In spite of the lack of answers to these important questions the current literature presents a wide variety of speculations about the behavioural significance of the theta state as the correlate of attention, motivation, voluntary movement, memory and learning. As a result, the dependence of the theta rhythm on the septo-hippocampal system implies the corresponding functional significance of these structures.

However, it seems to us that this kind of reasoning, implying a direct correlation between complex behavioural phenomena and particular rhythmic characteristics of relatively circumscribed neuronal populations, is basically wrong. If any kind of general EEG activity reflects the sum of activating and inhibitory influences on a given neuronal population, it is important to decide how the state of the neurons changes, whether their output increases or decreases in this state, or whether some other kind of reorganization, without a change in the level of the output signal (e.g. synchronization of discharge), takes place during a certain EEG pattern. Only then may we assess the significance of the active (or inactive) state of a given structure in the general machinery of the brain and reach conclusions about its function.

Working extracellularly in conscious animals (rabbits), we have pursued a very restricted aim: we simply wanted to understand what happens to the spontaneous and evoked (by sensory stimuli) activity of hippocampal and septal neurons deprived of some afferent influences. This approach does not help us to make a final decision about the functional significance of the septo-hippocampal interaction but, we hope, it brings us closer to understanding it.

GENERAL METHODS

All the experiments described here were performed in adult rabbits in standardized conditions.

One to two weeks before the experiments the surgery was performed under Nembutal (0.2 mg/kg, i.p.) anaesthesia and with local anaesthesia of pressure points in the stereotaxic apparatus by Novocaine. During the operation the directing plug for microelectrodes was placed on the skull, and the electrodes for electrical stimulation of various brain structures and for control records of the cortical and hippocampal EEG were inserted and cemented to the skull.

Before the experiments started, the animals were placed in the experimental chamber for one hour twice a day for 3–5 successive days for adaptation to the experimental environment. During the experiments awake rabbits were slightly restrained in a special box inside a screened sound-proof chamber. Stressful stimuli were carefully avoided.

The extracellular recording of single unit activity was made in fields CA1 and CA3 of the hippocampus and in the medial and lateral septal nuclei (in separate series). Tungsten microelectrodes with tip diameter 3–4 nm and resistance of 4–8 megohm in saline were driven by means of a distant hydraulic micromanipulator.

Neuronal activity was recorded on magnetic tape and later analysed both visually from the film and with the help of a computer, Electronics-100. In the latter case the medium frequency, interval histograms and autocorrelation functions were determined.

During the experiments a series (20–50) of sensory stimuli of different modalities were presented, and the types and dynamic changes of reactions were analysed. In some experimental series electrical stimulation of various brain structures was used for particular purposes. It was delivered from MSE-3 stimulators (Nihon Kohden) through bipolar Nichrom electrodes of diameter 100 nm. Rectangular pulses 0.8–1 ms in duration and of various frequencies (0.3–100 Hz) were used.

After the experiments the animals were killed and the loci of recording and stimulation as well as the extent of the lesions (described below) were analysed histologically on serial frozen brain slices.

RESULTS

Initial data on CA3 and CA1 hippocampal neurons

Data on 'normal' hippocampal activity were given by us previously (Vinogradova 1965; Vinogradova & Dudaeva 1971) and here we shall just list briefly its most important characteristics to facilitate comparisons with the deafferented hippocampus.

In field CA3 the majority of neurons recorded possessed medium to high spontaneous activity (10–15, 20–30 per second). Neuronal reactions to sensory stimuli are of a diffuse tonic type, greatly outlasting the stimulus (by 3–5 seconds and more). They consist of a general increase in the activity level (in 40% of units) or, more often, its prolonged decrease (in 60% of units) (Fig. 1A). Reactions of a given neuron to stimuli of different modalities are uniform. The repeated presentation of a stimulus brings about a gradual shortening and complete habituation of reactions after 8–30 presentations. Any change in the stimulus to which a reaction is habituated is followed by the reappearance of the initial effect.

In field CA1 the level of spontaneous activity is somewhat lower (5–15, usually not higher than 20 per second); some neurons are not spontaneously active.

About half the CA1 neurons respond to sensory stimuli in the same way as the CA3 neurons. The other neurons in this field have reactions of a 'specific' on-off patterned type, or of 'phasic' type (equal in duration to the stimulus) (Fig. 1B).

The specific feature of the dynamics of reactions in this field is that maximal well-developed responses usually appear after 2–3 presentations of a stimulus, while in CA3 the maximal response is evoked by the first stimulus in a series. The neuronal characteristics of fields CA1 and CA3 of the 'normal' hippocampus are summarized in graphical form in Fig. 6 (p. 157).

A rhythmic theta component can be detected in 12% of CA3 neurons and in 9% of CA1 neurons. During the application of sensory stimuli, theta modulation of the activity can occur concomitantly with activation or depression of the background level of discharge.

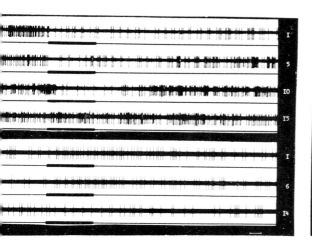
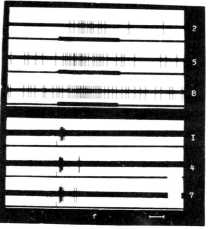

FIG. 1. Reactions of hippocampal neurons to sensory stimuli.

A. Tonic reactions of suppression *(above)* and increase *(below)* of spontaneous activity with gradual habituation typical of CA3 neurons;

B. Additional types of phasic *(above)* and 'specific' *(below)* responses in field CA1. Stimuli (pure tones and click) are indicated below the records of neuronal activity. On the right, the number of presentations of a stimulus. Time calibration, 200 ms.

Effects of septo-hippocampal disconnection on hippocampal neurons

In order to interrupt septo-hippocampal connections we introduced two bipolar electrodes at the level of the septo-fimbrial nucleus. All tissue was coagulated bilaterally in the narrowest 'bottleneck' between the hippocampus and septum (current 4.5 mA, for 20 s), and these structures were completely disconnected without significant lesion of either of them (Fig. 2A).

Usually experiments with recording of CA3 and CA1 neuronal activity started 10–14 days after the operation. Seventy neurons were recorded from in field CA3 and 60 in field CA1 in these series. In four animals registration of neuronal activity was started after longer (3–6 months) survival periods. Forty-three CA3 neurons were recorded from in these additional series.

Neither the level, nor the general, irregular type of spontaneous activity changed in the deseptalized hippocampus. The only effect on background activity was the complete absence of rhythmic theta bursts in neuronal activity, as well as of theta waves in the hippocampal EEG. Autocorrelograms of spontaneous activity and during the presentation of stimuli as a rule were of regular or exponential type. In some autocorrelograms a weak rhythmic component was observed, usually in the frequency range 2.0–3.2 Hz (delta range) (Fig. 4, p. 154).

There were obvious changes in neuronal sensory reactions. The most dramatic changes took place in field CA3. In the series done soon after the operation (10–14 days) the reactivity of the CA3 neurons to sensory stimuli was greatly decreased, from the normal value of 68% reactive neurons to 29%. In the long-term survival series, CA3 reactivity approached the normal level, but the characteristics of the reactions remained unusual for this field. Neurons with diffuse tonic reactions, normally dominant in CA3 (95%), now constituted only 57% of reactive units. Tonic effects preserved by these neurons become significantly shorter than in normal conditions (Fig. 5A, p. 155). Simultaneously the latencies of these reactions increased. While in the normal CA3 the latencies of responses to presentations of one-second pure tones varied from 25 to 100 ms (medium latency 87 ms), after deseptalization they shifted to 150–300 ms (medium latency 202 ms) (Fig. 5B). Besides the usual tonic effects in CA3 there appeared patterned responses of on-off type never observed by us in this field in normal conditions; the proportion of neurons with phasic responses increased (Fig. 2). The level of multimodal convergence was reduced (54%, compared to 93% in the control). Inhibitory reactions, dominating in the normal CA3, were observed only in 16% of neurons.

The changes in CA1 were in the same direction, but much less impressive. In the normal CA1 the proportion of multimodal neurons and units with

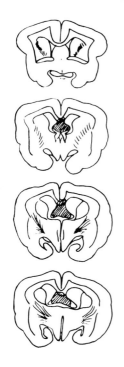
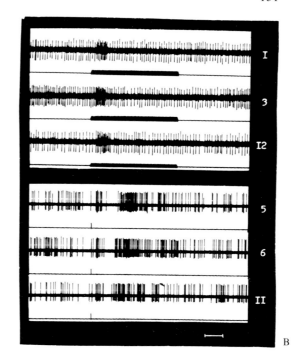

FIG. 2. Septo-hippocampal disconnection.
A. Schematic presentation of typical lesion at the level of septo-fimbrial nucleus.
B. Unusual 'specific' responses (to pure tone and click) appearing in field CA3 after interruption of connections with the septum.

tonic reactions is lower than in CA3. Tonic reactions are normally shorter, and inhibitory responses are observed less frequently (39%). Thus, further decreases in the level of convergence, in the proportion of tonic effects and in inhibitory reactions to sensory stimuli merely accentuated the initial characteristics of the CA1 field. The 'specific' and 'phasic' responses, normally present in this structure, only increased in proportion, without significant changes in latencies and patterns. Thus one can conclude that while septal deafferentation leads to drastic changes in field CA3, it does not greatly influence the principal characteristics of CA1. As a result of the transformation observed, the deseptalized CA3 resembles the normal (and deseptalized) CA1 field (Fig. 6, p. 157).

The dynamics of the reactions were seriously changed in both fields. With a short time of survival after septal disconnection, abnormal dynamics of diffuse reactions were observed in both fields, but especially in CA3. They were characterized by a very rapid disappearance of responses (after 3-5 repeated presentations of a stimulus). This does not resemble the normal gradual course of habituation and appears to be a sign of some neuronal abnormality. This is supported by the observation that after longer periods of survival such an abrupt disappearance of reactions is not usually observed. However, the normal process of habituation is nevertheless absent. A quite opposite course of dynamic changes with a gradual building up of reactions (by the 3rd-12th presentation of a stimulus) is observed, which is also more typical of normal CA1 and of the entorhinal cortex, as was shown in our laboratory (Stafekhina & Vinogradova 1976).

Thus, when septo-hippocampal connections have been interrupted the tonic reactions and inhibitory effects as well as the gradual habituation process suffer most. The neuronal reactions of the deseptalized hippocampus (patterned and phasic, with incremental dynamic changes) seem to reflect the characteristics of the remaining cortical input. It can also be concluded that in normal conditions this input mainly determines the type of reactions in the CA1 field, while septal input is functionally more important for CA3.

It is worth mentioning that in these experiments electrical stimulation of the midbrain reticular formation evoked weak diffuse activation or suppression in 10% of CA3 neurons and in 29% of CA1 neurons; there was no correlation between the characteristics of responses to reticular and sensory stimulation in the neurons of either field.

The effect of transection of the perforant path on hippocampal neurons

The interruption of the perforant path—the main pathway from the entorhinal cortex—was done surgically. A special curved spatula, 1.5 mm in width, connected to the ophthalmological coagulator DK-3, for prevention of bleeding, was introduced through the occipital neocortex into the lateral horn of the ventricle so that it passed over the curvature of the hippocampus, not damaging it. At a fixed depth it was brought into a vertical position and then rotated laterally through 90°. As a result, complete transection of all retrohippocampal structures and of the CA1 field was obtained at the level between the middle and lower third of the hippocampus. This operation was done bilaterally and, as the histological control showed, produced complete interruption of cortical afferents to the upper two-thirds of the hippocampus (Fig. 3). The experiments started 7-12 days after the operation. Sixty-eight neurons from CA3 and 135 from CA1 were recorded from in these series.

The level of spontaneous activity and its basic characteristics were not changed in either field. The only feature which distinguished the neuronal

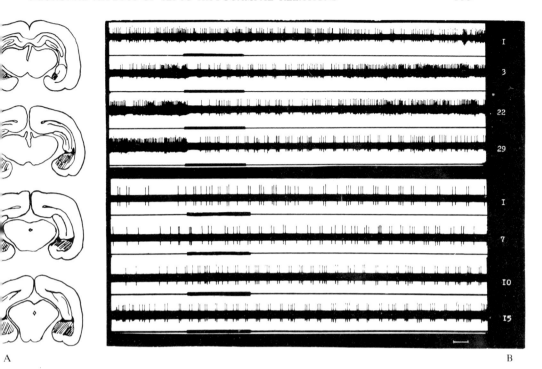

FIG. 3. The perforant path lesion.
A. Schematic presentation of typical lesion.
B. Tonic reactions of suppression *(above)* and increase *(below)* of activity of neurons in field CA3 with theta bursts and gradual incremental dynamics.

activity of the hippocampus without the perforant path from the normal one was that regular theta bursts were present in a large proportion of neurons. Autocorrelograms of spontaneous activity and during the presentation of sensory stimuli revealed a clear rhythmic component in the theta range of frequencies (4.8–5.9 Hz) (Fig. 4). The theta component was observed in 73% of CA3 neurons and in 62% of CA1 neurons. The hypersynchronized theta rhythm was observed in the EEG records.

The reactivity to sensory stimuli was slightly, but statistically reliably, increased in both fields (78% and 83% in CA3 and CA1 respectively). The majority of reactions in both fields were of tonic type and of longer duration than in the normal state (Figs. 3, 5B). The latencies of these tonic effects were normal (Fig. 5A). Many of these tonic reactions (though not all) contained theta bursts. The proportion of activating and inhibitory responses did not change much, and theta bursts could be observed against the background of

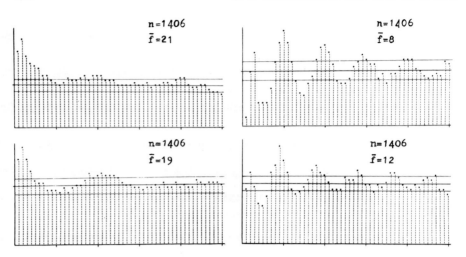

FIG. 4. Autocorrelograms of spontaneous neuronal spike activity. Left, after septal disconnection. Right, after perforant path lesion. Above, field CA3; below, field CA1. Bin, 20 ms.

both increased and decreased activity after a sensory stimulus was given; in some cases reactions consisted only in the appearance of theta patterning of neuronal discharges without any change in the level of activity.

Some neurons showed instability in the direction of their reactions. Sometimes the reactions had a stable diphasic form; they started with a short period of activation, followed by a prolonged tonic inhibitory phase. In other neurons this activatory component alone was observed during the early presentations of a stimulus, but with later presentations it was gradually replaced by a tonic inhibitory effect. There were also neurons in which reactions depended closely upon the level of previous background activity. In these cases the 'wave phenomenon' was observed: for example, the first stimulus evoked tonic activation of a neuron, then the second stimulus, presented during this increase in activity, evoked tonic inhibition, and so on. Such transitions of inhibitory and activating effects were not observed in the normal hippocampus.

The tonic responses were almost the only type of reaction observed in CA3 and constituted the majority of responses in CA1. But a small proportion of neurons in this field (10%), invariably observed in all experimental animals, had reactions of a different type. They were simple, relatively short-latency (14.5–31 ms) on-responses. As a rule they were observed in neurons without theta bursts in their activity.

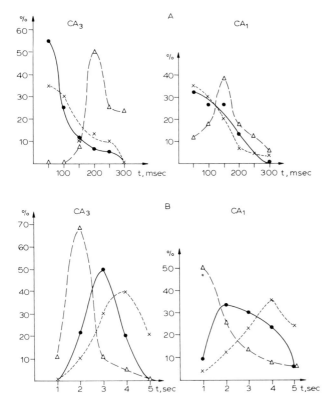

FIG. 5. Characteristics of tonic reactions of CA3 and CA1 neurons in different conditions. Distribution of latencies (A) and reaction durations (B) during presentation of auditory tonal stimuli (1 s long). Normal hippocampus, solid line. Without perforant path, dotted line. Without septum; broken line. Abscissa: time (A, in ms; B, in s). Ordinate: percentage of neurons.

The dynamics of reactions in the hippocampus lacking the perforant path during repeated presentations of sensory stimuli greatly differed from normal habituation. Usually, the initial reactions just slightly outlasted the duration of a stimulus (1 s for tones). But with each subsequent presentation of the same stimulus, reactions increased in duration, so that after 8–12 presentations they completely filled the intervals between stimuli (Fig. 3, p. 153). These reactions persisted in 96% of neurons of both fields even after 35–40 repeated presentations of a stimulus.

Electrical stimulation of the midbrain reticular formation produced reactions in 89% neurons in both fields. It is important to note that in the great majority of the neurons an extremely high correlation of reticular and sensory effects was observed. All characteristics of sensory responses (increase, or decrease of activity; presence or absence of theta bursts; dynamics of response changes with repeated stimulation) were reproduced by stimulating the reticular formation. It was surprising that the CA1 neurons with on-responses to sensory stimuli were the only group of units reacting to reticular formation stimulation by time-locked driving responses. Their latencies varied from 8.5 to 18 ms and usually were a few (4–12) milliseconds shorter than the latencies of the sensory on-responses (to click) of the same neurons.

Thus the elimination of the cortical input to the hippocampus leads to the exaggeration of characteristics depending on its reticulo-septal input. It seems that the desynchronizing influence of the cortical afferents, observed in the EEG of the hippocampus (Green & Adey 1956; Thomson & Laget 1966; Parmeggiani 1967), is also expressed in its neuronal spike activity. In the hippocampus without the perforant path, neuronal theta activity greatly increases in persistence and regularity and is present in a larger proportion of neurons.

It seems also that tonic reactions of hippocampal neurons are to a great extent independent of cortical input. That is why the lesion of the perforant path does not greatly alter the basic characteristics of neuronal reactions in CA3, where the tonic effects absolutely dominate in the normal state, but brings about obvious changes in the characteristics of CA1 sensory reactions (Fig. 6). Presumably the perforant path is essential for the patterned reactions of CA1 and without it CA1 starts to resemble CA3. Nevertheless, some on-effects remain in the CA1 lacking the cortical input. The data on reticular formation stimulation allow us to suggest that CA1 receives some direct afferents from the midbrain reticular formation, not relayed at the septal nuclei.

Cortical afferents seem not to be important for sensory inhibitory effects (tonic decrease of spontaneous activity), but the gradual habituation of responses depends considerably on their integrity. Without cortical input, an unlimited increase in reactions is observed. Thus, both main hippocampal afferent paths are necessary for the organization of the normal habituation of sensory reactions in the hippocampus.

Hippocampal neuronal activity without both septal and cortical afferents

Some data obtained in these conditions are also pertinent to the problem of the nature of theta activity in the hippocampus.

In these experiments, coagulation of the septo-hippocampal connection was first done. Then a month was allowed for the development of collateral

NEURONAL ASPECTS OF SEPTO-HIPPOCAMPAL RELATIONS 157

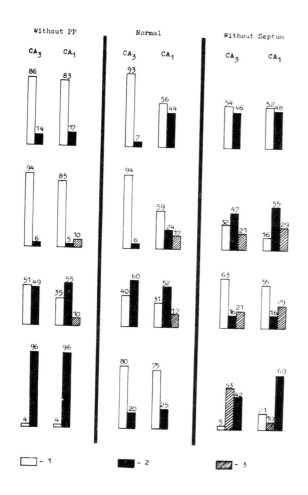

FIG. 6. Comparative characteristics of the CA3 and CA1 neurons in different conditions. Middle, normal hippocampus. Left, without perforant path. Right, without septal afferents. First row, the level of convergence of sensory modalities (1, multimodal; 2, unimodal units); Second row, types of reactions (1, tonic; 2, phasic; 3, patterned); Third row, direction of reactions (1, activation; 2, suppression of activity; 3, patterned reactions which are not evaluated on the basis of this parameter); Fourth row, dynamic characteristics (1, habituation; 2, stable reactions; 3, abrupt disappearance). Numbers above the columns are percentages of neurons with given characteristics. Note the increase in resemblance between the hippocampal fields without each of the afferent inputs; note also resemblance between the normal CA3 and hippocampus without perforant path, and between the normal CA1 and hippocampus without septum.

circulation, after which a bilateral lesion of the cortical path was made. The only difference from the operation described in the previous section was that the transection of the retrohippocampal structures and of the hippocampus was made at the level of the middle or upper third of the hippocampus. As a result, the dorsal hippocampus was completely isolated from the other brain structures except its contact with the retrosplenial cortex and stria Lancisii. The latter structure Cajal (Ramon y Cajal 1955) regarded as efferent for the hippocampus, and recent histological evidence shows that afferent fibres do not enter the hippocampus proper from the cingulum and cingulate cortex (Domesick 1969; Sotnichenko 1970). Thus, the dorsal part of the hippocampus being investigated could be regarded as almost completely deprived of extrinsic afferent connections.

The well-known resistance of hippocampal neurons to retrograde degeneration after their main efferent axons have been cut makes the investigation of neuronal activity in such a preparation possible. Even its 'specific activity' (the number of active units encountered in 1 mm^3 of the tissue) does not differ from that of the normal hippocampus. These experiments have been done so far only in field CA1 where recordings were made from 155 neurons. The recording of neuronal activity started one week after the second operation and continued for 5–11 weeks in different animals. The level and pattern of spontaneous activity again was not changed, except for the absence of theta bursts, as in the deseptalized hippocampus. Reactions to sensory stimuli were completely absent during the second week after the operation, though the cells preserved normal reactivity and thresholds of excitation to synaptic influences (electrical stimulation of Schaffer's collaterals). By the end of the second week after the operation amorphous long-latency activatory responses to various sensory stimuli appeared in a small proportion of neurons (12–20%). Later on the proportion of reactive units steadily increased, and by the end of 4–6 weeks it reached the normal level (65–70%). Reactions became more definite and stable, though the latencies remained long (106–468 ms for pure tones). Among these effects, by the end of the third week in all experimental animals there appeared complex on-responses with latencies of 47–177 ms for auditory stimuli, and 116–155 ms for visual ones. Habituation of all these reactions was absent, and a slow increment in responses with subsequent stabilization was observed instead.

In this situation an attempt was made to evoke intrinsic rhythmic activity in the hippocampus. First, we tried to reproduce the rhythmic synaptic bombardment using the intact inner connections of the hippocampus. Rhythmic five per second bursts of 100 Hz impulses, each burst lasting 100 ms, were delivered to Schaffer's collaterals or the remaining part of dorsal

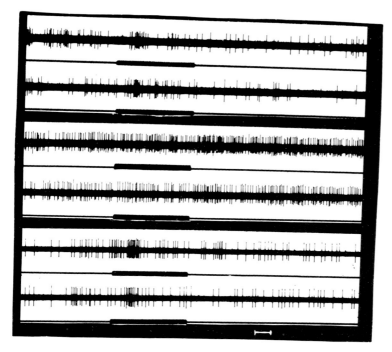

FIG. 7. Effect of physostigmine injection on neuronal activity of the isolated hippocampus. Above, on-response to pure tone before the injection. Middle, 25 min after injection. Note the increase of spontaneous activity and complete masking of the response. Below, 85 min after injection: reappearance of response.

subiculum. Complete inhibition of spontaneous activity for 1–2 s followed a series of such bursts, after which normal activity reappeared without any traces of rhythmic modulation. Prolongation of stimulation with bursts of impulses up to 15–20 s evoked epileptic discharges.

In other experiments we gave an intravenous injection of physostigmine (0.2 mg/kg), which provokes prominent theta rhythm in the intact hippocampus (Stumpf 1965). The injection was given after making the control record of spontaneous and evoked activity of a unit for about 10 min. During the first 10 min after the injection the level of spontaneous activity increased twofold and more; 40–50 min after injection it gradually declined to the initial control level. The sensory responses of the cells were usually completely masked by the background of augmented activity, but reappeared after its normalization (Fig. 7). The regularity of discharges increased without any signs of rhythmic modulation. It is well established that septal deafferentation chronically deprives the hippocampus of cholinergic afferents (Storm-Mathisen & Fonnum 1972; Mellgren & Srebro 1973; Kultas *et al.* 1974), so the action of physostigmine in the 'isolated' hippocampus can be explained as a result of denervation supersensitivity.

For our purposes it is important that the mere increase in activity of pyramidal neurons or their rhythmic synaptic activation through one of the intact inputs, involving the system of recurrent inhibition, is not sufficient to induce rhythmic activity. Septal input, with its spatial and temporal organization, seems to be indispensable for that.

The effect of septo-hippocampal disconnection on septal neuronal activity

Before a general discussion of the results it is necessary to look at the problem from another angle and to analyse the significance of hippocampal influences for septal neurons.

The normal characteristics of neurons in the medial and lateral septal nuclei (mS and lS) during the presentation of sensory stimuli have already been described by us (Vinogradova & Zolotukhina 1972, 1973), and will be only briefly summarized here.

In the normal septum neuronal activity in the mS and lS is very similar though the range of spontaneous activity is different. In the medial region neurons with low (3–6/s) and very high (up to 60–100/s) spontaneous activity were encountered. The last group was more usual in the lower parts of the mS, at the border with the nucleus of diagonal band and in its upper part. In the lS activity was more uniform and closer to the medium range (8–18/s).

In both nuclei about a third of the neurons (27–28%) were characterized by the presence of theta bursts in their spontaneous or evoked activity.

Reactions of nearly all neurons in the lS and of the majority (85%) of those in the mS were of tonic type with or without rhythmic bursts. Reactions of a neuron to stimuli of different modalities were uniform. In both nuclei tonic inhibitory responses with or without theta bursts were more numerous than activatory ones (48 and 37% in the mS; 58 and 42% in the lS respectively). In 15% of the mS neurons, usually in the lower part of the nucleus, reactions to auditory and visual stimuli had an initial on-component with prolonged secondary effects with or without theta bursts. Their latencies varied from short (18–30 ms) to very long (200–500 ms).

The main difference between the nuclei consisted in a different course of dynamic changes. While in the lS rapid and complete habituation developed after 8–12 repeated presentations of a stimulus (Fig. 8A), in 63% of the mS neurons habituation is absent, or there is only a partial habituation with

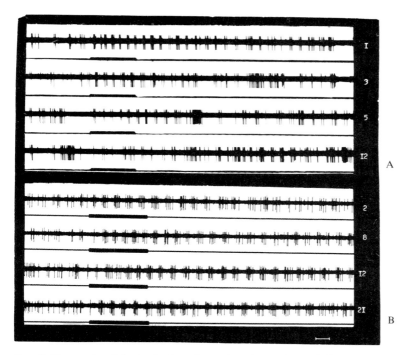

FIG. 8. Neuronal activity in the lateral septal nucleus. Above, before septo-hippocampal disconnection. Tonic activatory response (to pure tones) with theta bursts and gradual habituation. Below, after lesion of the afferent connections from the hippocampus. The same type of reaction with gradual increment.

shortening of the duration of the reactions (from tonic to phasic form) but without the complete disappearence of responses.

Short (1 s) low-frequency (15–30 Hz) tetanization of the hippocampus, besides oligosynaptic driving in some neurons, evoked diffuse effects of decreases or increases in activity. The inhibitory effects dominated in both nuclei (52 and 12% of responsive units respectively suppressed or activated in mS; 32 and 18% in lS). When theta bursts were present in the background activity of the neurons, stimulation of the hippocampus desynchronized them, or decreased their frequency, for 1–3 s after the stimulation.

Reactions to sensory stimuli applied after hippocampal stimulation were completely blocked or significantly decreased in 94% of mS neurons and in 58% of lS neurons. Stimulation of the midbrain reticular formation always produced opposite effects in the same neurons: appearance of reactions in unresponsive units; disinhibition of habituated responses and increases in the duration and intensity of responses.

The operation of septo-hippocampal disconnection, described above, leaves not only the hippocampus intact, but the septum as well. Thus it was used to analyse septo-hippocampal relations further. But it was found that in the mS a rapid pathological process develops, presumably due to damage to axons entering the hippocampus. That is why we started investigating neurons in the mS as early as 2–4 days after the operation. Nevertheless, the number of units with normal activity was low and many showed pathological traits (depolarizing bursts, low and unstable frequency of discharges, etc.). With longer periods of survival after operation the number of normal, active cells in the mS became minimal, but in the lS activity was quite indistinguishable from normal. After lesion of the hippocampo-septal afferents theta bursts were observed in a more restricted population of mS neurons (19%); in the lS their proportion significantly increased (48%). The theta burst activity observed in these conditions was more regular and persisted for longer periods in spontaneous activity.

The reactivity to sensory stimuli was normal. The proportion of inhibitory tonic effects was lowered (from 48% in mS and 58% in lS to 10% and 17% respectively). On-responses appeared in 13% of lS neurons, though in normal conditions they were never observed in this nucleus (Fig. 9).

Habituation was completely absent from both nuclei. In its place an unlimited increase in the duration of responses was observed with repeated presentations of sensory stimuli (Fig. 8B).

Stimulation of the midbrain reticular formation evoked responses in all neurons tested. The correlation of sensory and reticular effects, which was already high in the intact septum, further increased after septo-hippocampal disconnection. As in the hippocampus without cortical input, oligosynaptic driving by reticular formation stimulation was observed mainly in neurons with on-responses to sensory stimuli.

These data show that the hippocampus is necessary for inhibitory effects in the septum (suppression of activity, habituation), but does not take part in organizing its sensory reactions and is not necessary for the production of theta bursts.

DISCUSSION

Effects of lesions of hippocampal afferents

When one analyses the effects of lesions of the afferent paths to the hippocampus and septum it is important to take into account that they do not simply result in the elimination or subtraction of some afferent influences. In fact, such a procedure brings about a complete reorganization of the remaining

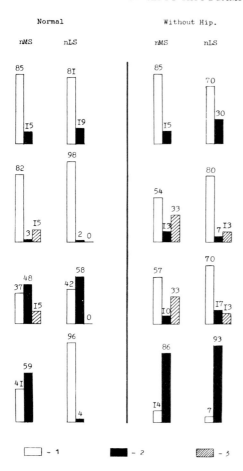

FIG. 9. Comparative characteristics of the mS and lS neurons in different conditions. Left, normal state. Right, without the hippocampus. First row, the level of convergence of sensory modalities (1, multimodal; 2, unimodal units); Second row, types of reactions (1, tonic; 2, phasic; 3, patterned). Third row, direction of reactions (1, activation; 2, suppression of activity; 3, patterned reactions, which are not evaluated on the basis of this parameter). Fourth row, dynamic characteristics (1, habituation; 2, stable reactions).

afferent systems with an increase in their influences and an exaggeration of their characteristics. The anatomical bases for these physiological effects in septum and hippocampus were conclusively shown by recent histological investigations on the sprouting of axonal collaterals (Raisman & Field 1973; Moore *et al.* 1971; Lynch *et al.* 1972; Storm-Mathisen & Fonnum 1972; Zimmer 1974). As the perforant path, and the fibres entering the hippocampus from the septal pole (through both the fimbria and fornix), are the two main known afferent sources for the hippocampus, elimination of one of them brings about a spread of contacts and increase of the influences of the remaining input.

From the facts described above it can be concluded that the level and general characteristics of spontaneous discharges of hippocampal neurons do not depend on its afferent systems. Rather, it represents a kind of 'carrier frequency', modulated by extrinsic synaptic influences. This suggestion is also supported by observations on hippocampal neurons in slices incubated *in vitro* (our unpublished data).

Sensory reactions of the hippocampal neurons of course depend on the afferent inputs, and various types of reaction depend to a great extent upon the different afferents. The tonic reactions, and especially, tonic inhibitory responses, depend closely on the reticulo-septal input. For the majority of 'specific' patterned and phasic responses the integrity of the cortical perforant path is necessary. The initial difference in the repertoire of reactions in CA3 and CA1 and their changes after lesion of each of these afferent paths suggest that functionally the direct cortical input has a stronger influence on the CA1 field, while reticulo-septal influences dominate in CA3. Without septal afferents CA1 does not differ much from normal and CA3 acquires unusual characteristics, resembling CA1. The opposite situation is seen after the perforant path lesion, where characteristics of CA3 are close to normal and CA1 more closely resembles CA3 (except for a small group of neurons with on-responses, depending on direct reticular influences).

It is also possible to suggest, on the basis of our experimental data, that, presumably through the retrosplenial cortex and upper part of the subiculum, some additional afferent path passes towards the hippocampus proper. Normally this path is very weak, so that it is difficult to reveal it by degeneration methods. But it seems that its sprouting after the lesion of both main hippocampal afferent paths results in a great increase in its functional significance. Unfortunately, we have not yet analysed the source of this path histologically, but both analysis of the sensory reactions and data on electrical stimulation indicate the posterior cingulate cortex as a possible source.

The types and latencies of sensory reactions in the 'isolated hippocampus' were close to those observed in the posterior cingulate cortex (Stafekhina & Vinogradova 1973). Its electrical stimulation evoked relatively short-latency (6–32 ms) driving in 79% of the CA1 neurons, which followed the stimulating impulses up to 60–90 Hz and more.

The origin of septal and hippocampal theta bursts

Our experimental data show beyond doubt that the integrity of hippocampal afferents to the septum is not important for neuronal theta activity in this structure. This crucial fact speaks against the suggestion made by McLennan & Miller (1974, 1976), who regard the closed loop between the septum and the hippocampus, including an inhibitory link, as a mechanism essential for the production of theta bursts in the septum. Quite to the contrary, stimulation of the hippocampus suppresses theta, or decreases its frequency (in the lS, where according to Raisman influences are orthodromic), and interruption of the hippocampal axons to the septum results in an increase in theta activity. Thus, at the septal level the hippocampus indeed acts as an antagonist of the activatory reticular formation, stimulation of which consistently evoked theta bursts in the septal neurons.

It is well known, and has also been shown in our laboratory (Kichigina & Vinogradova 1974), that the midbrain reticular areas, from which theta activity can be evoked in septum and hippocampus, do not show a rhythmic component in their neuronal activity. In fact, many reticular formation neurons are characterized by very regular, 'pacemaker'-like activity. The effect of a medial forebrain bundle lesion on the activity of septal units is now under study, but it seems that the septum indeed has the ability to transform the regular flow of reticular impulses into a quantal, burst-like form. Whether it is a unique property of single units in the septum, or (more probably) the result of intraseptal neuronal connectivity, is not clear, but certainly the septum may be regarded as a generator of the theta rhythm, as has been conclusively shown by many authors (Petsche & Stumpf 1960; Stumpf *et al.* 1962; Petsche *et al.* 1965; Macadar *et al.* 1970).

While neuronal theta activity is preserved in the septum without the hippocampus, it is absent in the hippocampus without the septum, in accordance with data on the hippocampal EEG. It seems that the preservation of an intrinsic system of recurrent inhibition alone is not sufficient for the production of repetitive rhythmic bursts. Perhaps the excitability of pyramidal neurons is lowered in the deafferented hippocampus and this prevents the rhythmic activity, but an increase in spontaneous frequency by physostigmine does not change the situation. It can also be suggested that an initial synchronous input is necessary for triggering theta bursts, but imitation of such rhythmic synchronized input through stimulation of remaining synaptic systems in the deseptalized hippocampus is also without effect in this respect.

Thus it seems that the specific spatial organization of septal afferents as well as the temporal patterning of their synaptic influences have the utmost importance for triggering theta bursts in hippocampal neuronal and EEG activity.

Functional significance of septal input to the hippocampus

Some histological investigations have indicated recently that septal afferents presumably terminate on the inhibitory basket interneurons of the hippocampus and dentate (Mosko *et al.* 1973; Kultas *et al.* 1974). The same suggestion has been made on electrophysiological grounds (Gogolák *et al.* 1967; Shaban 1970; Grantyn & Grantyn 1972). Some authors have even regarded all hippocampal neurons activated by septal input (Shaban 1970), or producing theta bursts (Fox & Ranck 1975), as basket interneurons. A decrease in the hippocampal output signal during theta states (e.g. during paradoxical sleep) was also reported (Mink *et al.* 1967; Belugou *et al.* 1968; Franzini *et al.* 1975). The drastic decrease in the proportion of general inhibitory effects in the deseptalized hippocampus makes this idea plausible, and it seems that direct afferent inhibition of septal origin in the hippocampus is a real possibility.

Nevertheless, we should not discard other histological data (Raisman *et al.* 1965; Rose *et al.* 1976) showing that septal afferents also terminate on basal and apical dendrites of pyramidal cells, especially in fields CA3-4. In this case, at least some of the 'theta neurons' in the hippocampus may belong to the pyramidal cells as well. Besides the fact that the spike activity of many 'theta cells' is indistinguishable from the main population of hippocampal neurons, this suggestion is strongly supported by the great increase in proportion of 'theta cells' in the hippocampus lacking cortical input. This may be explained as a result both of an increase in the synaptic weight of the septal afferents in the course of sprouting and of the elimination of the desynchronizing cortical influences.

We suggest that septal afferents, influencing both pyramidal and basket neurons in the hippocampus, evoke a complex reorganization of its neuronal activity. Being able to change the excitability of a pyramidal neuron as a whole, the septum may act as a synchronizing device, allowing the interaction of different inputs on the hippocampal neurons only if they come in strict temporal (phase) relations to the septal bursts. As we suggested earlier (Vinogradova 1965), the hippocampus may be regarded as the comparator for the memory system. 'Matching' the signals coming directly from the brainstem and those preprocessed in cortical structures, it determines the novelty of information, or its presence in the memory store, and accordingly regulates the neurophysiological conditions necessary for registering new information.

The septal 'bursting pacemaker' may increase the precision of the comparator operation, allowing the matching of just related events in both (septal and cortical) inputs.

It is important to remember that the integrity of both these inputs is necessary for the development of habituation of reactions to novel sensory stimuli, which is typical of the normal hippocampus. It is possible that the hippocampal processing of information is performed by only some of its active functional segments, 'selected' on the basis of the organized activation of cortical afferents. The septal input, involved in the widespread inhibition of other segments, may provide the background for this active process and increase the signal-to-noise ratio.

One more suggestion may be added. While we were comparing the neuronal repertoires of sensory reactions in the initial source of the cortical input to the hippocampus—the entorhinal cortex (Stafekhina & Vinogradova 1975, 1976)—and in its pre-CA3 relay, the dentate fascia (Vinogradova & Bragin 1975), it became obvious that some kind of secondary unification and simplification of signals takes place in the dentate. Differentiated, patterned reactions of entorhinal neurons are transformed into simple on-responses in the dentate. The fact that entorhinal signals reach CA1 only by a direct route, but the bulk of the cortical afferents are relayed on the dentate before entering CA3, seemed to be a possible explanation of the origin of more specific, more differentiated responses in CA1.

But exactly the same situation exists symmetrically in the reticulo-septal path to the hippocampus. Various sensory effects of reticular neurons are transformed into more uniform reactions at the septal pre-hippocampal relay. From the experiments with perforant path lesions one can conclude that CA3 again receives mainly relayed signals, while some direct oligosynaptic influences arrive at CA1. The small, but regularly observed population of CA1 neurons preserves on-responses in these conditions. These neurons never show theta bursts in their activity and as a rule are driven by reticular stimulation.

Thus, it seems that mS as well as dentate may be regarded as special 'mixers' of signals, predigesting the information before it enters CA3. This operation may be important for the special mode of operation of its neurons, which detect the novelty of stimuli, regardless of their specific characteristics.

Functional significance of descending hippocampal influences

The results of our neuronal investigations agree with anatomical data on relations between the hippocampus and septal nuclei described by Raisman (1966), and later confirmed by others both histologically (Segal & Landis 1974) and neurophysiologically (DeFrance *et al.* 1972). According to these authors mS may be regarded as an input relay nucleus for the hippocampus while the lS is its output relay structure. The obvious pathological state of the mS neurons after septo-hippocampal disconnection, with preservation of relatively normal activity in the lS, reflects the retrograde degeneration which develops in mS when main axons are lesioned, but the sparing of lS neurons when only synaptic hippocampal influences are eliminated.

In normal conditions the main difference between the lS and mS neurons is the rapid habituation of sensory responses in the former, but not in the latter. This may mean that a sensory message passes through the hippocampal gate to be transformed from a non-decremental form into a decremental one. This already implicates descending hippocampal influences in the organization of plastic decremental events. This conclusion is supported by the most important effect of hippocampal disconnection on lS neurons—complete disappearance of gradual habituation of their sensory responses. The opposite kind of dynamic change, with unlimited increase in the duration of reactions during repeated presentations of stimuli, takes place in the hippocampectomized septum.

The characteristics of sensory responses in the septum do not depend on the integrity of connections from the hippocampus. Quite to the contrary, the reactivity of lS neurons even increases after disconnection (the only exception being a decrease in the number of inhibitory effects), as well as the proportion of neurons with theta bursts. It seems that these characteristics of septal neurons are completely determined by ascending influences of the midbrain reticular formation, as the high correlation between the effects of sensory and reticular stimulation in the septum shows.

The changes in septal neuronal activity may reflect the sprouting of ascending reticular axons. This effect may be more pronounced in lS where, according to Raisman (1969), 56% of synapses are normally of hippocampal origin. An increase in reactivity, in the number of neurons with theta activity, and in the regularity and frequency of theta bursts, may also result from the elimination of hippocampal influences on brainstem structures.

Closely similar effects—suppression of sensory reactions during hippocampal electrical stimulation and incremental changes with complete loss of habituation after hippocampal lesion—were observed in our laboratory by V. F. Kichigina in the midbrain reticular formation. She showed that very complex relations exist between the hippocampus and non-specific midbrain structures (reticular tegmental nucleus, raphe nuclei). The hippocampus exerts an activatory influence on the raphe nuclei and, possibly through this structure, suppresses activity of the ascending midbrain reticular formation (Kichigina & Vinogradova 1974). The general suppressive influence of the hippocampus does not mean that its neurons function as inhibitory interneurons; rather it means that the hippocampus selectively activates some inhibitory structures, or elements.

In any case, the complete disappearance of habituation after lesion of the main hippocampal axons in the septum and midbrain reticular formation seems to be a close neuronal correlate of the defects in habituation of the orienting reaction which comprise one of the most consistently observed symptoms in animals with hippocampal dysfunction.

ACKNOWLEDGEMENTS

The authors wish to express their gratitude to A. M. Karanov, who did the computer analysis of the neuronal data, and to E. V. Lebedyeva for her invaluable help in the preparation of the manuscript and illustrations.

References

ANDERSEN, P., ECCLES, J. C. & LØYNING, P. (1964) Pathway of post-synaptic inhibition in the hippocampus. *J. Neurophysiol. 27*, 608–618
ARTEMENKO, D. P. (1972) Participation of the hippocampal neurones in generation of theta-waves. *Neurophysiologiya 4*, 531–539 (in Russian)
BEATTY, W. W. & SCHWARTZBAUM, J. S. (1968) Commonality and specificity of behavioral dysfunctions following septal and hippocampal lesions in rats. *J. Comp. Physiol. Psychol. 66*, 60–68
BELUGOU, J. L., BENOIT, O. & LEYGONIE, F. (1968) Décharges neuronales de l'hippocampe au cours de la veille et du sommeil. *J. Physiol. (Paris) 60*, 399
CLODY, D. E. & CARLTON, P. L. (1969) Behavioral effects of lesions of the medial septum of rats. *J. Comp. Physiol. Psychol. 67*, 344–352
DEFRANCE, J. F., SHIMONO, T. & KITAI, S. T. (1972) Hippocampal inputs to the lateral septal nucleus: patterns of facilitation and inhibition. *Brain Res. 37*, 333–339
DOMESICK, V. B. (1969) Projections from the cingulate cortex in the rat. *Brain Res. 12*, 296–320
DUNCAN, P. M. & DUNCAN, N. C. (1971) Free-operant and T-maze avoidance performance by septal and hippocampal-damaged rats. *Physiol. Behav. 7*, 687–693
FOX, S. E. & RANCK, J. B. (1975) Localization and anatomical identification of theta and complex spike cells in dorsal hippocampal formation of rats. *Exp. Neurol. 49*, 299–313
FRANZINI, C., CALASSO, M. & PARMEGGIANI, P. L. (1975) Attivazioni fasiche dei neuroni piramidali CA_1 dell'ippocampo durante il sonno a onde rapide dopo lesione del setto. *Boll. Soc. Ital. Biol. Sper. 51*, 981–986
GLICK, S. D., MARSANICO, R. G. & GREENSTEIN, S. (1974) Differential recovery of function following caudate, hippocampal, and septal lesions in mice. *J. Comp. Physiol. Psychol. 86*, 787–792
GOGOLÁK, G., KLINGBERG, F., OTSUKA, Y. & STUMPF, CH. (1967) Verhalten von Hippocampus-Pyramidenzellen bei retikulärer Reizung. *Experientia 23*, 190–191
GOGOLÁK, G., STUMPF, C., PETSCHE, H. & STERC, J. (1968) The firing pattern of septal neurons and the form of the hippocampal theta wave. *Brain Res. 7*, 201–205
GRANTYN, A. A. & GRANTYN, R. (1972) Postsynaptic responses of hippocampal neurons to mesencephalic stimulation: hyperpolarizing potentials. *Brain Res. 45*, 87–100
GREEN, J. D. & ADEY, W. R. (1956) Electrophysiological studies of hippocampal connections and excitability. *Electroencephalogr. Clin. Neurophysiol. 8*, 245–262
KICHIGINA, V. F. & VINOGRADOVA, O. S. (1974) The influence of the hippocampal stimulation upon the reticular formation units. *Fiziologicheskii Zhurnal 60*, 1648–1655 (in Russian)

Kultas, K. N., Smolikhina, T. I., Brazhnik, E. S. & Vinogradova, O. S. (1974) The effect of septal afferent lesions upon the acetylcholinesterase activity in the short-axon neurons of the hippocampus. *Doklady Academii Nauk.* 216, 462–463 (in Russian)

Lynch, G., Matthews, D. A., Mosko, S., Parks, T. & Cotman, C. (1972) Induced acetylcholinesterase-rich layer in rat dentate gyrus following entorhinal lesions. *Brain Res.* 42, 311–318

Macadar, O. J., Roig, J. A., Monti, J. M. & Budelli, R. (1970) The functional relationship between septal and hippocampal unit activity and hippocampal theta rhythm. *Physiol. Behav.* 5, 1443–1451

McLennan, H. & Miller, J. J. (1974) The hippocampal control of neuronal discharges in the septum of the rat. *J. Physiol. (Lond.)* 237, 607–624

McLennan, H. & Miller, J. J. (1976) Frequency-related inhibitory mechanisms controlling rhythmical activity in the septal area. *J. Physiol. (Lond.)* 254, 827–841

Mellgren, S. I. & Srebro, B. (1973) Changes in acetylcholinesterase and distribution of degenerating fibers in the hippocampal region after septal lesions in the rat. *Brain Res.* 52, 19–36

Mink, W. D., Best, P. J. & Olds, J. (1967) Neurons in paradoxical sleep and motivated behavior. *Science (Wash. D.C.)* 158, 1335–1336

Moore, K. Y., Björklund, A. & Stenevi, U. (1971) Plastic changes in the adrenergic innervation of the rat septal area in response to denervation. *Brain Res.* 33, 13–35

Morales, F. R., Roig, J. A., Monti, J. M., Macadar, O. & Budelli, T. (1971) Septal unit activity and hippocampal EEG during the sleep-wakefulness cycle of the rat. *Physiol. Behav.* 6, 563–568

Mosko, S., Lynch, G. & Cotman, C. W. (1973) The distribution of septal projections to the hippocampus of the rat. *J. Comp. Neurol.* 152, 163–174

Nikitina, G. M. & Boravova, A. I. (0972) Significance of the entorhinal input for development of the hippocampal EEG-correlate of the orienting reaction in ontogenesis. *Zhurnal Vysshei Nervnoi Deyatel'nosti* 22, 1023–1031 (in Russian)

Paxinos, G. (1975) The septum: neural systems involved in eating, drinking, irritability, muricide, copulation and activity in rats. *J. Comp. Physiol. Psychol.* 89, 1154–1168

Parmeggiani, P. L. (1967) On the functional significance of the hippocampal theta rhythm. *Prog. Brain Res.* 27, 413–441

Petsche, H. & Stumpf, C. (1960) Topographic and toposcopic study of origin and spread of the regular synchronised arousal pattern in the rabbit. *Electroencephalogr. Clin. Neurophysiol.* 12, 589

Petsche, H., Gogolák, G. & Van Zwieten, P. A. (1965) Rhythmicity of septal cell discharges at various levels of reticular excitation. *Electroencephalogr. Clin. Neurophysiol.* 19, 25–33

Poletaeva, I. I. (1968) The role of some subcortical structures in generation of the stress-rhythm in rabbit's electroencephalogram. *Biologicheskie Nauki,* N. 12, 31–34 (in Russian)

Raisman, G. (1966) The connexions of the septum. *Brain* 89, 317–348

Raisman, G. (1969) A comparison of the mode of termination of the hippocampal and hypothalamic afferents to the septal nuclei as revealed by electron microscopy of degeneration. *Exp. Brain Res.* 7, 317–343

Raisman, G. & Field, P. M. (1973) A quantitative investigation of the development of collateral reinnervation after partial deafferentation of the septal nuclei. *Brain Res.* 50, 241–264

Raisman, G., Cowan, W. M. & Powell, T. P. S. (1965) The extrinsic afferent, comissural and association fibres of the hippocampus. *Brain* 88, 963–996

Ramon y Cajal, S. (1955) *Studies on the Cerebral Cortex (Limbic Structures)*, Lloyd-Luke, London

Rose, A. M., Rattori, T. & Fibiger, H. C. (1976) Analysis of the septo-hippocampal pathway by light and electron microscopic autoradiography. *Brain Res.* 108, 170–174

Segal, M. & Landis, S. (1974) Afferents to the hippocampus of the rat studied with the method of retrograde transport of horseradish peroxidase. *Brain Res.* 78, 1–15

Shaban, V. M. (1970) Effect of the afferent paths lesions on hippocampal evoked potentials, theta-rhythm and neuronal activity. *Neurophysiologiya* 2, 439–447 (in Russian)

Sotnichenko, T. S. (1970) Experimental morphological investigation of the limbic cortex and

hippocampus in rodents and carnivora. *Zhurnal Evolyutsionnoi Biokhimii i Fiziologii 6*, 571–576 (in Russian)
SPENCER, W. A. & KANDEL, E. K. (1961) Hippocampal neuron responses to selective activation of recurrent collaterals of hippocampofugal axons. *Exp. Neurol. 4*, 149–156
STAFEKHINA, V. S. & VINOGRADOVA, O. S. (1973) Characteristics of sensory responses in neurons of the limbic (cingulate) cortex of the rabbit, in *The Limbic System of the Brain* (Cherkashin, A. N. & Kultas, K. N., eds.), pp. 191–201, Puschino-on-Oka (in Russian)
STAFEKHINA, V. S. & VINOGRADOVA, O. S. (1975) Sensory characteristics of the hippocampal cortical input. Entorhinal cortex. *Zhurnal Vysshei Nervnoi Deyatel'nosti 25*, 119–127 (in Russian)
STAFEKHINA, V. S. & VINOGRADOVA, O. S. (1976) Sensory characteristics of the hippocampal cortical input. Comparison between lateral and medial entorhinal cortex. *Zhurnal Vysshei Nervnoi Deyatel'nosti 226*, 1074–1081 (in Russian)
STORM-MATHISEN, J. & FONNUM, F. (1972) Localization of transmitter candidates in the hippocampal region. *Prog. Brain Res. 36*, 42–58
STUMPF, C. (1965) Drug action on the electrical activity of the hippocampus. *Int. Rev. Neurobiol. 8*, 77–138
STUMPF, C., PETSCHE, H. & GOGOLÁK, G. (1962) The significance of the rabbit's septum as a relay station between the midbrain and the hippocampus. *Electroencephalogr. Clin. Neurophysiol. 14*, 202–211
THOMSON, M. A. & LAGET, P. (1966) Influence des structures septales sur les rhythmes hippocampiques au cours de la maturation chez le lapin. *C.R. Séances Soc. Biol. 160*, 1775–1778
TURNER, B. H. (1970) Neural structures involved in the rage syndrome of the rat. *J. Comp. Physiol. Psychol. 71*, 103–113
URSIN, H. (1976) Inhibition and the septal nuclei: breakdown of the single concept model. *Acta Neurobiol. Exp. 36*, 91–115
VANDERWOLF, C. H. (1971) Limbic-diencephalic mechanisms of voluntary movements. *Psychol. Rev. 78*, 83–113
VINOGRADOVA, O. S. (1965) Dynamic classification of the hippocampal neurons. *Zhurnal Vysshei Nervnoi Deyatel'nosti 15*, 500–512 (in Russian)
VINOGRADOVA, O. S. & BRAGIN, A. G. (1975) Sensory characteristics of the hippocampal cortical input. Dentate fascia. *Zhurnal Vysshei Nervnoi Deyatel'nosti 25*, 410–420 (in Russian)
VINOGRADOVA, O. S. & DUDAEVA, K. I. (1971) Functional characteristics of the hippocampal field CA_1. *Zhurnal Vysshei Nervnoi Deyatel'nosti 21*, 577–585
VINOGRADOVA, O. S. & ZOLOTUKHINA, L. I. (1972) Sensory characteristics of the neurons in the medial and lateral septal nuclei. *Zhurnal Vysshei Nervnoi Deyatel'nosti 22*, 1260–1269
VINOGRADOVA, O. S. & ZOLOTUKHINA, L. I. (1973) The effect of electrical stimulation of the hippocampus and reticular formation upon the activity of neurons in the medial and lateral septal nuclei, in *The Limbic System of the Brain* (Cherkashin, A. N. & Kultas, K. N., eds.), pp. 161–173, Puschino-on-Oka (in Russian)
ZIMMER, J. (1974) Long term synaptic reorganization in rat fascia dentata deafferented at adolescent and adult stages: observations with the Timm method. *Brain Res. 76*, 336–342

Discussion

Weiskrantz: A number of issues arise: one is the question of the organization of theta; there is also the question of the characteristics of habituation in CA3 as a function of septal lesions, and the changes you reported. I wasn't clear whether this rapid disappearance of response showed the same specificity that habituation showed in a normal hippocampus.

Another issue is the question of novelty versus inhibition, because this leads us into two different kinds of interpretation of the septo-hippocampal influence. Inhibition will please some people greatly; novelty will please other people greatly!

Winocur: On this question of the characteristics in CA1 and CA3 in respect to the habituation pattern, you mentioned that the CA3 response habituates quite readily. Do CA1 responses habituate, and do CA1 cells respond to a variety of stimuli or are they unimodal?

Vinogradova: Habituation is almost linear in CA3; in CA1 it is different. There is a very short phase of building up of the response, and then it also gradually habituates.

The level of multimodality is lower in CA1 than in CA3. There are some unimodal units; the majority are bimodal. In CA3, units are responsive to all kinds of stimuli.

Winocur: Do septal lesions affect the habituation pattern in CA1 at all?

Vinogradova: Yes; normal habituation disappears from CA1, as well as from CA3, after septal lesions.

Gaffan: The question arose earlier of whether the long-lasting changes in the hippocampus are anything to do with memory, and this question seems highly relevant in the context of Dr Vinogradova's observations, because the detection of novelty must require a memory process (Gaffan 1976).

Vinogradova: This is already taking us into the realm of neurophilosophy! Let me answer *your* question, Dr Weiskrantz, on the rapid habituation in the deseptalized CA3. This was not selective at all. It occurs for the whole class of stimuli, so if we first use an auditory tone, there is a response which quickly disappears. Any other tone applied after that does not produce any response at all, which is quite impossible in the normal hippocampus.

Weiskrantz: I want to pursue Dr Gaffan's point, because it is important for our further discussion. As you describe the normal CA3 hippocampal units, they fit in very nicely to the Sokolov (1960) kind of model of habituation. Dishabituation occurs to a decrease in intensity, to a change in pattern—these features fit that general kind of scheme. I think Dr Gaffan's question is leading to the general point of whether that kind of approach, which entails the concept of a building up of an internal model with repetition of a particular external stimulus, helps us to understand the de-septalized hippocampus, or indeed the normal hippocampus. I don't think it *is* philosophy; it is a crucial question for understanding the implications of electrophysiological changes for memory.

Vinogradova: It seems that to have normal habituation in the hippocampus we need two inputs—both cortical and septal inputs. Only as a result of their interaction do we have this linear course of habituation. Without either input, we see severe derangement of the normal habituation process. We regard the hippocampus as a comparing device, which performs a match-mismatch operation. I think that habituation is a result of the interaction of septal and cortical input upon these units.

Ursin: There is an increasing amount of information showing that these septo-hippocampal structures are differentially involved in habituation and orienting responses (Köhler 1976 *a,b*). We also have additional data on lesions of the pyramidal cells of CA1 and CA3 in the dorsal hippocampus in the rat (T. F. Herrmann, H. Ursin & S. Levine, unpublished). Water-deprived, normal rats show a rise in corticosterone just before they expect to receive water; this rise is not present in the lesioned rats. Apart from that, we find no differences in the corticosterone response in these rats. Previously reported changes in diurnal rhythm and other hormone changes may be related to an expectancy function of these neurons, which seems to fit well with Dr Vinogradova's data.

Vanderwolf: Dr Vinogradova, when you make your lesions, you are also cutting off a piece of axon from the cells from which you record, or some of them. Do you know what the effect is?

Vinogradova: The hippocampus is very resistant to retrograde degeneration. We have discussed the possible number of collaterals of CA3 cells and it seems that even the system of Schaffer's collaterals alone is sufficient for survival of these neurons.

O'Keefe: Something that has bothered me for some time is the discrepancy between what people who record cells from the rabbit hippocampus are finding and what those who work on the rat are finding. We could not find these cell responses to simple sensory stimuli in CA1 in the rat, for example.

Vinogradova: Our data in rabbit are confirmed by other research workers (Dubrovinskaya 1971; Nikitina *et al.* 1975; Lidsky *et al.* 1974) who showed the same thing, so it is certainly true for rabbit.

O'Keefe: Dr Segal pointed out earlier that, in the rat at least, you can make a distinction, in both the CA fields and the dentate, between what appear to be two different types of cell: one which occasionally fires in a burst pattern where each spike is a different amplitude from the others (Dr Ranck's complex spike cell) and another type of cell with a more constant spike height which Dr Ranck calls a theta cell (Ranck 1973). You can make a lot of distinctions between these two classes in terms of 'spontaneous' firing rate, width of the action potential, and anatomical distribution within the hippocampus (Fox & Ranck 1975).

Vinogradova: We tried to find such correlations in the rabbit, and failed to find them. The so-called complex spike cells appear to me to be injured cells. I could not find these strong bursts with spike depolarization in normal rabbits.

Segal: On species differences, a recent paper by Miller & Groves (1977) shows inhibitory responses that are abolished by septal lesions. I found these same inhibitory responses in the rat hippocampus in response to a loud auditory stimulus and I found a correlation between locus coeruleus stimulation and inhibitory responsiveness (Segal & Bloom 1976). Your septal lesions would also interrupt the noradrenergic input; do you think these inhibitory responses might be mediated by the locus coeruleus?

Vinogradova: I agree completely, and I think that our further work will concern differentiation between the various components of the septal input.

Segal: It is also nice to see that the locus coeruleus innervation of CA3 is much heavier than that of CA1, and perhaps that is why you get more inhibitory responses in CA3.

Vinogradova: Yes; that is very possible.

Segal: To continue this same point, on your habituation curves; when I recorded from cells in the locus coeruleus I obtained rapid habituation (M. Segal, unpublished work 1977), so whether the hippocampus has to do with habituation or not I don't know, but if you find similar habituation curves in one of its afferents, this might be reflected in the hippocampus.

Vinogradova: This was the specific point of our studies. We have specifically investigated the question of whether there is true habituation in afferent systems of the hippocampus, to see if decremental effects in the hippocampus are just a passive mirror image of something that is transmitted to it. This seems *not* to be the situation.

Azmitia: However (to continue with this point of the role of the afferent systems in modulating habituation in the hippocampal system), you said that if a lesion is made in the perforant path the habituation response in CA3 is lost; furthermore, if a lesion is made between the hippocampus and the lateral septum the habituation response in the lateral septum is lost. Does this suggest, therefore, that the habituation in the lateral septum might also be a function of the perforant path? Have you investigated whether your perforant path lesions disrupt the habituation response in the lateral septum? The perforant path, or more properly the neurons whose axons comprise this pathway, might be a crucial centre in controlling the habituation response of the entire septo-hippocampal system.

Vinogradova: We haven't done this yet, but this is an important point. However, it is necessary to remember that habituation in the hippocampus is absent also after septal lesions with the perforant path intact. Thus, it cannot be a result of the perforant path action *per se*.

Zimmer: You said that you used two kinds of lesions for removing the perforant path input. In some experiments you denervated the upper third of the hippocampus and, in others, the middle third. Did you find any difference? By making the lesion corresponding to the upper third you will also damage some of the brainstem afferents curving around the cingulum; you wouldn't do that by a lesion corresponding to the lower third.

Vinogradova: The high level of lesions was used in our work on the so-called 'isolated hippocampus'.

Ranck: Do you have any data on the relation of the reactions of these cells to changes in neocortical EEG? Do you have any data on the slow wave hippocampal theta activity? And do you have any observations on the behaviour of the animal itself, and, if so, are there any correlations with the unit firing?

Vinogradova: The high level of lesions was used in our work on the so-called been done already, because the effects of perforant path or septal lesions on the hippocampal EEG are known. We made control EEG recordings in our rabbits at the beginning, middle and end of these lesion experiments and regularly observed desynchronization with delta components without the septum. This was a good control for the completeness of septal disconnection from the hippocampus. There was hypersynchronized theta without the perforant path, which may mean two things. It was interpreted at one time as the result of eliminating desynchronizing influences from the cortical input, but it might also be the result of plastic changes and an increase in the number of proliferating septal synapses.

Ranck: In the normal rabbit, is there any correlation with cellular activity during these slow waves?

Vinogradova: We did not use any special behavioural testing in these animals.

Lynch: Teyler has reported (Teyler & Alger 1976) that habituation can be elicited in the monosynaptic perforant path by electrical stimulation, in the rat. He showed that it matched some of the classical criteria of habituation. He found this in the perforant path but not in the Schaffer's collaterals. I wonder if this can be related to your thinking?

Vinogradova: I have some difficulty in interpreting Teyler's data, but influences of the perforant path on habituation of the sensory responses of pyramidal cells are indeed very important.

Gray: Could you summarize for us your views on the anatomical basis of the habituation occurring in CA3 and lateral septal neurons in the intact rabbit, on the basis of your studies?

Vinogradova: I am not certain what the exact anatomical basis is, but I can tell you our guesses about the physiology of the processes. In 1971 we investigated with Dr Dudaeva the influence of brief tetanic stimulation on the path ascending from the tegmental reticular nucleus, and the perforant path (Vinogradova & Dudaeva 1972). If we stimulated the reticular input, a lot of neurons in CA3 which were previously unresponsive became responsive, habituated reactions became dishabituated, weak reactions became stronger. After tetanization of the perforant path, responses to sensory stimuli were suppressed. Later, using the laminated structure of the hippocampus, we repeated this investigation, with stimulation of mossy fibres through the local electrode, which was implanted into the dentate fascia. The recording microelectrode was moved along the longitudinal axis of the hippocampus. It was obvious that only in the potentiated segment were CA3 reactions to sensory stimuli completely absent. In neighbouring segments there were some reactive neurons, and in distant segments there was normal reactivity. It was an experimental design that enabled us to see, in one and the same hippocampus, control and experimental effects of perforant path stimulation. Thus, in terms of our system, we can say that activation through the reticulo-septal input is equivalent to an increase in the novelty of stimuli, while tetanization of the perforant path is equivalent to an increase in the familiarity of stimuli. We tried to discover what kind of information is going to the hippocampus from the sources of these pathways. We haven't looked at locus coeruleus, but we have studied neuronal activity in the ventral tegmental nucleus, and dorsal and median raphe nuclei. We also investigated the lateral and medial entorhinal cortices and the strange symmetrical relay systems which exist for CA3 inputs on both sides (medial septal nucleus and the dentate fascia). We concluded that at the immediate relay systems of the hippocampus, as well as in the entorhinal cortex, there is no habituation, so one must assume that habituation is organized in the hippocampus itself, and is not just a passive reflection of any decremental changes in its inputs.

After this analysis of sensory inputs to the hippocampus, on the basis of experiments with lesions and stimulation, and with regard to the presence of long-latency potentiation in the perforant path and mossy fibres, it was possible to imagine the following sequence of events during habituation in the hippocampus. The first triggering stimulus which pushes the system out of a balanced state (that is, from a state of spontaneous activity) comes from brainstem structures through the septal route. In this state ('novelty', so to speak), the comparator is unbalanced; just one input comes to it and there is no matching input. The cortical signal to the structure builds up gradually, increasing in intensity, and long-lasting increased effectiveness of synapses is perhaps of importance for that. There is some critical point when the two inputs counterbalance; when the cortical signal becomes strong enough, it somehow blocks the reticular influence and the system returns to a balanced state (habituation, 'familiarity'). It is very difficult to explain from a neurophysiological point of view how an excitatory synaptic system (cortical) can block an activity which is coming from another source (reticulo-septal), and that is why we are now working on this problem in brain slices, trying to model this situation in a simple way (Vinogradova et al. 1976).

References

DUBROVINSKAYA, N. V. (1971) Phasic responses of hippocampal neurons and their possible functional significance. Zhurnal Vysshei Nervnoi Deyatel'nosti 21, 1084–1089

FOX, S. E. & RANCK, J. B., JR (1975) Localization and anatomical identification of theta and complex spike cells in dorsal hippocampal formation of rats. Exp. Neurol. 49, 299–313

GAFFAN, D. (1976) Recognition memory in animals, in Recall and Recognition (Brown, J., ed.), Wiley, London

KÖHLER, C. (1976a) Habituation of the orienting response after medial and lateral septal lesions in the albino rat. Behav. Biol. 16, 63–72

KÖHLER, C. (1976b) Habituation after dorsal hippocampal lesions: a test dependent phenomenon. Behav. Biol. 18, 89–110

LIDSKY, T. I., LEVINE, M. S. & MACGREGOR, S. (1974) Tonic and phasic effects, evoked concurrently by sensory stimuli in hippocampal units. Exp. Neurol. 44, 130–134

MILLER, S. W. & GROVES, P. M. (1977) Sensory evoked neuronal activity in the hippocampus before and after lesions of the medial septal nuclei. Physiol. Behav. 18, 141–146

NIKITINA, G. M., BORAVOVA, A. I., KRUCHKOVA, N. A. & POPOV, V. V. (1975) The role of sensory input in ontogenesis of neuronal activity in the hippocampus. Proceedings of XIIth Conference of the USSR Physiological Society, vol. 1, pp. 248–249 (Tbilisi)

RANCK, J. B., JR (1973) Studies on single neurons in dorsal hippocampal formation and septum in unrestrained rats. Exp. Neurol. 41, 461–555

SEGAL, M. & BLOOM, F. E. (1976) Norepinephrine in the rat hippocampus. IV. The effects of locus coeruleus stimulation on evoked hippocampal unit activity. Brain Res. 107, 513–525

SOKOLOV, YE. N. (1960) Neuronal models and the orienting reflex, in The Central Nervous System and Behavior (Brazier, M. A. B., ed.) (Transactions of the 3rd Conference), pp. 187–276, Josiah Macy, Jr Foundation, New York

TEYLER, T. J. & ALGER, B. E. (1976) Monosynaptic habituation in the vertebrate forebrain: the dentate gyrus examined in vitro. Brain Res. 115, 413–425

VINOGRADOVA, O. S. & DUDAEVA, K. I. (1972) On comparator function of the hippocampus. Doklady Academii Nauk SSSR 202, 486–489

VINOGRADOVA, O. S., BRAGIN, A. G., BRAŽNIK, E. S., KIČIGINA, V. F. & STAFECHINA, V. S. (1976) Auffassungen zur Funktion des Hippocampus und der mit ihm verbundenen Strukturen im Prozess der Informationsregistrierung. Z. Psychologie 184, 329–351

Single unit and lesion experiments on the sensory inputs to the hippocampal cognitive map

JOHN O'KEEFE and A. H. BLACK*

*Anatomy Department, University College London and Psychology Department, McMaster University, Hamilton, Ontario**

Abstract The hippocampal cognitive map theory states that the hippocampus calculates the animal's location in an environment and also the locations of objects such as rewards and threats. In this paper we report single cell experiments which explored how sensory inputs are used by the hippocampus to calculate spatial information and behavioural experiments which tested the sensory capabilities of fornix-lesioned rats. Both sets of experiments were done in cue-controlled enclosures which contained only a few distant cues by which the rat could locate itself and the goal. Other cues were eliminated by rotating the constellation of cues and the goal from trial to trial.

The results of the single cell experiment show that the place fields of hippocampal cells recorded in this environment are related to the controlled cues and, further, that some of these place cells maintain their fields after the removal of any two of four controlled cues. The lesion studies show that rats with damaged fornices can learn to approach distant cues behind and below the level of the goal but not ones behind and above the goal. A second study showed that the addition of redundant distant cues to the enclosure impairs the learning ability of the lesioned, but not the normal, animals.

In previous papers we have presented evidence in support of the view that the hippocampus functions as a cognitive map (e.g. O'Keefe 1976; Black *et al.* 1977; O'Keefe & Nadel 1978). In the present paper we wish to move on to a second stage of analysis and ask some detailed questions about the properties of the mapping system—in particular, the way in which it represents and transforms sensory information. In the first part of the paper we shall summarize the general theoretical framework but will give no evidence for it. Readers interested in the evidence should consult our earlier writings. The second part will deal with two questions. What sensory information is available to the mapping system? And how is that information represented in the mapping system? We shall describe methodological problems and discuss some preliminary data drawn from physiological studies of single unit activity in the hippocampus of the freely moving rat and behavioural studies of normal and lesioned rats. In the single cell studies we shall examine the hypothesis that hippocampal cells signal an animal's position in an environment and shall be asking how they use sensory information to do so. In the lesion studies, we shall consider the hypothesis that lesions to the hippocampus or fornix disrupt the spatial mapping system and force the animal to rely on other methods for solving problems.

PROBLEM SOLVING AND THE PLACE HYPOTHESIS

When faced with a problem such as finding food in a T-maze, rats usually do not act randomly but generate and test hypotheses about the solution. We have suggested that there are at least three different types of hypotheses: *place, or locale, hypotheses* which direct behaviour in relation to places in an environment (go to that part of the room, this part of the box is dangerous, this entire environment is dangerous: freeze); *guidance hypotheses* which specify approach to, or avoidance of, individual cues (approach the light, avoid the dark); and *orientation hypotheses* which involve movements referred to an egocentric body axis (turn right at the junction, keep the light in the upper left quadrant of the visual field) (see O'Keefe & Nadel 1978). Little is known about the factors which predispose an animal to select one hypothesis in preference to others or the mechanism for rejecting hypotheses. Clearly some environments and tasks elicit one class of hypothesis in preference to others. Closed mazes with a paucity of extra-maze information seem to bias the animal toward orientation (body turn) hypotheses while open, elevated mazes bias the animal toward place hypotheses. Also, overtraining and the use of non-correction procedures bias the animal towards orientation hypotheses while 'undertraining' and correction procedures bias the animal towards place hypotheses (A. H. Black & A. J. Dalrymple, unpublished manuscript). If the hypothesis originally selected does not lead to solution, the animal often switches to a different one. Reward contingencies clearly play a role here, but little is known of the mechanism.

One of the major tenets of cognitive map theory is that different hypotheses depend on information derived from different brain structures, and, in particular, that place hypotheses are dependent on the hippocampus. Furthermore, the information about the environment in these different areas of the brain is represented and manipulated in different ways. We have described some of these differences elsewhere (O'Keefe & Nadel 1978). Here we shall concentrate on the way in which sensory information is represented in the place system.

According to the cognitive map theory an environment is represented as a set of connected places. Each *object* in an environment to which an animal attends is represented as occurring in a place. Conversely, the primary way in which a *place* can be identified is by the cues or stimuli which are perceived when the animal is in that place. A place can also be identified in a secondary fashion, by its location relative to other places in an environment. Thus, theoretically, once an animal has located itself in an environment using the cues available to it at one location, it can use the mapping system to provide it with information about other places in the environment.

A main feature of the place system is that the central representations of objects and places in the environment are linked together in terms of their spatial and topographical relationships. This feature of the mapping system allows one to make specific predictions about the ways in which certain types of sensory information will be used when the animal employs a spatial hypothesis in problem-solving situations.

1. The mapping system allows the animal to 'act at a distance'. Consider the following experiment. An animal has formed a map of a particular situation in which two locations, A and B, can be identified. The stimuli and objects associated with location B cannot be perceived directly from location A, and vice versa. One places the animal directly into location B so that it is isolated from the rest of the situation (including A), and rewards that animal. Then, finally, one places the animal in location A and observes its behaviour. One would predict that the animal would move toward location B (even though it cannot perceive the stimuli and objects in location B that were paired with reward) because of the remembered relationship between the central representations of A and B in the animal's spatial map of the environment. We have suggested that this property of 'action at a distance' is unique to the hippocampal system, at least in the rat. Other hypotheses rely on cues which can be observed while the reward is being consumed or on highly specific chains of stimulus–response links.

2. In an environment that is rich in stimuli, animals using place hypotheses will not be disrupted by the removal of a particular cue in an environment. They will notice the change and investigate it but their location and the relative location of the goal will be calculated from the remaining cues. In contrast, the behaviour of animals using stimulus–response hypotheses should be more dependent on specific environmental cues, either close to the goal (guidance hypothesis), or at the start arm (orientation hypothesis).

3. Animals using place hypotheses will not need to rely on stimuli close to or behind the goal since the animal can use any available cue to locate its position and then calculate the direction of the reward using the mapping

system. Animals using stimulus–response hypotheses will rely on cues that are perceived just before or during reinforcement or on specific body turns or chains of body turns. Such cues will typically be located just beyond the choice-point, or close to or behind the goal.

4. As the number of cues in an environment increases, each place in that environment can be identified by a larger number of cues and more places can be distinguished from each other; the grain of the map gets finer. Thus increasing the number of cues in an environment should make place learning easier or at worst leave it unaffected. Stimulus–response hypotheses depend on the selection of the correct cue which will guide the rat to the goal or which will trigger off the appropriate motor chain. As the number of cues is increased many will be irrelevant and will thus increase the 'noise' from which a relevant cue must be selected and make it more difficult to select an appropriate stimulus–response hypothesis.

METHODOLOGICAL CONSIDERATIONS

Theories have a strong influence on one's experimental methods. Before we discuss some of the results directed at testing the above properties of the hippocampal mapping system we shall make several comments on the testing procedures.

1. Problems which can only be solved by place hypotheses

Most common laboratory tasks can be solved through the use of more than one hypothesis. Similarly, at least theoretically, the same sensory cue might be used by different brain areas to form the basis for different hypotheses. A light near the goal can be used to locate the animal's position in the apparatus and the relative position of the goal, or act as an approach cue, or it can elicit a stereotyped chain of motor movements. It is possible that we shall one day realize that the sensory cues for different hypotheses are in fact drawn from different classes of cues and that, for example, while the light as an object is used for guidance of approach responses, it is the shadows or light/dark gradients within the environment that it creates which are used by the mapping system. At this early stage in the investigation we do not know.

If the same cues can be used in different hypotheses, it follows that it is unlikely that a stimulus (environment) configuration can be devised which will force an animal to use one particular hypothesis in simple problem-solving tasks. There are more complex tasks, however, which do seem to require the use of a mapping system or its equivalent—for example, certain tasks which have been labelled 'reasoning problems' (Tolman & Gleitman 1949). Such tasks would employ the procedures outlined in the first of the predictions concerning the spatial mapping system that were listed above. Animals with damage to the hippocampal formation should fail to solve such problems.

2. Problems which can be solved through the use of place or stimulus–response hypotheses

Problems of the sort described above that can be solved only by place hypotheses are rare. In addition, they are complex, and failures among lesioned animals can usually be attributed to a variety of factors. For this reason, an alternative research strategy is often employed. It is to choose a problem which can be solved by several hypotheses and to look for the differences between normal and hippocampally damaged animals. The most appropriate procedure is to devise a task which is equally difficult for both groups, and then to develop probe techniques which permit one to determine the type of hypothesis that is being used by the subject—in particular, to determine whether changes in the properties of the cue configuration have differential effects.

We have developed an environment with a limited set of sensory cues which rats can use to solve problems through either place or stimulus–response hypotheses. We shall discuss four experiments that were done in the cue-controlled enclosure—one single unit study and three lesion studies. In the single unit study we asked which cues in the enclosure were responsible for the selective firing of the place cells in one part of the environment. The lesion studies were designed to answer the questions 'Can rats with damage to the fornix use distant cues to guide their behaviour?', 'Can they learn a task on the basis of distant cues', and 'What is the effect of increasing the number of cues within the enclosure?'

SINGLE UNIT STUDY

Some cells in the hippocampus of the freely moving rat fire when the animal occupies one part of an environment but not others (O'Keefe & Dostrovsky 1971; Ranck 1973; O'Keefe 1976; Olton et al. 1978). It was clear in our earliest studies that this selective unit firing in the *place field* was due to some environmental influence, since the behaviour of the animal was often irrelevant; many cells would fire when the animal was passively placed in the field as well as when the animal actively entered it. In these earlier studies of hippocampal units we found that eliminating the sensory information from different modalities by, for example, turning off the room lights or changing the intra-maze cues did not disrupt this selective unit firing in the place field. This might mean one of two things. Either the cells were receiving information along several different modalities and the elimination of any part of this input still left sufficient information to accurately identify the place field, or the cells received sensory information of a sort which was not disrupted by the manipulations we used; for example, they were responding to something like geomagnetic fields.

The unit studies were done at University College London in collaboration with D. Conway (O'Keefe & Conway 1978). Animals were trained inside a square enclosure, 215 cm on each side (Fig. 1). The walls were formed by floor-length black curtains. Within the enclosure, there were four cues by which the rat could locate itself: a dim light, a white card, a buzzer and a fan. Four male hooded rats were made hungry and trained to go to one of the arms of a T-maze to obtain reward. From trial to trial, the maze and the cues were randomly rotated by some multiple of 90° relative to the environment but maintained the same spatial relation to each other. In order to rule out other intra-maze cues as a means of solution, the arms were interchanged from trial to trial. Body turns were ruled out by randomly rotating the stem of the T-maze 180° relative to the cross-bar so that on one-half of the trials a right turn was required to reach the goal while on

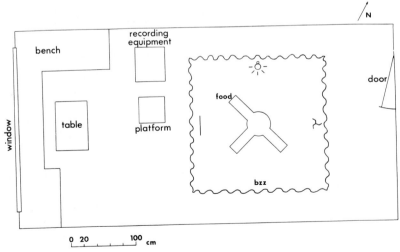

FIG. 1. Layout of the experimental room in the single cell experiment. The cue-controlled enclosure is the curtained-off area just to the right of centre. ☼, low wattage light; ⋋, fan; |, card; bzz, buzzer. (From O'Keefe & Conway 1978.)

the other half, a left turn was required. After the rats had learned the task, they were further trained so that after running to the goal arm and consuming the reward they should run to the non-goal arm and thence back to the start arm, where they received a second reward. Thus on each trial they made a complete circuit of the maze, giving the hippocampal units an equal opportunity to fire on all parts of the maze. The place where the units fired on the maze was recorded by pulsing a light-emitting diode on the rat's head whenever an action potential occurred and photographing these light flashes with an overhead camera.

Twelve units with place fields inside the enclosure were studied on sufficient trials to enable us to answer questions about their sensory inputs. We shall use one of these units as illustration (Fig. 2). The first finding was that the fields of all twelve units were related to the four controlled cues inside the enclosure and not to cues outside the enclosure or to the maze configuration itself (Fig. 2). This unanimity surprised us since we had not expected to so successfully eliminate the influence of the external world. It means that the hippocampal cells are under the influence of cues such as lights and cards and, at least in the cue-controlled enclosure, not under the influence of geomagnetic or other very distant sensory influences. Eight units were studied on probe trials in which we removed some of the spatial cues in order to test whether some cues were more important than others. For the unit illustrated, giving the rat either the card and the buzzer (Fig. 2H) or the light and the fan (Fig. 2I) was sufficient to maintain the field intact. Removal of all four cues resulted in a loss of the field and an increased firing over the total maze (Fig. 2G). It appears that the sensory stimuli act to block off the firing on parts of the maze outside the place field. Five of the eight units studied with these probes showed a similar behaviour.

For two other units the place field was dependent on one or two of the cues inside the enclosure and the light proved to be the most potent cue. Removal of the relevant cue(s) eliminated the firing in the place field, indicating that the influence was excitatory. The role of the wall cues was difficult to assess for the final unit since the field changed during this phase of testing.

We draw several conclusions from these single cell studies: *(a)* place cells in the rat hippocampus identify parts of an environment on the basis of common-or-garden cues, such as lights and sounds, and are not responding to some unidentified unusual stimulus such as geomagnetism. Notice that we have not ruled out the possibility that place cells could be influenced by such stimuli, only that they need not be and, in the present experiments, were not. *(b)* While some place cells fire in a part of an environment because they receive 'excitatory' sensory influences when the rat is there, others have

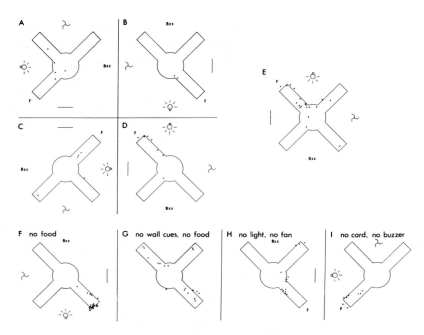

FIG. 2. A place cell whose field is dependent on 'inhibitory' sensory influences. A–D: four trials on which the controlled cues and the maze are rotated relative to the rest of the experimental room. The unit consistently fires in the goal arm. E: pictures A–D superimposed so that they all have the same orientation with respect to the controlled cues. The relationship between the cell firing and these cues can be seen clearly. F: a probe trial on which food is absent from the goal arm: the unit maintains place field but there is increased firing as animal sniffs around reward area. This firing is typical of *misplace* cells with place fields in the goal arm. G: probe trial on which all cues and the food are absent. Place field is lost and there is increased firing on maze. H: probe trial on which light and fan are absent. Place field maintained but increased firing. I: probe trial on which card and buzzer are absent. Place field maintained. (Modified from O'Keefe & Conway 1978.)

fields which are constructed, at least in part, by sensory influences which are 'inhibitory'. Removal of sensory cues increases the firing rate of one of these cells and in some cases increases the size of the place field. Removal of all sensory cues causes the cell to fire all over the environment. The place field where the cell fires is, in fact, a lacuna in a sea of suppression. *(c)* For these 'lacuna' place cells, no single stimulus is responsible for the selective firing in one part of an environment. Individual cues appear to 'inhibit' cell firing over large areas and there is a large degree of overlap amongst the inhibitory fields of the different cues. This redundancy ensures that the place field is independent of any particular cue.

We turn now to the results of experiments in the cue-controlled enclosure which examined the abilities of rats with damage to the hippocampal system.

LESION STUDIES

As pointed out above and in a previous paper (O'Keefe et al. 1975), the results of many behavioural lesion experiments could be explained by an inability of the rats to use spatial cues, but could also be explained by a failure to use distant cues. The first data that we shall present were directed, therefore, to the following question. Can rats with damage to the fornix use cues distant from the T-maze to guide their behaviour? The rationale for this study is straightforward. If rats with damage to the hippocampal formation cannot use distant cues, we can account for their deficit in spatial tasks in terms of some difficulty with the use of distant stimuli rather than an inability to use the spatial mapping system.

These studies were done at McMaster University in collaboration with H. Anchel and M. Hammer. They were carried out in a cylindrical cue-controlled enclosure (diameter 2.8 m). Both ceiling and floor could be rotated independently of each other; the walls were curtains suspended from the ceiling. A T-maze similar to the one used in the unit studies was located in the centre of the room. As in the unit studies, animals were trained to go to one of the arms of the cross-bar of the T in order to obtain reward, and the arm of the T was rotated 180° between trials in order to prevent the animal from solving the problem by turning in a given direction.

Six rats with lesions of the fornix and six control rats were trained to approach a stimulus placed just behind the goal at the end of one arm of the T-maze. The discriminative stimulus consisted of a white cardboard rectangle, with a 5 watt light protruding from it. The ceiling, floor, arms of the maze, etc. were rotated between trials in order to prevent the use of other cues. After the rats had reached a criterion of 17 out of 20 correct choices on a given day, the stimulus was moved away from the goal. For half of the animals it was moved 75 cm back from the end of the arm of the T-maze and 20 cm below the level of the arm (CUE-DOWN condition). In another group it was moved 75 cm back and 45 cm above the level of the arm (CUE-UP condition).* For both lesioned and control rats, there was slight deterioration

---

*Further animals have been trained in a CUE-UP condition in which the distance of the discriminative stimulus above the maze was equal to the distance of the discriminative stimulus below the maze (20 cm). The data for the two CUE-UP groups are much the same.

of performance (Fig. 3). But the important point is that there was no difference between animals with fornical lesions and controls. Both responded in the same way to the distant discriminative stimulus. Therefore, it would seem that rats with fornical lesions can use single distant stimuli.

One might ask whether the problem is not so much in the use of distant stimuli as in learning to use distant stimuli. In order to deal with this question, we trained animals with lesions of the fornix and controls using the same light-card discriminative stimulus as in the previous experiment under three conditions. In one condition, the compound light-card discriminative stimulus was placed directly behind and adjacent to the foodcup (CUE-NEAR condition). The second condition was the CUE-DOWN condition described above (i.e. the stimulus was moved back from and below the correct goal areas). The third condition was the CUE-UP condition (i.e. the stimulus was moved back from and above the correct goal area).

Data on speed of learning (number of blocks of 20 trials to criterion) are presented in Table 1. The control rats learned at the same speed under all conditions; the rats with lesions of the fornix learned slowly when the stimulus was in the CUE-UP position and more quickly when the stimulus was in the CUE-NEAR and CUE-DOWN positions. Individual comparisons were

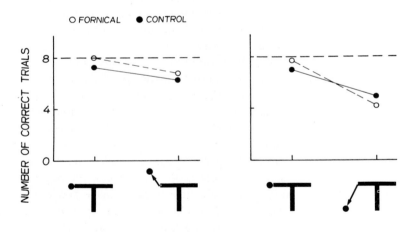

FIG. 3. Number of correct responses on eight trials for animals with lesions of the fornix and controls. The dot represents the position of the discriminative stimulus (SD) relative to the T-maze (viewed from the side). On the left-hand side of each graph the SD is adjacent to the correct goal arm of the T-maze. In one of the right-hand side conditions the SD was moved back from and above the correct goal arm; in the other right-hand side condition the SD was moved back from and below the correct goal arm.

TABLE 1

Speed of learning as a function of the location of the discriminative stimulus (SD)

	Blocks of trials to criterion		
	Distant SD (High)	Distant SD (Low)	Near SD
Fornical lesions	8.6	4.7	3.5
Normal controls	2.2	2.4	1.7

made among the six groups. Rats in the lesion group in the CUE-UP condition differed from all the other groups; the other groups did not differ from each other. It would seem then that rats with lesions of the fornix can learn to approach some distant stimuli (CUE-DOWN condition) as well as they do some nearby stimuli (CUE-NEAR condition).* Furthermore, at least in this situation, they do not differ significantly from controls.

One could suggest a number of hypotheses to account for these results. We shall describe only one which we found particularly interesting because it led to predictions which were supported by data of subsequent experiments. Animals with lesions of the fornix must be exposed to a stimulus, either just before or during reinforcement, for the stimulus to become a discriminative

---

*One can ask whether the rats really were responding to distant stimuli. Perhaps they were responding to within-apparatus stimuli; for example, perhaps one arm of the T-maze was illuminated more than the other arm because it was closer to the light-card discriminative stimulus. One might even suggest that the rats were not using the light-card cue at all, and were learning to approach the correct goal area on the basis of some unknown cue. In order to deal with these questions probe trials were interspersed among regular training trials after the criterion was met.

First, a second light-card cue identical to the first was placed in the same relationship to the other arm of the T-maze so that both of the goal areas were marked by the same light-card cues. We reasoned that the presence of two identical cues would confuse the rat if the light-card cue was being employed, and would not disrupt performance if some other stimulus was being employed. For the eight lesioned rats and eight control rats which were subjected to this probe, performance deteriorated to chance level. It seemed, therefore, that the rats were using the light-card cue.

The question as to whether some within-apparatus differences occurred as a consequence of the light-card cue's location nearer to one goal arm is more difficult to deal with. Foodcups were removed on some probe trials in case they reflected light differently. The performance of the rats was not disrupted by this probe. Also, the fact that the high-position cue group with fornical lesions differed from the middle- and low-position cue groups provides further support for the view that within-apparatus cues were not important. Presumably all three positions resulted in similar differences in within-apparatus illumination. Therefore, all three should have led to the same rate of learning. But, they did not. While it is difficult to rule out the use of within-apparatus cues, the probe data taken in conjunction with the speed of learning data indicate that it is unlikely that such cues were used. More data on this question are given in the footnote on p. 191.

stimulus. With stimuli near the apparatus (for example, on the floor of the correct goal arm) the physical arrangement of the stimuli ensures that this contiguity between stimulus and reinforcement will occur. But, for distant stimuli, this contiguity will only occur if the animal is perceiving the stimulus at the correct time. We would suggest that this occurred for stimuli placed below the floor of the runway because the rats were, in fact, paying attention to such stimuli before and during reinforcement. But it was much less likely to occur for stimuli placed above the floor because the rats were not paying attention to such stimuli at the crucial time.

We would argue further that normal animals are not constrained in this way—provided that they have formed cognitive maps of the experimental situation. That is, not only are the cues which are perceived just before or during reinforcement used as discriminative stimuli, but also stimuli which are related to these cues in the spatial map. In short, stimuli become part of the discriminative stimulus complex because of remembered spatial relationships among components of the spatial map.

If this hypothesis is correct one should be able to improve the performance of rats with fornical lesions by using distant stimuli that force the rat to attend to them at the crucial moment. Also, if the hypothesis is correct, the addition of further distant stimuli that are perceived at the crucial time but do not guide the animal to the goal area might act to confuse rats with fornical lesions. Suppose that a number of distinctive and different stimuli were scattered about the periphery of the room. If some of these were located so they would be perceived at the choice point but approaching them would lead the animal down the incorrect arm, animals with fornical lesions might become confused and learning would be slow. Control animals, on the other hand, should show no effect under these conditions since they would be aware of the spatial arrangements of the stimuli relative to each other and the correct goal. The following experiment was done in order to test this prediction.

In this experiment the discriminative stimulus was the same card-light stimulus that had been used in the previous experiment. It was placed in the CUE-DOWN position. (This was the position of the distant discriminative stimulus that led to more rapid learning among rats with lesions of the fornix.) In addition, four further stimuli were located about the periphery of the room—a string of Christmas tree lights, a cage in which the rat had been transported to and from the experimental room, a fan and a clicker.

Discrimination training procedures were identical in all respects to those used in the previous experiment. Six rats with lesions of the fornix and six control rats were trained on this discrimination.

The results were clear-cut. Rats with lesions of the fornix took longer than control rats to learn the discrimination (8.0 blocks of trials to criterion performance as against 2.8 blocks). An informal comparison of these results with the data from the previous experiment (Table 1) suggests that the effect of adding additional stimuli was to retard the speed of learning of animals with lesions of the fornix. There was no effect of additional stimuli for control rats. These results, then, indicate that our prediction was correct. The addition of stimuli produced slow learning in rats with lesions of the fornix but not in control rats.*

In conclusion, the results suggest that the animals with lesions of the fornix can use single distant stimuli once they have learned to attend to them, and can learn at least some discriminations based on distant stimuli with relative ease (CUE-DOWN condition). They do have trouble learning discriminations based on some distant single stimuli (CUE-UP condition), and in some multiple stimulus situations. We suggest that the latter difficulties arose because of the lack of a spatial mapping system which left the lesioned animals with learning strategies whose properties limited the animal to stimulus–response chain learning. We hope that experiments that we are now doing, in which stimuli are manipulated (removed, added and transposed) during or after learning, will resolve the question of the role of the hippocampus in spatial information processing in the near future.

ACKNOWLEDGEMENTS

The preparation of this paper and some of the research described in it was supported by the Wellcome Trust (U.K.), the National Research Council of Canada (APA 0042) and a NATO travel grant to A. H. Black.

---

*As in the previous experiment, we were worried whether the animals were responding to within-apparatus cues, in particular to cues that might be produced by differences in illumination. In order to check on this, an overhead light, which was considerably brighter than the lights used as distant stimuli, was placed directly above the choice point of the T-maze, and training continued after criterion was reached. There was a small disrupting effect of the introduction of the overhead light but there were no differences between lesioned and control groups in amount of disruption. This result supports the view that within-apparatus differences in illumination were not necessary for performance in lesioned rats, and did not contribute to difference between lesioned and control rats. It is important to note, however, that it took the animals with fornical lesions longer to re-reach the criterion of 17 out of 20 trials correct on a given day under this new condition than controls. That is, even though the disrupting effect was relatively small and the same in both groups at the beginning of training, it took the animals with fornical lesions longer to reach criterion again. The reason for this is not clear.

References

BLACK, A. H., NADEL, L. & O'KEEFE, J. (1977) Hippocampal function in avoidance learning and punishment. *Psychol. Bull. 84*, 1107–1129

O'KEEFE, J. (1976) Place units in the hippocampus of the freely moving rat. *Exp. Neurol. 51*, 78–109

O'KEEFE, J. & CONWAY, D. H. (1978) Hippocampal place units in the freely moving rat: why they fire where they fire. *Exp. Brain Res.*, in press

O'KEEFE, J. & DOSTROVSKY, J. (1971) The hippocampus as a spatial map. Preliminary evidence from unit activity in the freely-moving rat. *Brain Res. 34*, 171–175

O'KEEFE, J. & NADEL, L. (1978) *The Hippocampus as a Cognitive Map*, Oxford University Press, Oxford, in press

O'KEEFE, J., NADEL, L., KEIGHTLEY, S. & KILL, D. (1975) Fornix lesions selectively abolish place learning in the rat. *Exp. Neurol. 48*, 152–166

OLTON, D. S., BRANCH, M. & BEST, P. (1978) Spatial correlates of hippocampal unit activity. *Exp. Neurol. 58*, 387–409

RANCK, J. B., JR (1973) Studies on single neurons in dorsal hippocampal formation and septum in unrestrained rats. *Exp. Neurol. 41*, 461–555

TOLMAN, E. C. & GLEITMAN, H. (1949) Studies in learning and motivation: I. Equal reinforcements in both end-boxes followed by shock in one end-box. *J. Exp. Psychol. 39*, 810–819

Discussion

Weiskrantz: Two kinds of issue arise from this paper, firstly questions of detail; for example, where are these space units, how many are there, and do they habituate? A more general point is the implication that, among other things, the hippocampus is a device for coding space, which will take us to more highly theoretical issues.

Grossman: On a question of detail, how do your 'spatial' cells respond to sensory inputs which lack the complication of the spatial pattern of a presentation?

O'Keefe: When we started our unit work on the hippocampus we were very much aware of Dr Vinogradova's work and we spent a great deal of time and effort investigating the response of rat hippocampal cells to simple sensory inputs. We tried clicks, whistles and other noises. We waved striped cards, hands and other visual stimuli in front of the animal's eyes. We picked it up, manipulated its joints and rubbed it all over the body—the sorts of stimuli which we found worked fairly well in firing cells in the appropriate sensory neocortical areas. We did find changes in hippocampal units to these stimuli but they were always related to changes in the animal's state of arousal or alertness as determined by the rat's behaviour and its hippocampal EEG or, in the case of theta cells, to the animal's behavioural response to the stimulus. If we ensured that the rat was alert and attending before we presented the stimulus and the animal didn't move in response to the stimulus, we found no specifically sensory unit responses to the stimulus (O'Keefe & Dostrovsky 1971; cf. also Mays & Best 1975, who have found a similar dependence of hippocampal unit responsiveness to sensory stimuli on the rat's state of alertness). It was partly because we failed to find specific sensory responses in these hippocampal units that we changed to more complex spatial environments.

Molnár: If you do several trials, do you get the same results?

O'Keefe: We generally give the rats several ground trials, in which the animal is run in the standard situation, with the maze and the stimuli being rotated relative to the real world, to be sure that we have a stable field. Fig. 2 A-D (p. 186) shows four such trials.

Molnár: Are these well-trained rats?

O'Keefe: Yes. They are trained to a criterion of 9 out of 10 correct responses and given some over-training to teach them to do the circuit, but they are not over-trained. I am not sure that it makes much difference. We have no reason to believe that the place cells change if the rat is over-trained in a stable environment.

Segal: Where are you during the experiment?

O'Keefe: There are two of us doing the experiments, Dulcie Conway and myself. She is part of the maze and I am part of the external world! That is, she always puts the animals into the curtains from behind the start arm, and stays there, whereas I always stay in the same position in the experimental room where I can monitor the unit firing on the oscilloscope. So I stay as a stable part of the external environment and she varies with the start arm. Neither of us is any more, or less, a cue than any other part of the environment.

Rawlins: Professor Weiskrantz asked how many of these place cells there are, but if complex spike cells are inactive, except when the animal is located in the right place, perhaps it is not possible to assess the proportion of cells which behave in this way. It may be that there is a considerable population which is not seen, because it is composed of quiescent cells.

O'Keefe: That is right. I think that point is very relevant. Some place cells do not have a place field in every environment. In the experiments I have just described we recorded these place units in two environments. We looked for place fields in CA1 on the platform outside the maze as well as on the maze inside the cue-controlled enclosure. Some cells had fields only in one of the two environments, others had fields in both environments, and a third small number of cells did not have a field in either environment. So one cannot expect to find a place field for a place cell in every environment.

Of the cells which had place fields inside the cue-controlled enclosure, we selected the best cells which we thought we could record from for five hours or more, in order to run the rat on all of the various probe trials.

Weiskrantz: So the cells don't habituate?

O'Keefe: We have not seen any unprovoked changes in fields over long periods, provided that we treat the cells well.

Gray: That is not necessarily a discrepancy, because you are doing 30 trials, which may be too few in a complex situation for habituation to occur.

Gaffan: I am interested in the acquisition trials. Behaviourally, what hypothesis does the animal entertain before solution, and does it ever include an 'external world' hypothesis? I want to know whether the rat is conscious of the external world or not, and whether the cells' activities are determined by the rat's hypothesis. Do you record these cells' activity during the pre-solution period?

O'Keefe: There are about four questions there. The answer to the last one is no, we have not recorded from the place cells during the pre-solution period. Your first question concerned the hypotheses entertained in the pre-solution period. Some rats hit upon the correct place hypothesis right off. Those that don't, usually adopt a body-turn—or, as we call it, an orientation—hypothesis: the rat might turn left on a consecutive series of trials. Are the animals aware of the external world? We were rather surprised that none of our twelve place cells had fields related to the 'external world', because I would expect to find such cells. We have enough evidence from normal rats run in this environment that some animals are aware of the 'external world'; sometimes an animal adopts the hypothesis of going to, say, the arm close the oscilloscope even though this is rewarded only part of the time. This type of 'external world' hypothesis seems to occur more often when we train animals on a four-armed plus-shaped maze, where they have the opportunity to go to the same part of the external world on a greater percentage of trials than on a three-armed T-maze.

Gaffan: Have you ever tried to teach them to do that?

O'Keefe: No, but Dr Black has.

Black: We have attempted to train rats to respond to cues that were outside the enclosure in two experiments. In both we used a room which was cylindrical in shape, black on the inside, and whose ceiling and walls could be rotated as well as its floor. (This room is described in more detail in our paper.) In a first experiment, done in collaboration with M. Hammer and B. Ristow, we attempted to train rats to go to one arm of a T-maze. The correct arm was defined by its position relative to cues outside of the room. For example, the correct arm might always point to the north. There were no cues within the room, and the walls, maze, etc. were rotated between trials. All but one animal performed at chance level. One animal performed at above chance level on a few days, but it never maintained the performance for long. So, it would seem that most rats did not use cues outside the room to guide their behaviour in this situation.

In a second experiment, done in collaboration with G. Augerinos, we used an eight-arm maze. Food was placed at the end of each arm, and the rat was permitted to run until it had obtained all eight pellets of food. In this experiment, a series of stimuli was placed on the inner walls of the room. After the rat had learned to obtain the pellets of food without error, we carried out the following procedure. The rats were forced to make three different choices. Then the room was rotated 180°, and the rats were given five more choices. In making these five choices, the rat could orient with respect to the cues within the room that had been rotated, or with respect to cues outside of the room that had not been rotated. The rats made their choices on the basis of the cues within the room. Again, cues outside the room were not used by the rats to guide their behaviour.

Andersen: Many of us are happy about the fact that these cells seem to be sensitive to the animal's location in space. The presented data seem convincing, but my question is, how *important* is the relationship? In other studies of behavioural situations and their neural correlates the discharge rates are generally very much higher than yours. These are extremely low: about the same rate as the climbing fibre activation, in the cerebellum, where nobody has shown any clear relation to behaviour. So, is the relation so weak that it is not very interesting physiologically?

O'Keefe: One could only answer that if one could find a situation or stimulus which caused the cells to fire at a much greater rate. We don't know the dynamic range of the firing. The highest frequency we have seen in these cells is in the mismatch, or misplace, situation. This occurs when we are recording from a cell with a place field which includes or abuts on the reward area. On those trials when the rat runs to the goal but doesn't find the reward there, it sniffs vigorously in the goal area and there is a concomitant high frequency discharge in the cell—say, 10 spikes per second. Fig. 2E (p. 186) is a good example of this misplace firing.

Ranck: Complex spike cells rarely fire faster than 10 per second; my best might be 25/s, but only lasting for 1–2 s. In slow wave sleep, however, almost all fire at about 5/s, so sleep speeds them up to an intermediate range. The kind of rates that John O'Keefe sees are those I see too.

Andersen: 10 Hz is all right with me!

O'Keefe: Ten per second is rare, occurring only during some very specific behaviours in the place field.

Winocur: You said that you got no responses to external sensory stimuli. Have you looked for responses to stimuli that have previously been conditioned; that is, to familiar stimuli that have learned associations?

O'Keefe: We looked for unit responses to conditioned stimuli in our previous experiment (O'Keefe 1976) but didn't see any. There we placed the same small porcelain dish down in different arms of an elevated T-shaped maze and tapped it against the maze, making a clicking noise. The rats quickly learned to approach the dish (and the sound). They had become conditioned stimuli. In spite of this we never saw any unit responses to the presentation of these stimuli and a place unit only fired when the rat approached them if the animal ran across the place field of that unit in doing so.

Weiskrantz: There is a related point, namely, whether or not the animal has been trained, do these responses depend on some significant reward or some significant event being present, or are they pure place units? This is related to David Gaffan's point on the pre-solution period. Spatial responses in rats are extremely prominent and spontaneous spatial alternation is one clear indication of that. The question here is not whether these are place units, but whether they are jointly dependent on space *plus* the animal's having experienced a significant event in a particular place.

O'Keefe: It may be the case that if you found a unit which fired in a place and you then did something, in addition, to the animal in that place, you would increase the firing of the cell, but I haven't done this systematically.

Black: John O'Keefe has searched for place cells on a platform placed outside the enclosure in which the rats were reinforced. The animals were tested on this platform before they had been reinforced, and, I think, were not reinforced on the platform itself. He did find place cells in this situation. Therefore, the place cells don't seem to be related to training or to some significant reward, etc.

O'Keefe: For many years we have looked for these cells that way; in fact in this experiment we looked for place fields on that small platform outside the testing situation where the rat had never received a reward. It may have secondary reinforcing properties, because the rat is always taken from there and put into the cue-controlled enclosure in which he is rewarded. But we also recorded these cells in rats that have never been deprived of food or water, or rewarded or punished in the testing situation, except for the fact that they sit very close to me for some time! So we can find these cells in situations where there is no obvious reinforcement and the animals are not motivated.

Winocur: What about non-spatial tasks? What recording have you done from tasks where the primary cues were of a non-spatial nature?

O'Keefe: We haven't done anything on this, except what I mentioned in answer to your previous question.

Robinson: Dr O'Keefe, you said earlier that hippocampal cells in the rat did not respond to sensory input if you controlled for the animal's response. If you did *not* control for the animal's behaviour, did you find that you did get cellular responses to stimuli such as light flashes?

O'Keefe: That would pertain to the theta units. In the rat, these cells fire in relation to movements such as walking, sniffing and orienting; thus, if we present a stimulus and the animal orients towards it, the theta cells fire. If we present the same stimulus and the animal doesn't orient, we have not found any significant firing. However, let me add that I think that there is a real difference between the theta systems of the rabbit and rat; there is no reason to believe that theta cells in the rabbit would not respond to a sensory input in the absence of movement.

Gray: I want to take up the role of reinforcement again. As I understood your data, the cell which fired in the goal arm greatly increased its firing when you omitted the food, but did not do so when you omitted one of the stimuli associated with food. Is that correct?

O'Keefe: No. What we find so far is that when we get a misplace cell (one which increases its firing in the goal arm on a non-rewarded trial) those cells *also* increase their firing if you remove one of the spatial cues (cf. Fig. 2H, p. 186). If you remove all the cues they fire at much higher rates across the whole maze than if the cues were there (Fig. 2G). We tentatively conclude that all the cues in the situation, including the food, are acting to inhibit the firing of the misplace cell, and when the rat goes to the goal arm and doesn't find food an inhibitory influence is no longer there and the cell increases its firing. But that is in no way any different from the removal of any of the other spatial cues. But these are still early days and there may be a unique role for reward, or food, which we haven't yet found.

Vinogradova: Are the responses of the complex spike cells always larger in amplitude than those of all other cells?

O'Keefe: In our hands they tend to be, and in Dr Ranck's much more careful quantitative work they have a larger amplitude (Fox & Ranck 1975).

Vinogradova: This is difficult to see with extracellular recording because the amplitude of the spikes depends on the distance of the microelectrode from the cell, and this might mean that the complex spikes may occur in cells that are affected in some way by a closely placed microelectrode. These are very interesting data, but it is curious how we find in the brain what we are looking for! For example, in my earlier experiments I was specifically interested in *time* perception in the hippocampus and I have seen beautiful extrapolation in hippocampal units, when cells just after a few sensory stimuli given with equal intervals start to respond just before the stimulus, and can adjust to various time periods (Vinogradova 1965). The same was extensively shown by Dr Kopytova (Kopytova & Kulikova 1970). Dr Mering has shown that the hippocampus in dogs is especially important for time evaluation (Mering & Mukhin 1971). So I suspect that space and time and in fact *all* dimensions of the world are going through the hippocampus!

References

Fox, S. E. & Ranck, J. B., Jr (1975) Localization and anatomical identification of theta and complex spike cells in dorsal hippocampal formation of rats. *Exp. Neurol. 49*, 299–313

Kopytova, F. V. & Kulikova, L. K. (1970) Conditioned responses to time in neurons of dorsal hippocampus. *Zhurnal Vysshei Nervnoi Deyatel'nosti 20*, 1221–1230 (in Russian)

Mays, L. E. & Best, P. J. (1975) Hippocampal unit activity during arousal from sleep and in awake rats. *Exp. Neurol. 47*, 268–279

Mering, T. A. & Mukhin, E. I. (1971) Effect of hippocampal lesions on conditioned reflexes to time. *Zhurnal Vysshei Nervnoi Deyatel'nosti 21*, 1147–1153 (in Russian)

O'Keefe, J. (1976) Place units in the hippocampus of the freely moving rat. *Exp. Neurol. 51*, 89–109

O'Keefe, J. & Dostrovsky, J. (1971) The hippocampus as a spatial map. Preliminary evidence from unit activity in the freely moving rat. *Brain Res. 34*, 171–175

Vinogradova, O. S. (1965) Dynamic classification of the hippocampal neurons. *Zhurnal Vysshei Nervnoi Deyatel'nosti 15*, 500–512 (in Russian)

Hippocampal electrical activity during waking behaviour and sleep: analyses using centrally acting drugs

C. H. VANDERWOLF, R. KRAMIS and T. E. ROBINSON*

Department of Psychology, University of Western Ontario, London, Canada

Abstract Rhythmical slow activity (RSA) occurs in the hippocampus under many conditions including waking behaviour, active sleep and surgical anaesthesia. Under all these conditions RSA, apparently, is produced by the coupled operation of CA1 and dentate gyrus generators. Two ascending brainstem systems appear capable of initiating activity in these coupled generators. One system, ascending via the diagonal band and medial septal nucleus, may contain cholinergic synapses since it is blocked by atropine and stimulated by eserine. The RSA produced by this system usually has a frequency of 4–7 Hz and can occur during total immobility during the waking state, active sleep or anaesthesia. A second ascending system produces RSA of higher frequency (usually 7–12 Hz) and is active during waking if, and only if, movements such as walking occur. During active sleep this system is active only during phasic muscular twitches. Anaesthetics (ether, urethane) and morphine abolish activity in this second system but it is resistant to atropinic and nicotinic drugs. Amphetamine stimulates, and major tranquillizers depress the atropine-resistant system but these drugs do not abolish its normal relation to behaviour.

Neocortical activity appears to be controlled by two ascending systems which parallel closely those ascending to the hippocampus.

The slow wave activity of the hippocampus of a freely moving rat consists of several distinguishable wave patterns (Vanderwolf 1969; Whishaw & Vanderwolf 1973). These include: (1) rhythmical slow activity (RSA or 'theta rhythm') which consists of trains of approximately sinusoidal 6–12 Hz waves; (2) large amplitude irregular activity (LIA), a pattern which lacks the rhythmical character of RSA, contains waves of lower frequency than RSA, and includes irregularly occurring large amplitude spikes (50–100 ms duration, 2–5 times the background amplitude); (3) small amplitude irregular

---

Present address: Department of Psychobiology, University of California, Irvine, California 92717, USA.

activity (SIA), which appears as a sudden reduction of wave amplitude and normally lasts only 1–2 s, in contrast to RSA and LIA which may continue steadily for minutes or hours at a time; and (4) fast waves ranging from about 15 to 50 Hz.

Investigations in which moveable microelectrodes have been used to record hippocampal slow waves have helped to localize the structures which generate the different types of activity (Bland *et al.* 1975; Bland & Whishaw 1976; Winson 1974, 1976*a,b*; Fig. 1). Slow wave activity was recorded simultaneously from a reference microelectrode fixed in CA1 or the dentate gyrus and from a moveable microelectrode which was lowered through the hippocampus.

The results obtained vary somewhat depending on the species and type of preparation used, but typically the moving microelectrode encounters two distinct zones in which RSA of especially large amplitude can be recorded. One zone is located in the stratum oriens of CA1; the other is in stratum moleculare of the dentate granule cells just below the hippocampal fissure. Waveforms in these two zones are approximately 180° out of phase under a variety of conditions. Thus, when a positive-going wave appears in stratum oriens, a negative-going wave appears in stratum moleculare. Both zones produce LIA as well as RSA and probably produce SIA as well. Thus, all three patterns appear to be due to coupled activity in CA1 and dentate gyrus neurons. The mechanism of this coupling and the mechanism by which the slow waves are actually produced requires further investigation.

The hilus of the dentate gyrus is a focal source of fast activity in the rat (Fig. 1).

It is generally agreed that hippocampal activity (especially RSA) is related to behaviour in some way but the nature of this relation has, unfortunately, been the subject of much controversy. One of the sources of controversy has been a failure to appreciate the importance of electrode location and the necessity of obtaining a clear electrical signal from the hippocampus. Fig. 1 illustrates that an electrode may be placed in the hippocampus in such a way that it will reveal little or no RSA. In such a case, valid correlations of RSA and behaviour are not possible (Vanderwolf 1969; Whishaw & Vanderwolf 1973). Because of the reversed phase of the CA1 and dentate generator zones, a maximal slow wave signal can be obtained by taking difference recordings from a staggered bipolar electrode placed with one tip in stratum oriens and the other near the hippocampal fissure.

HIPPOCAMPAL SLOW WAVES AND BEHAVIOUR: TWO TYPES OF RSA

A normal waking rat behaving spontaneously in a recording cage always generates RSA during the performance of some motor patterns but produces

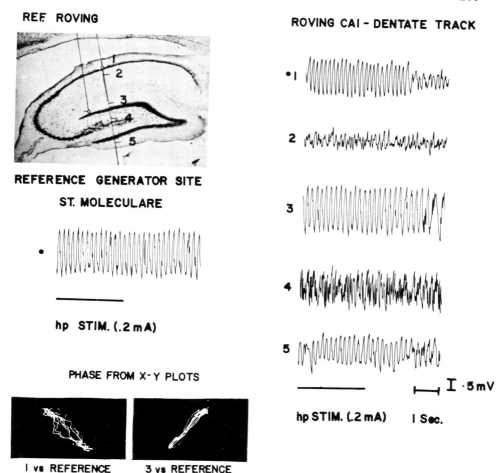

FIG. 1. Distribution of hippocampal RSA in a hippocampal profile obtained during muscular paralysis. Insert in upper left corner is a photomicrograph of a sagittal section of the hippocampal formation approximately 2.5 mm from the midline. The reference electrode is located in stratum moleculare and the lesion is indicated diagrammatically by the open circle. The penetration of the roving electrode is also indicated diagrammatically and the open circle at point 1 indicates the marker lesion made in stratum oriens. Records of the activity at each numbered location correspond to the numbered records to the right of the section. (1) stratum oriens; (2) null zone in stratum radiatum; (3) stratum moleculare; (4) fast activity in hilus; (5) stratum moleculare lower blade. The record below the section with the open circle next to it represents the activity recorded from the reference electrode in stratum moleculare. Solid bars indicate the duration of the electrical stimulation (0.3 mA) of the posterior hypothalamus. Phase is shown by the Lissajous patterns taken from X–Y plots. *Left:* RSA from point 1 in stratum oriens is approximately 180° out of phase compared with the reference in stratum moleculare. *Right:* RSA from point 3 in stratum moleculare is in phase with RSA recorded from the reference electrode in stratum moleculare. (From Whishaw *et al.* 1976, with permission.)

LIA instead during the performance of other motor patterns or during immobility (Fig. 2). Thus, RSA is always present whenever the rat walks but LIA is generally present when the rat is chewing food, washing its face or gnashing its teeth (a type of defensive or threat behaviour). Sensory stimuli elicit RSA only insofar as they are able to elicit the relevant motor activity. Somewhat different results are obtained in some other species. For example, rabbits exhibit RSA not only during locomotion, postural changes, etc. as rats do, but also during complete immobility. If a rabbit is startled by a sudden handclap it is likely to 'freeze'—that is, stand motionless with eyes wide open. Such freezing is often accompanied by a long train of very clear RSA. Further, if either rats or rabbits are anaesthetized with diethyl ether or urethane (ethyl carbamate) RSA will usually appear spontaneously even at a level of anaesthesia at which the corneal reflex is absent and there is only a minimal respiratory response to strong pinching. Other compounds used as anaesthetics, including chloroform, trichloroethylene, ethyl alcohol and chloral hydrate, yield similar results. RSA is also very prominent during active (paradoxical) sleep even though the animal is motionless or exhibits only slight twitching movements in this state.

It is not easy to conceive of some behavioural or psychological function which is equally present during waking, sleep and the coma produced by

NORMAL RELATION OF HIPPOCAMPAL ACTIVITY TO BEHAVIOR IN THE RAT

HIPPOCAMPUS	BEHAVIOR
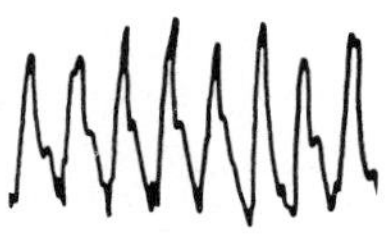 [RSA waveform]	Type 1. walking, running, swimming, rearing, jumping, digging, manipulation of objects with the forelimbs, isolated movements of the head or one limb, shifts of posture. Related terms: voluntary, appetitive, instrumental, purposive, operant, or "theta" behavior.
 [LIA waveform]	Type 2. a) alert immobility in any posture. b) licking, chewing, chattering the teeth, sneezing, startle response, vocalization, shivering, tremor, face-washing, scratching the fur, pelvic thrusting, ejaculation, defecation, urination, piloerection. Related terms: automatic, reflexive, consummatory, respondent, or "non-theta" behavior.

FIG. 2. Normal relation of hippocampal activity to behaviour in the rat. Upper record, RSA; lower record, LIA. (From Vanderwolf et al. 1975, with permission.)

anaesthetics. Perhaps RSA does not relate to any unitary function. In fact, functional diversity is suggested by recent evidence that there may be two pharmacologically distinct forms of RSA. Under some conditions RSA can be abolished by systemically administered anti-muscarinic drugs (atropine, atropine SO$_4$, scopolamine HBr) but under other conditions RSA is very resistant to these drugs. In both rats and rabbits, it is generally true that RSA occurring during behavioural immobility (in the waking state or during anaesthesia) is sensitive to atropinic drugs while the RSA accompanying locomotion and other forms of type 1 behaviour (Fig. 2) is resistant to them (Fig. 3). Atropine-sensitive RSA usually has a low frequency (4–7 Hz) and is resistant to anaesthetics, since it can occur during anaesthesia. Atropine-resistant RSA usually has a higher frequency (7–12 Hz) and is sensitive to anaesthetics since it, and the accompanying type 1 behaviour, are both abolished during anaesthesia. Thus, a combination of atropine plus ether or urethane abolishes all RSA. We are led to the concept of an atropine-resistant anaesthetic-sensitive form of RSA which is rigidly coupled to certain types of motor activity and an atropine-sensitive anaesthetic-resistant form of RSA which can be active during immobility under certain conditions (Kolb & Whishaw 1977; Kramis *et al.* 1975; Vanderwolf 1975; Vanderwolf *et al.* 1975; Whishaw 1976).

The concept that there are atropine-sensitive and anaesthetic-sensitive forms of RSA suggested a possible relation to the CA1 and dentate gyrus generator zones. It was conceivable that one zone would be sensitive to ether (or

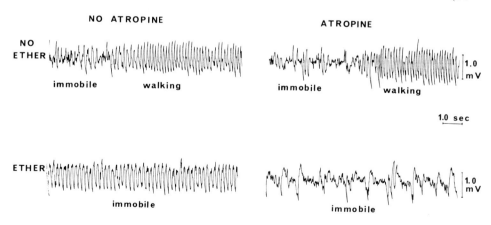

FIG. 3. Hippocampal slow wave activity under four conditions. (1) Undrugged state; (2) after atropine SO$_4$ (50 mg/kg, i.p.); (3) during ether anaesthesia; (4) during ether anaesthesia after atropine SO$_4$ (50 mg/kg, i.p.). Note: low frequency RSA during anaesthesia, higher frequency RSA during walking, and total absence of RSA if atropine is combined with anaesthesia.

urethane) while the other would be sensitive to atropine. Whishaw et al. (1976) recorded activity in the two generator zones in rats that were either anaesthetized with urethane or unanaesthetized and paralysed with d-tubocurarine. The results showed that RSA was present at both the CA1 zone and the dentate gyrus zone after either urethane or tubocurarine alone or tubocurarine plus atropine. No RSA was present in either generator zone after a combination of urethane and atropine. This suggests that atropine-resistant and anaesthetic-resistant forms of RSA are both produced by the same hippocampal mechanism involving both the CA1 and dentate gyrus generators. The pharmacological specificity of the two types of RSA probably resides in the afferent systems which initiate activity in the CA1 and dentate gyrus generator zones.

An interesting test of the hypothesis that there are two pharmacologically distinct mechanisms which can produce RSA is suggested by the phenomenon of active sleep. During active sleep, movement is strongly suppressed by combined descending postsynaptic and presynaptic inhibitory influences which act on the spinal motoneuron and its reflex afferents. Muscle tone is abolished. Despite this inhibition, phasic twitches of the somatic musculature occur periodically as a result of descending barrages which are excitatory to the spinal motoneurons (Jouvet 1967a; Pompeiano 1967). If hippocampal activity is monitored during active sleep, RSA is virtually continuous, but higher frequencies (about 8 Hz) occur during the phasic muscular twitches than during the intervals between twitches (6-7 Hz). When active sleep occurs in rats treated with atropine, RSA continues to accompany the muscular twitches but the RSA of the inter-twitch intervals is abolished (Robinson et al. 1977a; Fig. 4). Urethane has an opposite effect, abolishing both the muscular twitches and the higher frequency RSA which normally accompanies them. Lower frequency RSA, characteristic of the inter-twitch intervals, predominates after urethane treatment. Therefore, it appears that in active sleep, as in the waking state, atropine-resistant anaesthetic-sensitive RSA is an accompaniment of motor activity while atropine-sensitive anaesthetic-resistant RSA is present during immobility under certain conditions.

ATROPINE-SENSITIVE RETICULO-SEPTAL INFLUENCES ON HIPPOCAMPAL SLOW WAVES

Since the work of Green and Arduini it has been known that stimulation of the brainstem reticular formation will produce RSA in the hippocampus. It has been assumed that this effect is due to a pathway ascending from the reticular formation to the hippocampus through the medial septal nucleus

A. NO DRUG

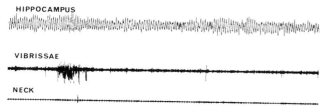

B. ATROPINE

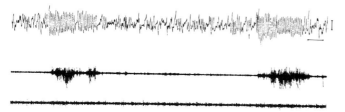

Fig. 4. Hippocampal slow wave activity during active sleep after deprivation of active sleep. A. Undrugged rat. Note higher frequency RSA during EMG burst in muscles under vibrissae, lower frequency RSA in interval after twitch. B. After atropine SO$_4$ (40 mg/kg, i.p.). Note that RSA occurs only in relation to EMG bursts. Neck muscle activity is not abolished after atropine.

and the diagonal band of Broca, since destruction or injection of procaine into the latter structures abolishes RSA (Brücke et al. 1959; Green & Arduini 1954). The hypothesis that two distinct types of afferents to the septum or hippocampus are capable of initiating RSA prompted us to study brainstem and septal influences on the hippocampus.

A number of facts suggest that the pathway responsible for producing RSA during behavioural immobility may be identical with an ascending cholinergic reticular system (Shute & Lewis 1967). An initial study (Vanderwolf 1975; Fig. 5) showed that electrical stimulation of the reticular formation could produce both atropine-sensitive RSA (present when the rat stood motionless during stimulation) and atropine-resistant RSA (present when stimulation resulted in running or jumping). A subsequent mapping study has shown that stimulation at sites scattered throughout the reticular formation, from the medulla to the upper midbrain, is capable of initiating RSA during behavioural immobility in undrugged or chlorpromazine-treated rats (Robinson & Vanderwolf 1978). With one exception (RSA elicited by

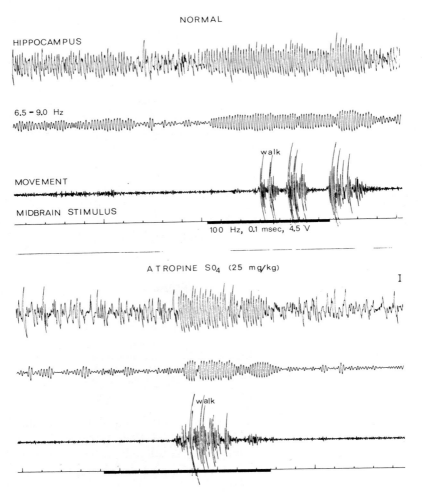

FIG. 5. Effects of atropine on hippocampal rhythmical slow activity (RSA) and behaviour elicited by stimulation of the midbrain reticular formation. (Hippocampal activity is shown as an unfiltered record and also after passage through a bandpass filter system.) *Top four tracings:* Undrugged rat. Note that *(a)* RSA accompanies small spontaneous movements (probably head movements); *(b)* RSA begins at onset of stimulation even though rat is motionless; walking begins only after a latent period of about 3 s; *(c)* RSA persists during the centrally elicited walking. *Bottom four tracings:* Atropinized rat. Note that *(a)* behaviour is essentially unchanged; walking begins after a latent period of several seconds; *(b)* RSA still accompanies walking behaviour; but *(c)* the RSA initially present during immobility appears to have been abolished. Reticular stimulation is constant throughout, 0.1 ms pulses at 100 Hz, and 4.5 V. Calibration: time marks 1.0 s, 500 μV. (From Vanderwolf 1975, with permission. Copyright 1975 American Psychological Association.)

stimulation of the region of subnucleus compactus nuclei pedunculopontini tegmenti of the midbrain; see Fifková & Maršala 1967), such RSA is sensitive to atropine. Atropine sensitivity, in itself, is not a proof that a cholinergic system is involved. The dose of atropine SO_4 necessary to block RSA during immobility is very large (25–50 mg/kg, i.v.), raising the possibility that its effects are non-specific. However, other facts support the concept of a cholinergic reticulo-septal or reticulo-hippocampal system which is active during immobility under certain conditions. Thus, rats treated with eserine (0.2–0.5 mg/kg, i.p.), an anticholinesterase, generate RSA during complete immobility (Vanderwolf 1975).

Further, it is worthy of note that atropine, atropine SO_4 and scopolamine HBr all possess the property of preventing RSA during immobility in a variety of tests. Atropine methyl nitrate does not have this effect, presumably because it fails to penetrate the CNS in an adequate concentration. Many other drugs which do have strong CNS effects (e.g., anaesthetics and major tranquillizers) also fail to block the RSA which occurs during immobility, suggesting that the effects of the atropine drugs are specific to a cholinergic system. Further, the cholinergic synapses in this system may be muscarinic since both nicotine and mecamylamine fail to block RSA in urethane-anaesthetized rats (Whishaw et al. 1978). Iontophoretic studies have also suggested the existence of muscarinic cholinergic receptors in the hippocampus (Biscoe & Straughan 1966).

Lewis & Shute (1967) and other investigators (see Kuhar 1975, for a review) have shown that the medial septal nucleus and diagonal band are prominent

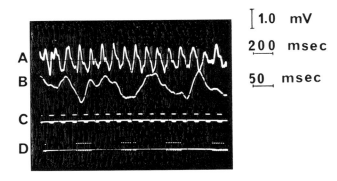

FIG. 6. Hippocampal slow wave activity during grouped pulse stimulation of the diagonal band of Broca. A and B show slow waves; C and D show associated stimuli. A and C show slow sweep; B and D show first 500 ms of A and C at faster sweep. Stimulus parameters: 60 μA; 0.4 ms pulse duration; 25 ms train duration; 10 trains/s; 7–9 pulses/train. Note that hippocampal waves are synchronized with the pulse trains.

sources of acetylcholinesterase-containing projections to the hippocampus. Since these same structures have been shown to play a role in the production of RSA it seemed worthwhile to stimulate them in freely moving rats while recording from the hippocampus (Kramis & Vanderwolf 1977). Stimulation was delivered in rhythmically recurring bursts (Fig. 6) of rectangular pulses in order to mimic the pattern of activity of the diagonal band and medial septal burst cells described by Petsche *et al.* (1962). These cells have often been regarded as pacemakers of the RSA rhythm. Such burst stimulation of the diagonal band produces a hippocampal waveform resembling spontaneous RSA (Kramis & Routtenberg 1977) but having a frequency which is determined by the repetition rate of the bursts of stimuli (frequency-specific driving). Pulses presented at 5–12 bursts/s are an exceptionally effective method of

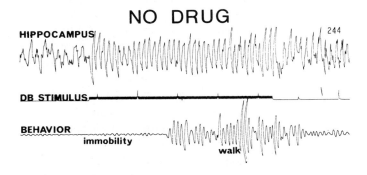

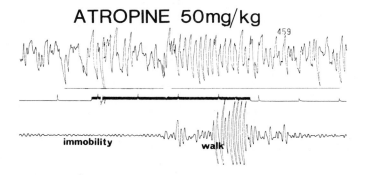

Fig. 7. Hippocampal slow wave activity in relation to behaviour during grouped pulse stimulation of the diagonal band of Broca. No drug: diagonal band (DB) stimulus drives hippocampal rhythm during both immobility and walking. After atropine, driving occurs only during head movement and walking. Stimulus parameters as in Fig. 6.

driving slow wave activity in the hippocampus. Currents as low as 5–30 μA often produce good driving if grouped pulses are used but steady trains of pulses (e.g. 100 Hz) produce little or no effect until current levels reach 100–300 μA, and then produce flattening of the record rather than RSA. Single pulses at 5–12 Hz are also less effective than grouped pulses.

The frequency-specific hippocampal driving elicited by septal stimulation in the normal waking rat is largely independent of behaviour. Regardless of whether the rat stands motionless or walks about, good driving can be maintained for long periods at 6–12 Hz. However, after an injection of atropine (50 mg/kg, i.p.) driving cannot be obtained during immobility even though driving occurs immediately whenever the rat walks or moves its head (Fig. 7). This suggests that the frequency-specific driving elicited during immobility in the undrugged rat depends on atropine-sensitive (probably cholinergic) afferents from the diagonal band to the hippocampus. The data also indicate that a second system, which can be activated only if the rat moves appropriately, is resistant to atropine. It is puzzling that, in the presence of atropine, septal stimulation can still determine the frequency of the RSA waveform occurring during walking but cannot initiate RSA by itself. Walking and other similar movements appear to open a 'gate' which enables the septal stimulus to exert an effect on RSA frequency. The means by which behaviour (or the mechanisms that produce it) can influence septo-hippocampal transmission are completely unknown.

Frequency-specific driving of the hippocampus remains possible in rats deeply anaesthetized with urethane. Effective current levels are similar to those required in the unanaesthetized rat. However, if atropine is administered, all driving is abolished, indicating that only the atropine-sensitive system is active in the urethane-drugged rat. Thus the RSA-like hippocampal waves produced by burst stimulation of the diagonal band have a pharmacology and relation to behaviour which is similar to those of spontaneously occurring RSA. One system, which permits diagonal band-hippocampal driving during immobility, is sensitive to atropine but resistant to urethane anaesthesia. A second system, which permits diagonal band-hippocampal driving during walking, struggling, and so on, is sensitive to urethane but resistant to atropine. If atropine and urethane are administered together, diagonal band-hippocampal driving is completely abolished.

TWO TYPES OF NEOCORTICAL LOW VOLTAGE FAST WAVES

Relations between slow wave activity and behaviour, of the type described in this paper, can be demonstrated in the neocortex as well as in the hippocampus. However, as is well known, the morphology of the waveforms in the two structures differs. Electrical stimulation of the reticular formation produces low voltage fast activity (LVFA) in the neocortex rather than the RSA which is characteristic of the hippocampus. Despite the difference in wave morphology there are many parallels in the relations of RSA and LVFA to behaviour. In the case of LVFA these relations have been obscured for many years as a result of the prevailing opinion that LVFA is a correlate of arousal, alertness, or wakefulness. Critical examination of the available published data shows that this 'correlation' is largely non-existent (Vanderwolf 1978). However, by means of centrally acting drugs, clear relations can be demonstrated between the slow wave activity of the neocortex and concurrent motor activity.

In a normal freely moving rat there is no obvious correlation between motor activity and the spontaneous slow wave activity of the neocortex. In most cases, LVFA of the same general appearance is present during walking, face-washing and immobility in a posture in which the head is held up against gravity and the eyes are fully open. (Large amplitude slow waves may appear on some occasions but, for the purposes of this paper, they can be ignored). If atropine SO_4 (25–150 mg/kg, i.p.) is administered, a strong correlation between neocortical activity and behaviour appears (Vanderwolf 1975; Fig. 8). Large amplitude irregular slow waves occur during immobility and during other type 2 behaviours such as shivering, face-washing or gnashing the teeth. A form of low voltage fast activity occurs during walking, head movement and other type 1 behaviours. Such atropine-resistant LVFA can also be produced by direct electrical stimulation of the hypothalamus or brainstem reticular formation in unanaesthetized rats. However, passive movement or sensory stimuli which do not result in movement are ineffective. These observations suggest a parallel between reticular control of neocortical activity and reticular control of hippocampal activity. There appears to be an atropine-sensitive type of LVFA which, like atropine-sensitive RSA, can occur during immobility and other type 2 behaviour. In addition, there is an atropine-resistant type of LVFA which, like atropine-resistant RSA, occurs only during the performance of such motor acts as walking or head movement. Atropine-sensitive LVFA differs from atropine-sensitive RSA in its level of spontaneous occurrence. In rats, LVFA is very common during immobility whereas RSA is absent in this condition unless special experimental procedures are applied (e.g., anaesthetics, reticular stimulation, etc.). Nonetheless, atropine SO_4 prevents the occurrence of both waveforms during immobility.

There are further parallels between LVFA and RSA. It has been recognized

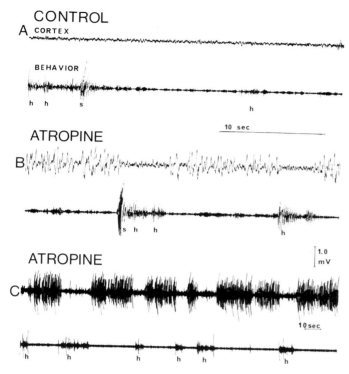

FIG. 8. Neocortical slow wave activity in relation to behaviour in a rat. A: control record after trifluoperazine HCl (5 mg/kg, i.p.); B: after the addition of atropine SO₄ (50 mg/kg, i.p.); C: low speed record after trifluoperazine and atropine. h, head movement; s, stepping. Note relation between slow waves and motor activity in B and C. Trifluoperazine counteracts the excessive motor activity produced by atropine.

for many years that LVFA is prominent during surgical anaesthesia induced by diethyl ether, chloroform and similar agents. Such LVFA is atropine-sensitive, resembling the LVFA which occurs during waking immobility (Vanderwolf et al. 1975). Atropine-resistant LVFA is abolished during surgical anaesthesia. Further, during active sleep, atropine-resistant LVFA accompanies the phasic twitches of the somatic musculature while atropine-sensitive LVFA occurs in the intervals between twitches (Robinson et al. 1977a).

In sum, it is likely that both the neocortex and hippocampus receive two types of input from the reticular formation. One input, which may be active during behavioural immobility, is sensitive to atropinic drugs but is resistant to many anaesthetics and may be spontaneously active during surgical anaesthesia. The other input is active during, and only during, the performance of certain types of motor activity, such as locomotion and spontaneous head movement. This input is resistant to atropinic drugs but is sensitive to anaesthetics. The atropine-sensitive input to the neocortex may be cholinergic, paralleling the atropine-sensitive input to the hippocampus. Many experiments suggest the existence of an ascending cholinergic reticular system which projects to the thalamus and neocortex (Shute & Lewis 1967; Krnjević 1974; Phillis 1976).

ATROPINE-RESISTANT RETICULAR INFLUENCE ON THE HIPPOCAMPUS AND NEOCORTEX

Although there has been a good deal of evidence favouring the view that the ascending reticular projections to the hippocampus and neocortex are cholinergic there have also been suggestions that a catecholaminergic component is involved. Early findings (Dell 1958; Rothballer 1956) suggested that adrenaline or noradrenaline might be a transmitter in the reticular formation. Later work has not favoured this hypothesis (Baust et al. 1963; Mantegazzini et al. 1959) but it has received new life with the discovery of the diffuse noradrenergic projections from the locus coeruleus to the hippocampus and neocortex. Various other compounds, including serotonin (5-hydroxytryptamine) (Jouvet 1967b) and dopamine (Mantegazzini & Glässer 1960), have also been suggested as playing a role in the control of hippocampal or neocortical slow wave activity.

Keeping this background in mind, we have investigated the mechanisms that produce movement-related atropine-resistant RSA and LVFA in the hippocampus and neocortex, respectively. One of the methods used involved the following steps. *(a)* Hippocampal and neocortical activity were recorded during spontaneous behaviour on a small-movement-sensing platform and also during struggling induced by picking the rat up briefly and during electrical stimulation of the midbrain reticular formation. *(b)* A test drug was administered (i.p.) and the foregoing procedures were repeated, once or several times, after delays ranging from 30 min to several days. *(c)* Atropine SO_4 was administered (25–50 mg/kg, i.p.) and the procedures in *(a)* were repeated again. Alternatively, atropine was administered first and the test drug second. Sometimes additional procedures were carried out.

The effect of ethyl ether in such experiments has already been discussed. In summary, ether permits atropine-sensitive RSA and LVFA but blocks both patterns in their atropine-resistant form.

Ethyl urethane is more selective than ether in that, although it also affects the atropine-resistant system, it blocks RSA only, sparing LVFA. Thus, after an atropine injection, neither RSA or LVFA occur during immobility although both wave patterns continue to be found during locomotion, struggling, etc. If an atropinized rat is anaesthetized with ether, RSA and LVFA both disappear and cannot be elicited even by noxious peripheral stimuli or strong stimulation of the reticular formation. If an atropinized rat is anaesthetized with urethane, however, RSA is abolished selectively. LVFA can still be elicited readily by a tail pinch or by stimulation of the reticular formation even though the rat is deeply anaesthetized and immobile (Fig. 9). Therefore it appears that urethane abolishes atropine-resistant RSA but not atropine-resistant LVFA (Whishaw *et al.* 1976).

Reserpine, in contrast, acts on the atropine-resistant system in a manner opposite to urethane, abolishing LVFA instead of RSA. Rats treated with reserpine (5–10 mg/kg) pass, after several hours, into a state characterized by three properties. (1) Behaviourally, the rats are severely cataleptic and forward locomotion is almost impossible even during a severe tail pinch. The rats squeal readily, can back up, and a tail pinch easily elicits a flexion-turning movement which ends in biting. (2) RSA of 5–6 Hz occurs during immobility,

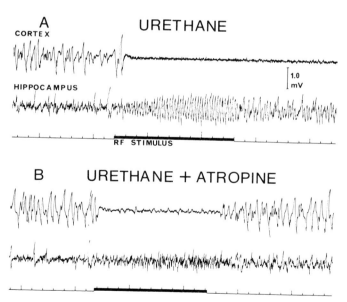

FIG. 9. Effects of urethane and atropine on the neocortical and hippocampal response to stimulation of the midbrain reticular formation in a rat. A: record following urethane (1.5 g/kg, i.p.). Rat in surgical anaesthesia. B: record following addition of atropine SO$_4$ (50 mg/kg, i.p.). Note that hippocampal RSA response is abolished by atropine, but modified LVFA persists in the neocortex. Reticular formation (RF) stimulus: 100 Hz, 0.1 ms pulse duration, 250 μA.

either spontaneously or after sensory stimuli. (Thus, reserpine resembles urethane, ethyl alcohol and the volatile anaesthetics to the extent that it releases atropine-sensitive RSA.) Higher frequency RSA occurs during movement. (3) Neocortical activity consists mainly of large amplitude slow waves or 6–10 Hz spindles if the rat is not disturbed. Sensory stimuli or stimulation of the reticular formation provoke good LVFA during either movement or immobility.

When atropine is administered to rats in this state, all RSA occurring during behavioural immobility is abolished but the RSA accompanying movement tends to persist. However, the morphology of the RSA waves is somewhat altered (Fig. 10). In contrast, atropine abolishes all LVFA in reserpinized rats. The neocortical record consists of continuous large amplitude irregular slow waves which cannot be blocked even by strong reticular stimulation or by noxious sensory stimulation, which continue to elicit movement as they did before atropinization. Thus, reserpine does not abolish atropine-resistant RSA but does, apparently, abolish atropine-resistant LVFA.

The contrasting effects of reserpine (abolition of atropine-resistant LVFA) and of urethane (abolition of atropine-resistant RSA) suggest a neurochemical differentiation of the atropine-resistant inputs to the hippocampus and the neocortex. The nature of this differentiation is unknown, however.

It is possible that some of the effects of reserpine are due to a depletion of noradrenaline or other brain monoamines. Fluorometric assays indicated that 10 mg/kg of reserpine reduced the noradrenaline content of the neocortex and hippocampus to 5.7% of the normal value. Dopamine was reduced to 17.4% and serotonin to 2.0% of the normal values. Therefore, we did additional experiments to examine the possibility that specific monoamines play a role in atropine-resistant RSA and LVFA.

Since the noradrenaline-containing projections to the neocortex and hippocampus originate in the locus coeruleus, particular attention was devoted to it. We found that electrical stimulation of the locus coeruleus in freely moving rats produces both atropine-sensitive and atropine-resistant hippocampal RSA and neocortical LVFA, but the current required is about five times what is necessary to achieve the same effects by stimulating reticular structures ventral to the locus coeruleus. Therefore, the locus coeruleus effect is probably due to the spread of current to these ventral structures (Robinson et al. 1977b). Further, neither electrolytic destruction of the locus coeruleus (Kolb & Whishaw 1977), nor depletion of forebrain noradrenaline (by acute or chronic administration of α-methyl-p-tyrosine or FLA-63 or by systemic injections of 6-hydroxydopamine in the first few days of life) result in a loss of atropine-resistant RSA or atropine-resistant LVFA.

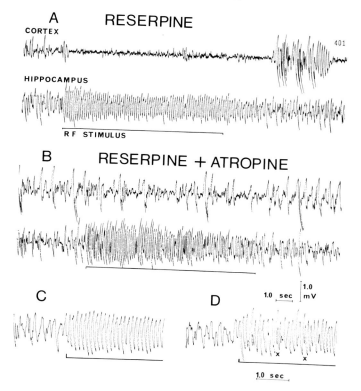

FIG. 10. Effects of reserpine and atropine on the neocortical and hippocampal response to stimulation of the midbrain reticular formation in a rat. A. record after reserpine (10 mg/kg, i.p.). B: record after addition of atropine SO₄ (25 mg/kg, i.p.). Note that LVFA is abolished but RSA persists in modified form. C: higher speed record showing hippocampal RSA after reserpine alone. D: higher speed record showing hippocampal RSA after reserpine plus atropine. Note abnormal deeply notched RSA waves, ×. Reticular formation (RF) stimulus (100 Hz, 0.1 ms pulse duration, 7.0 V) produced elevation of the head and rotation of the trunk both before and after atropine treatment.

Atropine-sensitive RSA and LVFA are also preserved (Robinson et al. 1977b). Finally, Jones et al. (1977) have been unable to demonstrate that lesions of the locus coeruleus alter electrocortical slow wave activity. Consequently, it seems unlikely that ascending noradrenergic projections play an important role in the control of RSA and LVFA.

The role of serotonin was investigated by treating rats with DL-*p*-chlorophenylalanine, an effective inhibitor of the synthesis of serotonin (Koe & Weissman 1966). This drug failed to block either atropine-sensitive or atropine-resistant RSA and LVFA over a period of three days after an injection of 500 mg/kg. Thus, serotonin is probably not crucial to the mechanisms discussed here.

The possible role of dopamine in the production of RSA and LVFA is difficult to evaluate. The drug α-methyl-*p*-tyrosine produced a depletion of forebrain dopamine to 23% of the normal level without abolishing atropine-resistant RSA or atropine-resistant LVFA (Robinson *et al.* 1977*b*). A further test of the role of dopamine consisted of treating rats with various phenothiazines, butyrophenones, or diphenylbutylamines, drugs which are believed to block dopaminergic transmission (Andén *et al.* 1970). Drugs that were tested include chlorpromazine, perphenazine, trifluoperazine, haloperidol, and pimozide. If these drugs are combined with atropine in rats, RSA and LVFA occur very infrequently. This effect is a correlate of the immobility and catalepsy produced by the drugs, since RSA and LVFA occur and remain well correlated with struggling and other movements whenever they can be elicited. Thus, dopaminergic blocking drugs change the probability of occurrence of atropine-resistant RSA and LVFA but do not abolish these wave patterns, even in large doses. For example, haloperidol in doses of 1, 10 or 25 mg/kg does not abolish the atropine-resistant RSA accompanying struggling behaviour or even change its frequency, which remains at 7–8 Hz at all dose levels. α-Methyl-*p*-tyrosine has a similar effect.

These data suggest that dopamine plays a role in a system which regulates the probability of occurrence of atropine-resistant RSA and LVFA, together with the associated type 1 behaviour, without being directly involved in generating the wave patterns. The effects of amphetamine appear to be consistent with such a hypothesis. Low doses of D-amphetamine (1–2 mg/kg) appear to act as a selective stimulant of type 1 behaviour, increasing walking, rearing and head movement at the expense of immobility, grooming or feeding (Vanderwolf 1975). Consequently, due to the strong correlation between type 1 behaviour and atropine-resistant RSA and LVFA, amphetamine increases the likelihood of the occurrence of these wave patterns in a given time; it does not, however, alter the normal relation of the wave patterns to behaviour. According to the data of Kelly *et al.* (1975) the locomotion-enhancing effect of low doses of amphetamine is dependent on the dopaminergic projection from the midbrain to the nucleus accumbens septi.

More direct effects were observed in experiments using morphine. Of all the compounds tested, morphine SO_4 was the only one capable of abolishing atropine-resistant RSA and LVFA without producing general anaesthesia. In rats, doses of morphine SO_4 of 15 mg/kg or more produce a severe catalepsy with greater muscular rigidity and immobility than is seen after treatment with phenothiazines, haloperidol or pimozide. Despite this, loud sounds or light touches on the back and sides produce sudden lunges forward in which the rat may advance several steps. This 'locomotion' is not under normal environmental control since the rats will walk off a table without hesitation. Stimulation of the midbrain reticular formation also elicits turning or sudden lunging. During the cataleptic state the neocortex exhibits either large amplitude slow waves or LVFA, the two patterns alternating

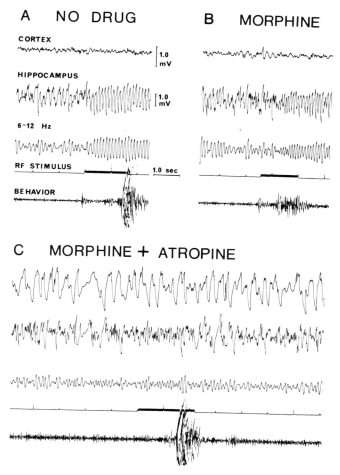

FIG. 11. Effects of morphine and atropine on the neocortical and hippocampal response to stimulation of the midbrain reticular formation in a rat. A: record in the undrugged state. B: record after morphine SO_4 (60 mg/kg, i.p.). C: record after addition of atropine SO_4 (50 mg/kg, i.p.). Note that RSA and LVFA are both abolished. Reticular formation (RF) stimulus (100 Hz, 0.1 ms pulse duration, 100 μA) elicited 'freeze' response followed by walking. 6–12 Hz, bandpass filtered record of hippocampus.

spontaneously at short intervals. RSA is usually absent but can be induced (less readily than normal) by electrical stimulation of the reticular formation or by stroking and manipulating the rat to induce movements. In some morphine-treated rats, RSA of 6–7 Hz appears spontaneously during behavioural immobility. When atropine is administered during the morphine-induced catalepsy it is no longer possible to elicit either LVFA or RSA even though movement is still possible, as before (Fig. 11). This indicates that the atropine-resistant RSA and LVFA normally present during type 1 movement had already been abolished by the morphine. The RSA and LVFA which remains after morphine treatment alone is apparently entirely of the atropine-sensitive type. Morphine had little or no effect on atropine-resistant RSA and LVFA at 5 mg/kg but produced a substantial degradation of these wave patterns at 15 mg/kg. Larger doses (30 or 60 mg/kg) produced a total or near total loss of atropine-resistant RSA and LVFA in all rats. If naloxone (0.4–0.8 mg/rat) was given in addition to morphine and atropine, atropine-resistant RSA and LVFA returned rapidly as the cataleptic state disappeared. It appears that morphine shares some property with anaesthetics such as diethyl ether since, combined with atropine, it will totally abolish both RSA and LVFA. It is conceivable that this shared property is related to analgesia in some way, since both morphine and ether are good analgesics.

The physiological basis of these effects of morphine remains obscure. Morphine has been shown to interact with many brain systems including those involving acetylcholine, serotonin and dopamine (Hockman & Bieger 1976). Recently it has been suggested that morphine combines with receptors whose normal ligands are endorphins or enkephalins, polypeptides which occur naturally in the pituitary gland and in the brain (Hughes 1975; Hughes et al. 1975). Morphine receptors are widely distributed in the brain (Kuhar et al. 1973; Pert et al. 1976; Atweh & Kuhar 1977) and it is possible that endorphins play a role in the production of atropine-resistant RSA and LVFA.

In conclusion, data from a variety of experiments are consistent with the hypothesis that two distinct types of input from the reticular formation are capable of generating RSA in the hippocampus and LVFA in the neocortex. One of these inputs, blocked by atropine and stimulated by eserine, may be identical with the cholinergic reticular system of Lewis and Shute. The functions and behavioural correlates of activity in this system are largely unknown.

The second ascending system is active if, and only if, movements of a certain class are also in progress. This system is stimulated by D-amphetamine and depressed by dopamine-blocking drugs. Its activity is largely or totally abolished by morphine and such anaesthetics as ether or chloroform. This system may play an important role in the generation of spontaneous behaviour and may be a substrate for the action of several major classes of drugs that influence behaviour.

ACKNOWLEDGEMENTS

This research was supported by grant AO-118 from the National Research Council of Canada.
Drugs used in the investigations were kindly supplied by: CIBA-GEIGY Ltd (reserpine); Health Protection Branch, Health and Welfare Canada (morphine SO_4); McNeil Laboratories (haloperidol, pimozide); and Poulenc Ltd (chlorpromazine, trifluoperazine).
We are indebted to Dr Bruce Pappas of Carleton University who performed the assays of brain monoamines.

References

ANDÉN, N. E., BUTCHER, S. G., CORRODI, H., FUXE, K. & UNGERSTEDT, V. (1970) Receptor activity and turnover of dopamine and noradrenaline after neuroleptics. *Eur. J. Pharmacol. 11*, 303–314
ATWEH, S. F. & KUHAR, M. J. (1977) Autoradiographic localization of opiate receptors in rat brain. II. The brain stem. *Brain Res. 129*, 1–12
BAUST, W., NIEMCZYK, H. & VIETH, J. (1963) The action of blood pressure on the ascending reticular activating system with special reference to adrenaline-induced EEG arousal. *Electroencephalogr. Clin. Neurophysiol. 15*, 63–72
BISCOE, T. J. & STRAUGHAN, D. W. (1966) Microelectrophoretic studies of neurons in the cat hippocampus. *J. Physiol. (Lond.) 183*, 341–359
BLAND, B. H. & WHISHAW, I. Q. (1976) Generators and topography of hippocampal theta (RSA) in the anaesthetized and freely moving rat. *Brain Res. 118*, 259–280
BLAND, B. H., ANDERSEN, P. & GANES, T. (1975) Two generators of hippocampal theta activity in rabbits. *Brain Res. 94*, 199–218
BRÜCKE, F., PETSCHE, H., PILLAT, B. & DEISENHAMMER, E. (1959) Über Veränderung des Hippocampus-Elektrencephalogrammes beim Kaninchen nach Novocain-injektion in die Septumregion. *Naunyn-Schmiedeberg's Arch. Exp. Pathol. Pharmakol. 237*, 276–284
DELL, P. C. (1958) Humoral effects on the brain stem reticular formations, in *Reticular Formation of the Brain* (Jasper, H. H., Proctor, L. D., Knighton, R. S., Noshay, W. C. & Costello, R. T., eds.), pp. 365–379, Little, Brown, Boston
FIFKOVÁ, E. & MARŠALA, J. (1967) Stereotaxic atlases for the cat, rabbit and rat, in *Electrophysiological Methods in Biological Research*, 3rd edn. (Bureš, J., Petrán, M. & Zachar, J., eds.), pp. 653–731, Academic Press, New York
GREEN, J. D. & ARDUINI, A. A. (1954) Hippocampal electrical activity in arousal. *J. Neurophysiol. 17*, 533–557
HOCKMAN, C. H. & BIEGER, D. (eds.) (1976) *Chemical Transmission in the Mammalian Central Nervous System*, University Park Press, Baltimore
HUGHES, J. (1975) Isolation of an endogenous compound from the brain with pharmacological properties similar to morphine. *Brain Res. 88*, 295–308
HUGHES, J., SMITH, T. W., KOSTERLITZ, H. W., FOTHERGILL, L. A., MORGAN, B. A. & MORRIS, H. R. (1975) Identification of two related pentapeptides from the brain with potent opiate agonist activity. *Nature (Lond.) 258*, 577–579

JONES, B. E., HARPER, S. T. & HALARIS, A. E. (1977) Effects of locus coeruleus lesions upon cerebral monoamine content, sleep-wakefulness states and the response to amphetamine in the cat. *Brain Res. 124*, 473–496

JOUVET, M. (1967a) Neurophysiology of the states of sleep. *Physiol. Rev. 47*, 117–177

JOUVET, M. (1967b) Mechanisms of the states of sleep: a neuropharmacological approach. *Res. Publ. Assoc. Nerv. Ment. Dis. 45*, 86–126

KELLY, P. H., SEVIOUR, P. W. & IVERSEN, S. D. (1975) Amphetamine and apomorphine responses in the rat following 6-OHDA lesions of the nucleus accumbens septi and corpus striatum. *Brain Res. 94*, 507–522

KOE, K. B. & WEISSMAN, A. (1966) P-chlorophenylalanine: a specific depletor of brain serotonin. *J. Pharmacol. Exp. Ther. 154*, 499–516

KOLB, B. & WHISHAW, I. Q. (1977) The effect of brain lesions and atropine on hippocampal and neocortical EEG in the freely moving rat. *Exp. Neurol. 56*, 1–22

KRAMIS, R. C. & ROUTTENBERG, A. (1977) Dissociation of hippocampal EEG from its behavioral correlates by septal and hippocampal electrical stimulation. *Brain Res. 125*, 37–49

KRAMIS, R. C. & VANDERWOLF, C. H. (1977) Two-RSA concept supported: effects of atropine, urethane and septal stimulation. *Neurosci. Abstr. 3*, 201

KRAMIS, R., VANDERWOLF, C. H. & BLAND, B. H. (1975) Two types of hippocampal rhythmical slow activity in both the rabbit and the rat: relations to behavior and effects of atropine, diethyl ether, urethane, and pentobarbital. *Exp. Neurol. 49*, 58–85

KRNJEVIĆ, K. (1974) Chemical nature of synaptic transmission in vertebrates. *Physiol. Rev. 54*, 418–540

KUHAR, M. J. (1975) Cholinergic neurons: septal-hippocampal relationships, in *The Hippocampus*, vol. 1: *Structure and Development* (Isaacson, R. L. & Pribram, K. H., eds.), pp. 269–283, Plenum Press, New York

KUHAR, M. J., PERT, C. B. & SNYDER, S. H. (1973) Regional distribution of opiate receptor binding in monkey and human brain. *Nature (Lond.) 245*, 447–450

LEWIS, P. R. & SHUTE, C. C. D. (1967) The cholinergic limbic system: projection to hippocampal formation, medial cortex, nuclei of the ascending cholinergic reticular system, and the subfornical organ and supra-optic crest. *Brain 90*, 521–540

MANTEGAZZINI, P. & GLÄSSER, A. (1960) Action de la DL-3-4-dioxyphénylalanine (DOPA) et la dopamine sur l'activité électrique du chat 'cerveau isolé'. *Arch. Ital. Biol. 98*, 351–366

MANTEGAZZINI, P., POECK, K. & SANTIBAÑEZ, G. (1959) The action of adrenaline and noradrenaline on the cortical electrical activity of the 'encéphale isolé' cat. *Arch. Ital. Biol. 97*, 222–242

PERT, C. B., KUHAR, M. J. & SNYDER, S. H. (1976) Opiate receptor: autoradiographic localization in rat brain. *Proc. Natl. Acad. Sci. U.S.A. 73*, 3729–3733

PETSCHE, H., STUMPF, CH. & GOGOLÁK, G. (1962) The significance of the rabbit's septum as a relay station between the midbrain and the hippocampus. I. The control of hippocampal arousal activity by the septum cells. *Electroencephalogr. Clin. Neurophysiol. 14*, 202–211

PHILLIS, J. W. (1976) Acetylcholine and synaptic transmission in the central nervous system, in *Chemical Transmission in the Mammalian Central Nervous System* (Hockman, C. H. & Bieger, D., eds.), pp. 159–213, University Park Press, Baltimore

POMPEIANO, O. (1967) The neurophysiological mechanisms of the postural and motor events during desynchronized sleep. *Res. Publ. Assoc. Nerv. Ment. Dis. 45*, 351–423

ROBINSON, T. E. & VANDERWOLF, C. H. (1978) Electrical stimulation of the brain stem in freely moving rats. II. Effects on hippocampal and neocortical electrical activity, and relations to behavior. *Exp. Neurol.*, in press

ROBINSON, T. E., KRAMIS, R. C. & VANDERWOLF, C. H. (1977a) Two types of cerebral activation during active sleep: relations to behavior. *Brain Res. 124*, 544–549

ROBINSON, T. E., VANDERWOLF, C. H. & PAPPAS, B. A. (1977b) Are the dorsal noradrenergic bundle projections from the locus coeruleus important for neocortical or hippocampal activation? *Brain Res. 8*, 77–138

ROTHBALLER, A. B. (1956) Studies on the adrenaline-sensitive component of the reticular activating system. *Electroencephalogr. Clin. Neurophysiol. 8*, 603–621

SHUTE, C. C. D. & LEWIS, P. R. (1967) The ascending cholinergic reticular system: neocortical, olfactory and subcortical projections. *Brain 90*, 497–520

VANDERWOLF, C. H. (1969) Hippocampal electrical activity and voluntary movement in the rat. *Electroencephalogr. Clin. Neurophysiol. 26*, 407–418

VANDERWOLF, C. H. (1975) Neocortical and hippocampal activation in relation to behavior: effects of atropine, eserine, phenothiazines and amphetamine. *J. Comp. Physiol. Psychol. 88*, 300–323

VANDERWOLF, C. H. (1978) Role of the cerebral cortex and ascending activating systems in the control of behavior, in *Handbook of Behavioral Neurobiology*, Plenum Press, New York, in press

VANDERWOLF, C. H., KRAMIS, R. C., GILLESPIE, L. A. & BLAND, B. H. (1975) Hippocampal rhythmical slow activity and neocortical low voltage fast activity: relations to behavior, in *The Hippocampus*, vol. 2: *Neurophysiology and Behavior* (Isaacson, R. L. & Pribram, K. H., eds.), pp. 101–128, Plenum Press, New York

WHISHAW, I. Q. (1976) The effects of alcohol and atropine on EEG and behavior in the rabbit. *Psychopharmacologia 48*, 83–90

WHISHAW, I. Q. & VANDERWOLF, C. H. (1973) Hippocampal EEG and behavior: changes in amplitude and frequency of RSA (theta rhythm) associated with spontaneous and learned movement patterns in rats and cats. *Behav. Biol. 8*, 461–484

WHISHAW, I. Q., BLAND, B. H. & BAYER, S. (1978) Hippocampal granule cell X-irradiation: effects on rhythmical slow wave activity (RSA) and responses to muscarinic and nicotinic antagonists in anaesthetized and freely moving rats. *Brain Res.*, in press

WHISHAW, I. Q., BLAND, B. H., ROBINSON, T. E. & VANDERWOLF, C. H. (1976) Neuromuscular blockade: the effects on two hippocampal RSA (theta) systems and neocortical desynchronization. *Brain Res. Bull. 1*, 573–581

WINSON, J. (1974) Patterns of hippocampal theta rhythm in the freely moving rat. *Electroencephalogr. Clin. Neurophysiol. 36*, 291–301

WINSON, J. (1976a) Hippocampal theta rhythm. I. Depth profiles in the curarized rat. *Brain Res. 103*, 57–70

WINSON, J. (1976b) Hippocampal theta rhythm. II. Depth profiles in the freely moving rabbit. *Brain Res. 103*, 71–79

Discussion

Grossman: Among the best predictors of EEG sleep are complex spikes in the hippocampus. During aroused wakefulness and during paradoxical sleep these spikes are only rarely seen. They become quite common just before (as well as during) slow wave sleep and quite infrequent again just before the onset of paradoxical sleep (Hartse *et al.* 1975). Have you noted this spike activity and, if so, have you observed any correlation with hippocampal theta?

Robinson: We often saw large-amplitude spikes in the hippocampal record during slow wave sleep. In rats deprived of active (paradoxical) sleep the frequency of occurrence of these spikes often increased before the animal entered active sleep.

Vanderwolf: I don't think they predict theta. We see thick spikes, with a duration of around 100 ms. These spikes happen in a lot of circumstances, when an animal is just standing motionless. Some animals show them and others don't. They become more frequent in slow wave sleep, and they occur after atropine too. Atropine doesn't abolish them.

DeFrance: I am a little worried about the effects you obtained with morphine. Did you deliver something such as naloxone to reverse the effects?

Vanderwolf: Yes; naloxone has a dramatic effect.

DeFrance: What about lower doses of morphine?

Vanderwolf: I tested 5, 15, 30 and 60 mg/kg and gave a few animals 100 mg/kg. We get a fullblown effect at around 30 mg/kg, a lesser one at 15, and nothing at 5 mg/kg. Naloxone brings the animals right back; before the drug they are cataleptic but after treatment with 1–2 mg/kg of it they move around in a few minutes and there is a recovery of atropine-resistant RSA (theta rhythm) in the hippocampus and of atropine-resistant low voltage fast activity in the neocortex.

Koella: When you tried to deplete catecholamines you gave reserpine in rather large doses. In our experience and in that of many others, reserpine alone is not enough to eliminate completely the active pool of catecholamines in the terminals. You get better depletion when you pretreat with reserpine about 16–24 hours previously and follow that with α-methyl-p-tyrosine an hour before testing. If you stimulate locus coeruleus in rats and look for inhibitory effects on single cells in the cortex or the cerebellum, reserpine alone never eliminates this inhibition. You are able to knock out this coeruleus-induced cell depression only if you use the combination. You only get enough lowering of the active pool of noradrenaline to eliminate the transmission through adrenergic information channels when you use reserpine and α-methyl-p-tyrosine. Have you tried that combination?

Vanderwolf: No. Noradrenaline in the cortex and the hippocampus were reduced to 5.7% after reserpine.

Koella: That does not seem to be enough.

Vanderwolf: It isn't a complete abolition. After α-methyl-p-tyrosine alone we got depletion of forebrain (including diencephalon) noradrenaline to 14.5% of normal values.

Robinson: In adult rats which were treated neonatally with 6-hydroxydopamine we found a permanent 83% depletion of cerebral noradrenaline (Robinson *et al.* 1977). To try to ensure a more complete depletion of noradrenaline we gave these neonatally treated animals an additional injection of 2.5 mg/kg of reserpine (when they were adults) and recorded hippocampal electrical activity periodically for the next 12 hours. This 'double treatment' failed to abolish either atropine-sensitive or atropine-resistant forms of hippocampal rhythmical slow activity (Robinson *et al.* 1977).

Andersen: From a reductionist point of view your findings, with the exception of the effect of volatile anaesthetics, could be explained not as two different types but as a continuum, ranging from weak activation to more intense activation. It shouldn't be too difficult to find a way to explain the discrepant effect of ether. One system is surely easier than two as an explanation?

Vanderwolf: These two systems are of course hypothetical, because we don't have an anatomical base for them. One thing in favour of two systems is that there have been an enormous number of puzzling observations of EEG, behaviour and the effects of centrally acting drugs for 40 years, and they all fall into place with this scheme.

O'Keefe: It is important to consider the effects of urethane, which I believe you find abolishes the movement-related theta but does not affect the atropine-sensitive theta. If this is so, then experiments done on the theta activity in the urethane-anaesthetized rat, such as those by McLennan & Miller (1974), are only dealing with one of the two theta systems.

Vanderwolf: Urethane is not the same as ether here, because it has a different effect on the neocortex, but the three volatile anaesthetics that I tried are all exactly alike.

Lynch: In your septal stimulation studies, was it the case that the best spots for recording one kind of theta were the best spots for the other kind, or was there a dissociation? Are they anatomically driven from the same places?

Vanderwolf: Yes, but I don't know to what extent one can say that septal stimulation drives atropine-resistant theta. You can control its frequency, but you can't turn it on by a septal stimulus. A septal stimulus does not make the animal walk around if you stimulate at 10 Hz; he may walk or he may not. The atropine-resistant system only comes on *when* he walks. The sites in the medial septum and diagonal band that drive atropine-sensitive theta in a frequency-specific way also drive atropine-resistant theta, however.

Lynch: But are they the *best* spots for driving atropine-resistant theta?

Vanderwolf: We never noticed any difference.

Lynch: You never see one without the other?

Vanderwolf: No. They have different thresholds, for one thing, particularly in the brainstem; if you stimulate at different points throughout the reticular formation, as Terry Robinson did, you find on the whole that the atropine-sensitive effect has a lower threshold; it is easier to get atropine-sensitive theta than atropine-resistant theta. You can get atropine-sensitive electrographic changes without making the animal walk around, but when he walks the atropine-resistant theta comes in.

Lynch: Doesn't that make you wonder about the pharmacology—the fact that the two systems are anatomically intertwined?

Vanderwolf: No; we had hoped that we could separate them, and that was the purpose of going through the reticular formation looking for sites that would give one or the other. We didn't find a great deal, with one exception; Terry Robinson showed that if you stimulate in the nucleus compactus, which is just under the cuneiform nucleus in the midbrain, approximately the region where the A8 dopamine fibres seem to come from, you can drive atropine-resistant theta without the animal walking. That is the only site in the brainstem that seems to do that. So there is something special about that site.

Vinogradova: Have you tried to stimulate the reticular formation with increasing currents? Old data by Stumpf (1965) and Petsche et al. (1965) show that there is a linear relation between theta frequency and the intensity or frequency of midbrain reticular formation stimulation. Thus, the theta range of frequencies can be regarded as a continuum. But of course this is a somewhat artificial situation and perhaps in normal conditions, as shown for example by Donald Lindsley, several places in the brainstem participate in triggering theta activity (Lindsley & Wilson 1975). Among them, some may be more or less sensitive to the effect of anaesthetics, and only some of them (it seems to be the rostral reticular brainstem) are cholinergic. The differences in theta that you see should have some anatomical basis. Have you analysed the effects of electrical stimulation from various places in the brainstem?

Vanderwolf: We stimulated at sites from the hypothalamus back into the medulla. Generally speaking you can elicit both kinds of theta from all these sites, the atropine-sensitive theta having a lower threshold as a rule. After you give atropine, if you try to get theta in an immobile rat by raising the current, it doesn't happen. The rule seems to be that for most sites (the exception is the nucleus compactus site) after atropine it is impossible to produce theta in the rat by stimulating the reticular formation unless it walks around or makes some other type 1 movement.

Vinogradova: How do you explain this possibility of having various frequencies of theta from the same point?

Vanderwolf: If you get a high frequency train of RSA by stimulating the reticular formation, and the animal is motionless at the same time, it will be sensitive to atropine, regardless of the frequency. If you startle a rabbit, without its moving, except for the brief startle response itself, you can produce theta that can go as high as 15 Hz for half a second. That theta is atropine-sensitive. Therefore, both the atropine-sensitive and atropine-resistant systems are capable of driving theta at high frequencies, although atropine-sensitive theta *usually* has a lower frequency than atropine-resistant theta.

Segal: I have a problem with your interpretation. You find that RSA (theta) is associated with walking and a number of drugs which affect walking. Could you dissociate the RSA from the walking?

Vanderwolf: Morphine does that.

Segal: But for the rest of the drugs, you have different theta frequencies depending on the speed of walking. Could you show that these drugs affect walking and the theta frequency differentially? Did you do any frequency analysis to see if the effects are more specific for RSA than for the walking? For example, if amphetamine causes the rats to walk and then you get more RSA, it is not that it affects RSA but that it affects walking and therefore it affects RSA. If you were to find that amphetamine did not cause the rats to walk faster but did cause the RSA to grow faster, you would have a dissociation between the two.

Vanderwolf: We don't find that. I tested that very carefully, after amphetamine, in fact. The frequency during walking after amphetamine is exactly the same as the frequency during walking without amphetamine.

Another factor which I should mention is that in making comparisons of specific frequencies it is important to control the body temperature of the rat. If you give rats a drug or anything else that heats them up, the theta frequency rises. That control isn't always done and it is essential.

Weiskrantz: The general question here is what effects are specific and what effects are indirect via the behavioural changes that are induced.

Vanderwolf: The effects of drugs like amphetamine and the phenothiazines are non-specific in that sense. The RSA waves seem to be normal, as far as I can tell, provided that you make measurements during the same behaviours before and after the administration of the drug. I think that these drugs act by increasing or decreasing activity in the circuits that generate type 1 behaviour but do not alter the mode of operation of the circuits.

Winocur: May I extend this subject of functional significance? Can you try to reconcile your results with RSA and long-term activity and some of the data on lesions? There are two reliable findings here that can perhaps be related. You found that theta activity is clearly correlated with walking and voluntary movement generally, and also in the rat, data on the effects of lesions reliably show that the hippocampus is necessary for situations in which conditioned stimuli take on a new significance and animals have to learn new responses. In your observations in natural settings, when the animal is moving about, he is moving between stimuli; he constantly finds himself in a situation where he has to suppress old responses to stimuli which he has previously encountered, or which may still be there, as he encounters new stimuli. So he is responding to changing stimulus conditions. Perhaps one could then speculate that your findings reflect the process involved in the animal changing his responses to the changing stimuli that he encounters as a result of moving through the environment.

Vanderwolf: It is hard to know what role stimuli play. A few years ago (L. A. Gillespie, unpublished work 1972) we tried removing sensory inputs; we had rats that were blinded, anosmic, had the vibrissae shaved off, and had the inner ear destroyed. None of this seemed to do very much to theta waves. When the animals move about after all this, they still show theta waves in the same relation to behaviour as before.

O'Keefe: Do they show both types of theta?

Vanderwolf: We did that experiment before we knew about the two kinds of theta, and we haven't done it again, so I don't know.

Gray: I would like to try to re-state Gordon Winocur's question in a less theoretical way. The key point is that there are beautiful correlations between theta and movement on the one hand but, on the other, many of us know that a variety of lesions—destruction of the hippocampus; destruction of the septal area in such a way that there is no theta; severing the connections between the septal area and the hippocampus so that theta is abolished—all leave us with animals which perform the behaviours which Dr Vanderwolf called type 1, or voluntary, with perfect ease and no observable impairment. That is a big problem for the interpretation of what theta means and what these data mean.

Weiskrantz: This raises the whole question of correlation or causation, of course, which we shall take up again (see General Discussion II, pp. 309–326).

References

HARTSE, K. M., EISENHART, S. F., BERGMAN, B. & RECHTSCHAFFEN, A. (1975) Hippocampal spikes during sleep, wakefulness, and arousal in the cat, in *Sleep Research*, vol. 4 (Chase, M. H., Stern, W. C. & Walter, P. L., eds.), p. 30, *Brain Information Service*, Los Angeles

LINDSLEY, D. B. & WILSON, C. L. (1975) Brain-stem–hypothalamic systems influencing hippocampal activity and behavior, in *The Hippocampus*, vol. 2: *Neurophysiology and Behavior* (Isaacson, R. L. & Pribram, K. H., eds.), pp. 247–278, Plenum Press, New York

MCLENNAN, H. & MILLER, J. J. (1974) The hippocampal control of neuronal discharges in the septum of the rat. *J. Physiol. (Lond.) 237*, 607–624

PETSCHE, H., GOGOLÁK, G. & VAN ZWIETEN, P. A. (1965) Rhythmicity of septal cell discharges at various levels of reticular excitation. *Electroencephalogr. Clin. Neurophysiol. 19*, 25–33

ROBINSON, T. E., VANDERWOLF, C. H. & PAPPAS, B. A. (1977) Are the dorsal noradrenergic bundle projections from the locus coeruleus important for hippocampal or neocortical activation? *Brain Res. 138*, 75–98

STUMPF, C. (1965) Drug action on the electrical activity of the hippocampus. *Int. Rev. Neurobiol. 8*, 77–138

An experimental 'dissection' of the septal syndrome

SEBASTIAN P. GROSSMAN

Department of Psychology, University of Chicago

Abstract I present evidence from several different lines of investigation (including partial lesions, intraseptal and intrahippocampal drug injections, and surgical transections of the major fibre systems that enter, leave, or traverse the area) which indicates that individual components of the septal syndrome in the rat and cat reflect an interruption of different neural elements. The 'disinhibitory' effects of septal lesions on behaviour that is suppressed as a consequence of non-reward are due to an interruption of septo-hippocampal (or hippocampo-septal) connections that do not have a cholinergic synapse in the septum but do have one in the hippocampus. The inhibitory effects of punishment are not mediated by the same pathways but involve amygdalo-septal (no directionality intended) projections that have a cholinergic synapse in the septal area. This component of the septum may communicate with the hippocampus via entorhinal and periamygdaloid projections. This pathway does not appear to have a cholinergic synapse in the hippocampus. The effects of septal lesions on active avoidance are related to an interruption of at least two pathways. The facilitated shuttle box conditioned avoidance response acquisition is related mainly to an interruption of ventral connections of the septum with the lower brainstem. The impaired acquisition of most other avoidance problems seems to be due to an interruption of components of the stria medullaris. Both pathways have a cholinergic synapse in the septal area. The hyperdipsia, finickiness, and sudden weight loss are related to an interruption of pathways that interconnect the septum with the lower brainstem.

The septum and, more generally, the septo-hippocampal system have been the subject of a great deal of scientific interest and investigation ever since stereotaxic surgery became feasible forty years ago. Lesions in this relatively primitive area of the brain produce dramatic effects on various behaviours. Partly because of historical accident it has been fashionable to interpret the behavioural influences of the septum in terms of global functions.

Papez started this tradition when he proposed in 1937, largely on the basis of anatomical considerations, that the septum might be a nodal point in the circuit which he postulated to be the substrate of 'emotion' or 'affect'. Twenty-five years later, McCleary (1961, 1966) expressed strong reservations about the mentalistic concepts Papez used and proposed to replace them with the more mechanistic if no less global notion of 'response inhibition'. McCleary's hypothesis provided a tailor-made description of the 'passive avoidance' deficit which McCleary (1961) and Kaada and his associates (1962) had observed, and accounted adequately for the transient rage and loss of conditioned emotional response that Brady & Nauta (1953) had described earlier. A general loss of control over response inhibition did not readily account for King's classic observation (1958) that rats with septal lesions learned a shuttle box conditioned avoidance response (CAR) faster than controls. McCleary (1966) steered around this potentially troublesome shoal by proposing an imaginative explanation based on the observation that cats with septal lesions learned conditioned avoidance responses no better or even worse than controls when the animal was taught to avoid painful shock by always escaping from compartment A rather than by jumping back and forth between the two compartments of a shuttle box (Lubar 1964; Zucker 1965). According to this interpretation, normal animals find it difficult to learn the standard shuttle box CAR because it involves returning to the compartment where they have most recently been shocked or threatened with shock. Animals which have a 'passive avoidance' deficit (i.e., loss of 'response inhibition') do not experience this difficulty and consequently learn the avoidance response faster.

In spite of the generality of the concept and McCleary's inventiveness in stretching it to accommodate the rapidly growing literature on the many behavioural effects of septal lesions, the heuristic value of his parsimonious hypothesis decreased with every new publication on the subject. Various investigators and reviewers of the voluminous literature on the behavioural functions of the septo-hippocampal system have proposed alternatives or additions to the response-inhibition hypothesis, such as 'incentive motivation' (Harvey & Hunt 1965) and related concepts (Fried 1970, 1972). It has been clear for some years that the growing list of behavioural dysfunctions reported after septal lesions has become so long and so diverse that theoretical constructs used to account for them in a unitary manner are so amorphous and general as to be of doubtful heuristic usefulness.

SEPTAL LESIONS

I have reviewed the salient features of the experimental literature elsewhere (Grossman 1976) and do not want to duplicate this effort here. It may be instructive, however, to briefly survey the basic behavioural consequences of septal lesions (without attempting to cover the numerous reports which attempt to define their boundary conditions). The information summarized in Table 1 is self-explanatory for investigators in the field. I hope that a few comments will make this material comprehensible to colleagues from allied disciplines.

Under the heading of 'aggressive behaviours' we find the classic, if perhaps a bit misleadingly named, symptom of 'rage'. More specifically, this refers to the fact that many rats with septal lesions strenuously object to normal handling for a few weeks after surgery. This irritability disappears, often within a few days, when the animals are handled (Brady & Nauta 1955), when they are housed in pairs (Ahmad & Harvey 1968), or even when nothing at all is done to

TABLE 1

Principal behavioural effects of septal lesions

Aggressive behaviours
    (a) Increased resistance to handling, footshock, loud noises, and bright lights
    (b) Increased shock-induced 'fighting' and muricide but decreased dominance and fighting in food competition situation

Avoidance behaviours
    (a) Faster acquisition of shuttle-box CAR (even under discriminated 'Sidman' contingencies)
    (b) Slower acquisition of most or all other CARs

Conditioned emotional responses
    (a) Unavoidable shock or related CS less disruptive of ingestive or instrumental behaviour

Punishment ('passive avoidance')
    (a) Ingestive as well as operant responding less affected by punishment

Inhibitory control of responding
    (a) FI, DRL, 'Sidman' avoidance, and extinction responding increased (i.e., inefficient)
    (b) Reversal learning and go/no-go discrimination learning poor

Other operants
    (a) CRF and FR responding increased

Food, water and weight
    (a) Water intake increased
    (b) Food intake normal
    (c) Sharp initial weight loss

Taste preferences
    (a) Intake of palatable (sweet or salty) solutions increased
    (b) Intake of unpalatable (quinine, very salty) solutions decreased

Locomotor activity
    (a) Activity increased in novel/or complex environments
    (b) Activity decreased in wheels, tilt boxes, home cages

promote recovery (Yutzey *et al.* 1964). Although the septal 'rage' syndrome represents the cornerstone of any hypothesis that relates the septum to 'affect', 'emotion' or similar concepts, its relation to the many other behavioural symptoms of septal lesions is, at best, tenuous. Not only does it disappear within a few weeks whereas most other behavioural changes appear to be permanent or at least much more persistent (e.g., Miczek *et al.* 1972), but a number of comparative studies (e.g., Votaw 1960; Moore 1964; Bunnell & Smith 1966; Sodetz *et al.* 1967) have shown little evidence of septal 'rage' in species other than the rat.

In the rat, where septal 'rage' is not uncommon, it is accompanied by an apparently general hyperreactivity to strong and potentially noxious stimuli such as footshock (Lints & Harvey 1969); loud noises (Brady & Nauta 1953); or bright lights (Nielson *et al.* 1965). This general hypersensitivity appears to be as transient as the septal 'rage' itself (Harvey & Lints 1971; Miczek *et al.* 1972).

It is not entirely clear that the increased response to handling which characterizes many rats with septal lesions during the first two or three weeks after surgery should be interpreted as an increase in 'aggressiveness', as early investigators did rather uncritically. Our own extensive experience with rats with septal lesions indicates that they will indeed bite (i.e., 'aggress') more readily when cornered, but prefer to retreat whenever the experimental situation permits it. We (Miczek & Grossman 1972) as well as others (Ahmad & Harvey 1968; Blanchard & Blanchard 1968) have observed an increase in shock-induced or 'reflexive' fighting after septal lesions. However, we share the opinion of many others in the field that the threatening postures counted as 'aggressive' behaviours in this test might, more properly, be classified as exaggerated defensive reactions, due, perhaps, to general irritability. That this may be a correct interpretation is suggested by observations on interspecific aggressive behaviours. Mouse killing, for instance, is not increased by septal lesions when the potential antagonists are permitted to live together for some time before surgery (Karli *et al.* 1969) but is greatly increased when the mouse is introduced into the rat's cage for the first time shortly after surgery (Miczek & Grossman 1972). Also compatible with this hypothesis is our observation (Miczek & Grossman 1972) that the effects of septal lesions on intra- as well as interspecific aggressive behaviours show a transient time course that is similar to that described for exaggerated responses to bright lights (Nielson *et al.* 1965) or painful footshock (Lints & Harvey 1969).

Perhaps the most perplexing behavioural effects of septal lesions are seen in active avoidance situations. King's classic observation of facilitated CAR acquisition in the shuttle box (King 1958) has been replicated in numerous laboratories and species, including the rat (Miczek *et al.* 1972); mouse (Carlson 1970); hamster (Matalka & Bunnell 1968); and cat (McCleary 1961). However, sensible interpretations of this effect have been hard to come by because it appears to be peculiar to the shuttle box paradigm. Septal lesions *retard* CAR acquisition in T-mazes (Nielson *et al.* 1965); multi-chambered alleys (McNew & Thompson 1966); shelf-jumping (Hamilton 1969) or lever-pressing situations (Hamilton 1969); and even in the shuttle box apparatus (Kenyon & Krieckhaus 1965; Zucker 1965) when the conditions are altered slightly so that the animal always escapes from one distinct compartment into another rather than by shuttling back and forth between the two. We (Hamilton *et al.* 1968; Hamilton 1969; Kelsey & Grossman 1969) have compared the effects of septal lesions in a number of experimental paradigms and have reached the conclusion that it is the ubiquitous shuttle box that may give 'atypical' results for some as yet unknown reason.

The third major category of behaviours that appear to be affected by septal lesions are ingestive or operant responses that are suppressed as a result of punishment or non-reward. Brady & Nauta (1953, 1955) first reported that rats with septal lesions failed to display a conditioned emotional response or CER (i.e., freezing and interruption of ongoing operant responding) during the presentation of a signal for unavoidable painful shock. McCleary (1961) subsequently demonstrated that septal lesions also severely impaired the normal inhibitory response to punishment in a 'passive avoidance' situation (i.e., animals continue to feed even though each contact with the food results in painful shock to the mouth). McCleary's observation has been replicated by numerous investigators (e.g., Fox *et al.* 1964; Kaada *et al.* 1962). Impaired responses to punishment have also been seen in a variety of other experimental paradigms, including food- or water-rewarded lever pressing (Schwartzbaum *et al.* 1964; Harvey *et al.* 1965) and situations where the punished response is not appetitively reinforced (McNew & Thompson 1966; McCleary 1966). The 'disinhibitory' effects of septal lesions on punished responding depend, to some extent, on the animal's experience with unpunished responding both before and after the introduction of the punishment contingencies (Middaugh & Lubar 1970; Fried 1970) but constitute one of the most 'robust' components of the septal lesion syndrome under most experimental conditions.

McCleary (1961, 1966) suggested some years ago that the impaired response to punishment might reflect a general tendency to 'perseverate' in previously rewarded behaviours because he observed that cats with septal lesions also found it difficult to reverse a position habit or inhibit a recently acquired avoidance response. McCleary's proposal that a general loss of 'response inhibition' might account for most or all of the effects of septal lesions has been the subject of much discussion and dispute. The more limited thesis that the inhibitory effects of punishment and non-reward are mediated by common, septal, mechanisms has been widely accepted.

The results of our own investigations as well as related work by others (below) suggests that even this hypothesis may not be true, even though large septal lesions do, indeed, 'disinhibit' non-rewarded as well as punished behaviour in a variety of experimental situations. The two experimental paradigms that have been used most extensively to examine the effects of septal lesions on an animal's ability to inhibit non-rewarded responding use fixed interval (FI) or differential reinforcement of low rate (DRL) schedules of reinforcement. In the former, only those lever presses are rewarded which occur a pre-determined time (usually 30–60 seconds) after the last reward. In the latter, a similar schedule of reinforcement is in effect except that responses made before the required waiting period has elapsed not only fail to procure reinforcements but are, in effect, punished because they re-set the clock that determines the delay between successive rewards. Under both test conditions normal rats reasonably quickly learn not to respond for some time after the last reward (i.e., withhold potentially rewarded behaviour).

In the fixed interval paradigm, a well-trained animal begins to respond in the second half of the FI and emits more and more responses as the time of the next reward approaches, thus producing the classic 'scallop' in a cumulative record of its behaviour. Rats with septal lesions overrespond in such a paradigm (Ellen & Powell 1962a,b; Beatty & Schwartzbaum 1968a) and it has been suggested that the principal effect of the lesion might be an increase in 'perseverative' responding during the normally quiet first half of the FI (Beatty & Schwartzbaum 1968a). Our own, recent experiments (Ross & Grossman 1975) have demonstrated a marked but transient effect of the lesion on responding during the first half of the FI and permanent 'disinhibitory' effects only with respect to terminal rate (i.e., the burst of activity that occurs towards the end of the second half of the FI). In conjunction with observations from a DRL (differential reward of low rate of responding) paradigm (below), we proposed that the rat with septal lesion may not 'perseverate' at all but, instead, find it difficult to withhold the next, potentially rewarded response.

In the DRL paradigm, a rat is penalized for responses that occur before the inter-reward delay has elapsed and receives no rewards at all unless it learns to space successive responses so that they occur no more frequently than once per delay interval. Needless to say, this is a difficult assignment for a rat but most normal animals eventually become surprisingly adept at 'timing' their responses accurately (possibly aided by organizing chains of intervening behaviours that help bridge the delay). Rats with septal lesions cannot space their responses in this situation. Instead, they overrespond and receive few, if any, reinforcements (Ellen *et al.* 1964; Carey 1967; MacDougall *et al.* 1969). It has been suggested (Caplan & Stamm 1967) that the effects of septal lesions may be ameliorated when the shift from continuous reinforcement to the DRL contingencies is made gradually. However, we (Kelsey & Grossman 1971) have found that rats with septal lesions learn and perform a DRL problem poorly even in a situation where the DRL contingencies are present from the start of training and 'perseveration' in previously rewarded behaviour patterns is thus impossible. Indeed, rats that were required to alternate responses to two levers and wait a specified time between successive responses, made not more but fewer 'perseverative' responses to the most recently correct lever than controls and 'overresponded' exclusively by failing to withhold the next, potentially correct response. This, of course, is in agreement with our observation (above) that these animals overrespond mainly in the terminal portion of the FI, rather than during the initial post-reinforcement phase (Ross & Grossman 1975).

Septal lesions also increase non-rewarded responding during extinction (Schwartzbaum *et al.* 1964) and in a variety of situations where previously correct responses are no longer rewarded, as during reversal learning (McCleary 1966; Gittelson & Donovick 1968), 'go/no-go' training (Schwartzbaum *et al.* 1964), or the acquisition of problems requiring alternation (Schwartzbaum & Donovick 1968; Dallard 1970). Increases in rate of responding have also been reported when continuous reinforcement schedules (Lorens & Kondo 1969; Neill *et al.* 1974; Ross & Grossman 1975), or schedules (such as fixed ratio) that do not differentially reward the withholding of potentially correct responses (Carey 1969; Hothersall *et al.* 1970), are used.

Lastly, there is a class of effects that may be related to metabolic consequences of septal lesions. Most conspicuous of these is an often pronounced hyperdipsia (Harvey & Hunt 1965) which does not appear to be related to prandial needs (Blass & Hanson 1970) or reflect a primary polyuria (Lubar *et al.* 1968) as some investigators (Besch & van Dyne 1969) have suggested. The effects of septal lesions on water intake appear to reflect a selective disinhibition of reactions to hypovolaemic conditions (Blass & Hanson 1970) and, perhaps, angiotensin in particular (Blass *et al.* 1974). Food intake tends to be normal after septal lesions (Kelsey & Grossman 1969; Lorens & Kondo 1969) although very small transient increases have been reported by some investigators (Singh & Meyer 1968). Indeed, we (Ross *et al.* 1975a) as well as others (Beatty & Schwarzbaum 1968b) have noted a sharp drop in body weight immediately after septal lesions which is not recovered in spite of normal or slightly above normal food intake.

The effects of septal lesions on food intake are potentially complicated by a marked increase in reactivity to palatability after the lesion. Rats with septal lesions consume significantly more than controls of sucrose- or saccharin-sweetened water or milk (Beatty & Schwartzbaum 1967, 1968b; Ross et al. 1975a) and saline solutions (Negro-Vilar et al. 1967) but significantly less of quinine-adulterated fluids (Donovick et al. 1970). These effects appear to be independent of the hyperdipsia (Beatty & Schwarzbaum 1968b; Chiaraviglio 1969).

ANATOMICAL CONSIDERATIONS

When one considers the diversity of the many behavioural effects of septal lesions, it becomes obvious why it has been difficult to conceptualize a function or even multiple interrelated functions for the septal area. Even the most global of the theoretical constructs (e.g. 'emotionality', 'incentive motivation', 'behavioural inhibition') that have been proposed by various investigators of septal functions have invariably been found wanting, unless one defines them in such general terms as to render the concepts practically useless.

It is, of course, possible that this nodal point in the limbic–midbrain circuit may interrelate information from many diverse sources in such complex ways that we have simply failed to grasp the nature of its influence. It has, however, seemed increasingly likely that this structure, like other major subdivisions of the brain, does not exercise a single unitary influence on behaviour but contains a diversity of neural elements that are concerned with different behaviours and possibly unrelated psychological processes.

Experimental support for this conclusion is abundant but it has been difficult to construct a coherent picture from the often contradictory literature. Although the septum is far from structurally homogeneous, the boundaries between adjacent subdivisions are not as clear as they are in adjacent subcortical structures like the thalamus and hypothalamus and there is considerable confusion and disagreement about the associated terminology (Andy & Stephan 1964; Swanson & Cowan 1976). Using Swanson & Cowan's (1976) nomenclature, one can distinguish a medial division that includes the medial septal nucleus dorsally and the nucleus of the diagonal band of Broca ventrally; a lateral division consisting of the lateral nucleus; a posterior division including the septofimbrial and triangular nuclei; and a ventral division consisting of the bed nucleus of the stria terminalis. Some investigators also distinguish a septo-hippocampal nucleus near the dorsal midline and a bed nucleus of the anterior commissure near the origin of the post-commissural fornix. Most if not all of the nuclear masses of the septum can be subdivided further on the grounds of ultrastructural considerations but no attempt has as yet been made to relate these smaller subdivisions to specific functions.

The task of the investigator of septal functions is complicated by the profusion of fibre systems that originate, terminate, or course through the septal region (Raisman 1966; Powell & Hines 1975; Swanson & Cowan 1976). The medial septal region projects to the hippocampus, dentate gyrus, and entorhinal cortex via the fornix. The lateral region receives inputs from the subiculum and hippocampus via pre-commissural components of the fornix, and projects, in turn, to the medial region. Ventrally, the medial and lateral septal nuclei and bed nucleus of the stria terminalis project to the preoptic region and hypothalamus, particularly the ventromedial mamillary region. The hypothalamus and lower brainstem, in turn, project profusely to most aspects of the septum via the medial forebrain bundle. The bed nucleus of the stria terminalis receives projections from most nuclei of the amygdala and projects back to the central and medial regions of that structure. The bed nucleus also projects to and receives connections from the hypothalamus and lower portions of the brainstem. Since the hippocampus proper has no direct projections to the hypothalamus and the amygdala no direct connections with caudal portions of the hypothalamus, these relays in the septum may be of great functional significance. Projections from the septo-fimbrial and triangular nuclei of the septum to the habenula similarly provide a link between the hippocampus and hypothalamus on the one hand and the habenular-interpeduncular system and anterior thalamus on the other. In their course through the septal region, many of these fibre systems are diffuse and extensively interdigitated, and this has made it difficult to identify functionally distinct components of the septum.

PARTIAL LESIONS

Numerous investigators have attempted to provide relevant information by examining the behavioural effects of small lesions in some of the major subdivisions of the septum. With some notable exceptions, the success of this approach has been limited. The most consistent evidence for localization of function in the septal area concerns its apparently general inhibitory influences on instrumental behaviours. There is some consensus that behaviour suppressed by non-reward (e.g., responding on DRL or FI schedules of reinforcement or reversal learning) appears to be 'disinhibited' quite specifically by damage to the anterior aspects of the septum that are traversed by the dorsal fornix (McCleary 1966; Carey 1968; Donovick 1968; Hamilton 1970), or by near-complete destruction of the fornix (including the fimbria) itself (Mac-

Dougall *et al.* 1969). Behaviour suppressed by punishment (i.e., the classic 'passive avoidance' paradigm) surprisingly does not seem to be affected by such lesions (McCleary 1966; Van Hoesen *et al.* 1969; Hamilton *et al.* 1970). Indeed, when we added antero-ventral septal lesions (that maximally impaired the rat's response to punishment) the effectiveness of the dorsal lesion was reduced or abolished (Hamilton *et al.* 1970).

Other components of the septal lesion syndrome have been less amenable to isolation and localization by means of partial lesions. We (Hamilton *et al.* 1970) as well as others (e.g., Donovick 1968; Van Hoesen *et al.* 1969) have consistently found that destruction of any aspect of the septum or adjacent tissues such as the nucleus accumbens septi (Lorens *et al.* 1970) facilitates CAR acquisition in a shuttle box. Septal 'rage' similarly has been observed after damage to various portions of the septum (Harrison & Lyon 1957) although lesions in the ventral segment appear to be most effective in eliciting the hyperreactivity syndrome (Stark & Henderson 1966; Turner 1970). Turner (1970) has therefore suggested that the stria terminalis might be responsible, at least in part, for this component of the septal syndrome. Perhaps because of strain differences we have seen little evidence of 'rage' even after extensive damage to the septal area and have excluded this measure from our test battery because of the unreliability of subjective assessments of 'irritability' under these circumstances. We (Miczek *et al.* 1974) have investigated the effects of lesions in the bed nucleus of the stria terminalis on various aggressive behaviours that are influenced by septal lesions and have the same transient time course that characterizes septal 'rage'. In these studies, we found only partial support for Turner's hypothesis. Extensive damage to the bed nucleus of the stria terminalis reproduced the loss of dominance and fighting in food-competition tests which are typically seen in rats with septal lesions but not the sharp increase in shock-induced aggression and muricide which we (Miczek & Grossman 1972) and others (Ahmad & Harvey 1968; Blanchard & Blanchard 1968) have observed after damage to the septal area. A very similar pattern of effects was seen in our studies (Miczek *et al.* 1974) when we examined the effects of lesions in the amygdaloid complex, suggesting that the effects of septal lesions on fighting in food-competition situations but not the changes seen in other aggressive behaviours may reflect, at least in part, an amygdalo-septal interaction mediated by the stria terminalis. The effects of septal lesions on aggressive behaviours also probably reflect an interference with hypothalamo-septal (no directionality intended) connections since lesions in the ventromedial hypothalamus (VMH) reproduce the effects of septal lesions in both food-competition and shock-induced fighting situations (Grossman 1972). Transection of the rostral connections of that region also resulted in a loss of dominance and fighting in our food-competition test but coronal cuts caudal or lateral to the VMH did not affect aggressive behaviour (Grossman & Grossman 1970; Grossman 1970*a*).

The effects of septal lesions on ingestive behaviour (hyperdipsia and 'finickiness') have been the subject of considerable controversy. Some investigators have reported that the hyperdipsia that is seen after some septal lesions could not be localized (Carey 1969; Sorenson & Harvey 1971). Others (Lubar *et al.* 1968, 1969; Besch & van Dyne 1969) have indicated that the effect is seen most reliably after damage to posterior aspects of the septum. Finickiness (i.e., increased reactivity to positive or negative taste qualities) also appears to be seen preferentially after posterior lesions but the relationship to the hyperdipsia is in doubt (Carey 1969; Chiaraviglio 1969).

Even this necessarily brief and selective review indicates that a different approach is needed if we are to understand the behavioural functions of the septum. Several alternative approaches are suggested by the prevalence of fibre systems in the ultrastructure of the septum. Lesions anywhere in this region interrupt components of major fibre systems, including several which do not terminate or originate in the septum (even though collaterals may exist). One cannot hope to understand the functions of the septum until one can distinguish between the effects of lesions and/or stimulation of perikarya endogenous to the septum and those that are the result of an activation or destruction of fibres of passage.

DIFFERENTIAL RECOVERY OF FUNCTION

It is commonly observed that the physiological and behavioural effects of brain lesions change, often dramatically, as a function of the surgery–test interval. In some instances, nearly complete recovery of function seems to occur within a few days or weeks. In others, there is little evidence of a return of function. Even progressive deterioration of behavioural capabilities has been observed in a number of experiments.

The anatomical complexity of the septal area (as well as the diversity of the effects of septal lesions) suggested to us, some years ago, that different behavioural effects of lesions in this region might have different recovery functions. That this might be a fruitful approach was indicated by a number of studies (e.g. Brady & Nauta 1953, 1955; Yutzey *et al.* 1964) which had demonstrated that the 'rage' response to normal handling that characterizes the early postoperative period disappears within two or three weeks whereas other behavioural effects of septal lesions appear to be permanent.

We first replicated Ahmad & Harvey's (1968) observation that septal lesions increase shock-induced or 'reflexive' fighting in the rat only when the animals are tested shortly after surgery. It is tempting to relate this apparent recovery to the transient time course of the effects of septal lesions on shock-sensitivity but a closer look at the available data does not support this hypothesis. Ahmad & Harvey (1968) reported only slightly different recovery functions for shock-sensitivity and shock-induced fighting but argued nonetheless that the two tests reflected disruptions of different neural functions. We (Miczek & Grossman 1972) investigated the time course of the effects of septal lesions in some detail and succeeded in demonstrating a sharp increase in reactivity to foot-shock at a time (10 days after surgery) when shock-induced fighting had returned to normal. Other tests of intra- and interspecific aggressive behaviours revealed recovery functions that were very similar to that observed in the shock-induced fighting test. Dominant male rats became submissive and refused to fight conspecifics in a food-competition situation when tested 5-8 days after surgery but not when tested (or re-tested) 15 days after the lesion. All of the animals with septal lesions also killed mice introduced into their home cage during the first week but not 15 days after surgery. It is interesting to note that animals that killed a mouse during the initial postoperative period continued to kill on all subsequent tests whereas animals that were not tested until the 10th day after surgery did not kill on this or subsequent tests. These observations suggest that the effects of septal lesions on various aggressive behaviours may have a similar recovery function and may thus reflect a disruption of a common neural mechanism. We have, however, observed quite selective effects of lesions in the bed nucleus of the stria terminalis on food-competition fighting (Miczek *et al.* 1974) and these observations indicate that more specific influences on some types of antagonistic behaviours may also originate in the septal region.

We (Miczek *et al.* 1972) have also investigated the time course of the effects of large septal lesions on active and passive avoidance behaviour. The results of this investigation demonstrated that the marked facilitation of CAR acquisition in the shuttle box that is such a prominent component of the septal lesion syndrome is not present two days after surgery. It appeared, fully developed, when animals with septal lesions were first tested five days after surgery. The passive avoidance deficit, on the other hand, was most pronounced two days after surgery and significantly smaller when the animals were tested five or 10 days after the operation. The clearly different and, indeed, opposite time course of the lesion effects on active and passive avoidance behaviours is of special interest in view of McCleary's (1966) suggestion that the passive avoidance deficit might be responsible for the facilitated shuttle box CAR acquisition. Such an interpretation is not tenable in view of our data.

Since both active and passive avoidance behaviours are maintained by electric shock, it appeared important to have a closer look at the development (and eventual decline) of the hypersensitivity to footshock which numerous investigators (e.g. Lints & Harvey 1969) have reported. We (Miczek et al. 1972) were intrigued to find that the jump threshold to footshock was clearly elevated five and 10 days after surgery (when CAR acquisition was facilitated) but not two days after the lesion (when CAR acquisition was essentially normal but passive avoidance of shock most severely impaired). This pattern of results suggests that differential sensitivity to shock may contribute to the time course of CAR acquisition or passive avoidance responding. One must remember, however, that the effects of septal lesions on active as well as passive avoidance behaviour persist long after the increased sensitivity to footshock has disappeared (e.g. Ross et al. 1975a).

We typically find little evidence of 'rage' reactions to normal handling in our albino rats and thus cannot relate this component of the septal lesion syndrome directly to the results of our tests of intra- and inter-specific aggressive behaviours. The time course of the 'rage' reactions which other investigators (e.g. Brady & Nauta 1955; Yutzey et al. (1964) have reported, is, however, sufficiently similar to those observed in our tests of shock-induced and food-competition fighting (as well as muricide) to suggest that the septal area may exert a common influence on aggressive behaviours. The very different time courses of the effects of septal lesions on active and passive avoidance acquisition indicate that a disruption of different neural mechanisms may be responsible for the disturbances in aggressive and avoidance behaviours. Our results further suggest that the effects of septal lesions on active and passive avoidance responses also do not reflect a common dysfunction.

PHARMACOLOGICAL EXPERIMENTS

The only practically viable experimental approach to the problem described above (p. 237) has been the use of microinjection techniques which permit the direct, intracerebral administration of compounds which selectively facilitate or inhibit synaptic transmission. It is generally accepted that not all chemical synapses in the brain use the same neurotransmitter and the clinical and experimental psychopharmacological literature indicates that functionally related neurons may use a common transmitter, perhaps to assure concurrent activation and prevent cross-talk in areas of extensive anatomical overlap with other neural systems. This raises the possibility that only neurons which subserve a common function might be affected by microinjections of a particular neurotransmitter or compounds which affect its metabolism. The techniques available for the delivery of drugs to specific portions of the brains of freely moving animals are still crude, and perplexing questions concerning appropriate doses, diffusion parameters, and possibly occurring non-specific effects of microinjected compounds have been raised (Rech 1968; Marczynski 1967; Singer & Montgomery 1973). Quite selective behavioural effects have nonetheless been obtained with this technique (see Grossman 1964a, 1967, 1970b, 1972 for review), including some of interest to this discussion.

We have consistently found that only cholinomimetic and cholinolytic compounds produce major changes in the types of behaviours that are typically affected by septal lesions and have confined our systematic investigation to these compounds. Our conclusions about the possible contribution of cellular components of the septal area to various behaviours are thus limited to soma and dendrites that have cholinergic receptor sites. The role of other neurohumours unquestionably deserves further investigation in view of the demonstrated projection of catecholaminergic pathways to the septum. The initial focus on cholinergic components of the area appears nonetheless reasonable in view of a large and generally consistent literature indicating that the behaviours affected by septal lesions are modified, in many instances quite selectively, by systemic injections of cholinergic and anticholinergic compounds (see Carlton & Markiewicz 1971; Bignami *et al.* 1975 for review).

The picture is clearest when one looks at avoidance behaviour. When I first made microinjections into the septal area of the rat, I (Grossman 1964b) observed that intraseptal injections of carbachol, a powerful cholinomimetic compound that produces prolonged excitatory effects at cholinergic synapses, totally blocked the acquisition of a simple shuttle box CAR and interfered, to a lesser extent, with the performance of a previously acquired avoidance response. The anticholinergic compound atropine which blocks transmission at cholinergic synapses produced opposite effects which were similar to those seen after septal lesions. In subsequent experiments, my associates and I demonstrated the similarity between the effects of septal lesions and intraseptal atropine injections in a number of different avoidance paradigms. Injections of the anticholinergic compound scopolamine into the septum of cats or rats consistently facilitated the acquisition of conditioned avoidance responses in a shuttle box (Hamilton & Grossman 1969; Kelsey & Grossman 1969) but the acquisition of a one-way (shelf-jump) avoidance problem was significantly retarded (Hamilton & Grossman 1969) and the learning of an unusual shuttle box avoidance problem which others had shown to be unaffected by septal lesions (Lubar 1964) was normal after intraseptal atropine injections (Hamilton *et al.* 1968).

It may be interesting to note that intraseptal injections of anticholinergics produce no signs of irritability or photophobia, both commonly seen in rats with septal lesions and believed by some to contribute to their effects on active avoidance learning. The consistent observation that intraseptal injections of anticholinergics facilitate, inhibit, or have no effect on CAR acquisition, depending on the nature of the test situation, in exactly the same fashion as septal lesions do, provides evidence of the anatomical and behavioural specificity of the observed drug effects. Our observation that the drug treatments did not produce all of the effects of septal lesions (see also below) constitutes evidence for a selective drug action on some components of the septal area.

Our investigation of the effects of intraseptal injections of anticholinergic compounds on behaviours known to be 'disinhibited' by septal lesions provided further evidence for the distinction between the area's influence on reactions to punishment and non-reward which we had made earlier on the basis of differential lesion effects (p. 232 above). In the traditional passive avoidance paradigm where feeding or drinking is punished by electric shock, and in a so-called 'forced-extinction' paradigm where the execution of a previously learned avoidance response is punished, we (Hamilton et al. 1968) observed clear evidence of disinhibitory effects after intraseptal injections of atropine. However, in classic tests of 'perseverative' responding, atropine-treated cats learned a reversal of a position habit (Hamilton et al. 1968) and atropine-treated rats learned the reversal of a shuttle box CAR (without punishment) as rapidly as controls (Kelsey & Grossman 1969).

Intraseptal injections of a cholinomimetic (eserine) or anticholinergic (scopolamine) in the rat also failed to affect rate of responding in a modified DRL paradigm (Kelsey & Grossman 1975) which had shown reliable facilitatory effects of septal lesions (Kelsey & Grossman 1971). This observation provides a particularly nice demonstration of the usefulness of the intracerebral injection technique, because systemic injections of cholinomimetic or anticholinergic compounds did affect DRL responding in this experiment and these drug effects were eliminated by septal lesions (Kelsey & Grossman 1975). The overall pattern of effects suggests that DRL responding may be controlled, at least in part, by a neural pathway which courses through the septum without synapse (or, at least, without cholinergic synapse) but relies on cholinergic synapses elsewhere. Since we have observed disinhibitory effects of intrahippocampal injections of scopolamine and atropine in the same DRL paradigm (Ross & Grossman 1974) it seems possible that this is the site of action (or at least one of the sites of action) that mediate the effects of systemically administered anticholinergic compounds on behaviour in this situation.

That this conclusion may apply more generally to behaviour suppressed as a consequence of non-reward is suggested by our additional observation (Ross et al. 1975b) that intrahippocampal (as well as intrahypothalamic) injections of anticholinergic compounds facilitated responding during the extinction component of a multiple reinforcement schedule without affecting the rate of responding during a punished component of the same schedule. Punished ingestive behaviour in a traditional 'passive avoidance' paradigm also was not affected by intrahippocampal or intrahypothalamic injections of atropine. This interesting differentiation between intraseptal and intrahippocampal drug effects suggests that the pathways that mediate, at least in part, the response to punishment may follow much the same course as the pathways that mediate the effects of non-reward (since hippocampal as well as hypothalamic lesions do affect the response to punishment) but differ by having a cholinergic synapse in the septal area, and no cholinergic synapse in the hippocampus or hypothalamus.

The complexity of the central pathways that mediate reactions to punishment or non-reward is undoubtedly far greater than this simplistic description implies. We obtained some indication of this when we examined the effects of intraseptal injections of cholinomimetic and cholinolytic compounds on behaviour in an adaptation of the free-operant (Sidman) avoidance paradigm to the shuttle box (Kelsey & Grossman 1975). In our version of this test, the rat has a choice of dealing with the problem in terms of a conventional shuttle box CAR (by emitting avoidance responses to a periodically presented conditioned stimulus, CS) or in terms of the free-operant contingency which permits avoidance not only of the shock itself but also of the signal that threatens shock by the occasional spontaneous performance of the same behaviour. We (Kelsey & Grossman 1971) had found earlier that rats with septal lesions learned to avoid faster than controls in this situation and did so preferentially by spontaneously emitting the CAR, in contrast to controls which preferred to wait until the CS was presented. The behaviour of the rat with septal lesions was, however, grossly inefficient—it emitted many more responses than would be required for perfect avoidance performance. The results of our transection studies (below, p. 247) support our general conclusion that this is still another instance of the disinhibitory effects of septal lesions on previously (or potentially) rewarded behaviour. The results of our drug studies are in general agreement with this conclusion but indicate that pathways outside the septal area play an important role in the control of behaviour in this situation. As in our DRL test, intraseptal injections of anticholinergics did not reproduce the effects of septal lesions on avoidance behaviour but systemic administrations of the same compounds did. However, large lesions in the septal area failed to interfere with the effects of systemically administered cholinergic or anticholinergic compounds on responding in the modified free-operant situation, in contrast to the DRL paradigm where the same lesions abolished all systemic drug effects (Kelsey & Grossman 1975).

The contribution of extra-septal pathways to inhibitory influences over behaviour that is suppressed as a result of non-reward obviously requires additional study. It would appear at this time that the efficient execution of free-operant avoidance behaviour (which requires inhibitory control over potentially reinforced responding) depends, at least in part, on a pathway that does not pass through the septal area and appears to have a cholinergic synapse elsewhere in the brain. The pathway that provides inhibitory control over responding in the DRL situation, on the other hand, appears to pass through the septum although it too relies on a cholinergic synapse elsewhere in the brain. The effects of our intrahippocampal drug injections (Ross & Grossman 1974; Ross et al. 1975b) suggest that the hippocampus may be at least one of the places where one might look for these synapses.

A discussion of the effects of intraseptal drug injections would be incomplete without mention of the fact that the application of carbachol, eserine, or other cholinomimetics elicits drinking in sated rats (Grossman 1964b). Since similar effects can be obtained from various other subcortical injection sites (Fisher & Coury 1962), one might dismiss the matter, were it not for the fact that some septal lesions produce marked hyperdipsia (Harvey & Hunt 1965; Blass & Hanson 1970) and electrical stimulation of certain septal sites appears to inhibit the drinking response to extracellular thirst stimuli (Moran & Blass 1976) or lateral hypothalamic stimulation (Sibole et al. 1971). The direction of the drug effect is paradoxical unless the powerful cholinomimetic compounds used in these experiments exert disruptive effects. Such an interpretation is consistent with Sorenson & Harvey's (1971) observation that only those septal lesions which resulted in a decline in brain acetylcholine produced hyperdipsia. It seems likely that cholinergic components of the brain are involved in the regulation of water intake but the specific role of the septum remains to be elucidated.

LESIONS IN RELATED STRUCTURES

Because of questions which have been raised about the physiological nature of direct intracerebral injections of neurotransmitters or compounds that affect their metabolism (e.g., Rech 1968; Singer & Montgomery 1973), it is important to verify and supplement the information we have obtained by this technique.

The most logical alternative would be an investigation of the effects of selective interruption of the fibre systems that pass through the septum without synapse. Unfortunately there is, at this time, no practical way to obtain an answer to this straightforward question. The closest we have been able to come to the ideal experiment is to selectively transect the principal fibre systems that are, to some extent, interrupted by septal lesions.

To develop a meaningful plan of attack on this project, one must examine the literature that relates the structures where these pathways originate or terminate to the behaviours which we know to be affected by septal lesions. A review of this literature is beyond the scope of our discussion. It must suffice here to remind the reader of some of the principal experimental observations that seem particularly relevant to an understanding of the influences some of the major limbic system structures may exert on septal functions.

The septum is the target of the principal caudal efferents from the hippocampus and the source of many reciprocal afferents to that structure (Raisman 1966; Swanson & Cowan 1976; Siegel & Edinger 1976; Powell & Hines 1975). There is, moreover, considerable evidence (Stumpf 1965; Vinogradova 1975) that neurons in the septal area exercise some control of the neural activity that gives rise to the prominent theta rhythm which characterizes hippocampal EEG records under certain conditions.

We have already discussed the results from some of our drug studies which suggest that the pathways that mediate the response to non-reward may synapse in the hippocampus. The hippocampal lesion literature amply supports this conclusion. Damage to this structure increases responding during extinction and during reward schedules (such as FI or DRL) which normally maintain low rates of responding much as septal lesions do (e.g., Peretz 1965; Kimble & Kimble 1965; Clark & Isaacson 1965; Beatty & Schwartzbaum 1968a). Our own (Hamilton *et al.* 1970) observation that lesions in the medial septum (which projects to the hippocampus via the fimbria and dorsal fornix) but not lesions in the lateral septum (which has no direct projections to that structure) produce 'disinhibitory' effects in various test paradigms also supports this interpretation. One review of the literature on the behavioural effects of hippocampal lesions which has not lost its relevance although it is nearly ten years old (Kimble 1968) concluded that many of the deficits seen in rats with hippocampal lesions could be traced to an inability to inhibit previously rewarded behaviours and to adapt, in general, to changing reward contingencies. This interpretation was based on experiments demonstrating that hippocampal lesions result in *(a)* greater resistance to extinction in both appetitive (Jarrard *et al.* 1964) and aversive (Isaacson *et al.* 1961) instrumental test paradigms; *(b)* impaired responses to changes in lesions. In the first experiment of this series (Ross *et al.* 1975*a*), we found, somewhat to our consternation, that neither fornicotomy nor transection of the ventral connections of the septum produced the passive avoidance deficit that is such a prominent component of the septal lesion syndrome. We were perplexed by this observation, in view of the many studies which have shown passive avoidance deficits after hypothalamic (Kaada *et al.* 1962; Sclafani & Grossman 1971) as well as hippocampal (Kimura 1958) lesions, but our observation has held up in a number of subsequent replications.

When we examined animals which had sustained stria terminalis or stria medullaris transections (Ross & Grossman 1977), we once again failed to see any reliable effects of stria medullaris transection but finally obtained passive avoidance deficits after cuts across the stria terminalis. The latter observation is in excellent agreement with Ursin's (1965) report that stria terminalis lesions in the cat resulted in passive avoidance deficits and many reports of similar deficits after damage to the amygdala or adjacent peri-amygdaloid cortex (Pellegrino 1968; Grossman *et al.* 1975). Our observation that stria medullaris cuts failed to affect passive avoidance behaviour is not entirely congruent with the report by Van Hoesen *et al.* (1969) that habenula lesions produced not only shuttle box CAR facilitation but also impaired passive avoidance behaviour. It appears possible that both effects may be due to an interruption of habenulo-interpeduncular connections since damage to the interpeduncular nucleus produces passive avoidance deficits as well as CAR facilitation (Wilson *et al.* 1972).

The results of our pharmacological studies (above) consistently indicate that the inhibitory effects of punishment (as observed in the traditional passive avoidance paradigm) may be mediated by different central mechanisms from the inhibitory effects of non-reward. The results of our knife-cut studies confirm this interpretation. Whereas fornicotomy failed to have the anticipated disinhibitory effects on punished responding in a passive avoidance situation, a very clear effect was seen when we tested the same animals in a DRL paradigm. Rats with fornicotomies or septal lesions overresponded and received fewer reinforcements than controls when required to space (and alternate) their responses to two levers. Rats which received cuts ventral to the septum performed like controls in this situation. A closer look at the distribution of responses revealed the interesting fact that both the experimental groups that overresponded in the DRL test did so by making anticipatory errors (i.e., pressing the lever that resulted in the next reinforcement if the required delay had elapsed) rather than by making 'perseverative' errors (i.e., responding to the lever that had most recently provided reinforcement). This provides important evidence against theoretical interpretations (e.g.,

McCleary 1966) which suggest that the 'disinhibitory' effects of septal lesions may reflect a tendency to 'perseverate' in the most recently rewarded behaviour. Our observation that cuts below the septum did not result in any increase in responding in this situation argues against the suggestion (Beatty & Schwartzbaum 1968a,b) that the 'disinhibitory' effects of septal lesions might reflect an increase in incentive- or motivation-related processes mediated by septo-diencephalic connections.

Because stria medullaris/habenula lesions have been reported to result in disinhibitory effects in a DRL paradigm, we (Ross & Grossman 1977) tested rats with stria medullaris transections in our DRL situation. In general agreement with the earlier report by MacDougall et al. (1969), we found that transection of this pathway does not reproduce the extensive and persistent 'disinhibitory' effects of septal lesions (or fornicotomies) but produced minor and transient disruptions of the normal pattern of responding.

To confirm our suspicion that the 'disinhibitory' effects of septal lesions on non-rewarded behaviour may be mediated specifically by its connections with the hippocampus, we (Ross & Grossman 1975) compared the performance of rats with septal lesions and cuts above or below the septum in a standard lever-pressing situation, employing a fixed interval (FI) schedule of reinforcement. Immediately after surgery, all three experimental groups showed a transient increase in responding during the normally quiet first half of the FI. Only fornicotomy and septal lesions produced persisting disinhibitory effects on responding that were confined mainly to the second half of the FI and may thus reflect 'anticipatory' rather than 'perseverative' errors.

The results of our tests of operant responding in appetitively reinforced test paradigms dovetail perfectly with the effects of our cuts on behaviour in the free-operant avoidance situation (above). Like the FI and DRL paradigm, the free-operant avoidance contingencies differentially reinforce low frequency behaviour. The comparison is clearest when one compares FI and free-operant avoidance paradigms where only one response per unit time (usually 30 or 60 seconds in our experiments) optimizes the reward/effort ratio. Additional responses are neither rewarded nor punished (except in the limited sense that they do not result in reward or the avoidance of shock). It is of considerable theoretical interest that fornicotomies, like septal lesions, produced significant overresponding in both of these paradigms even though quite different motivational mechanisms must be activated. We thus have rather strong evidence for the conclusion that the 'disinhibitory' effects of septal lesions may be due, specifically, to an interruption of septo-hippocampal connections.

The final section of our investigation of the effects of surgical transections of septal connections was devoted to the changes in ingestive behaviour that have been seen after some septal lesions. In the first part of the investigation we (Ross *et al.* 1975*a*) found that transections of the ventral connections of the septum but not fornicotomy produced hyperdipsia and the sudden and apparently irretrievable weight loss seen after large septal lesions. *Ad libitum* food intake was not affected by any of our procedures but the intake of palatable sweetened milk was increased by ventral as well as dorsal cuts and septal lesions themselves. That the 'finickiness' is not merely another instance of the general 'disinhibitory' effects of septal lesions (which appear to be mediated by hippocampal connections) is further suggested by our observation that ventral as well as dorsal cuts (and, of course, septal lesions) increased responding when the animals were permitted to obtain palatable rewards (on a continuous reinforcement schedule) by alternating responses to two levers placed at opposite ends of a runway. The effectiveness of the ventral cuts in this paradigm is particularly noteworthy in view of the fact that it is identical to our DRL test (which did not show any effects of cuts ventral to the septum) except that a 30-second delay requirement was imposed in the latter. In more recent parts of this investigation (Ross & Grossman 1977), we have ascertained that transections of the stria medullaris or stria terminalis do not affect ingestive behaviour or body weight.

CONCLUSIONS

Where then do we stand in our understanding of the many and complex behavioural functions of the septum and their relation to the major fibre systems that enter or leave the area?

One of the most prominent and pervasive consequences of septal lesions—their 'disinhibitory' effect on behaviour suppressed as a consequence of non-reward—is almost certainly due to an interruption of septo-hippocampal (directionality is not intended by the word-order) connections. This conclusion is supported by a variety of experimental findings, including the apparently preferential effects of anterior septal lesions on reversal learning or responding maintained by FI or DRL schedules of reinforcement; the selective effects of fornicotomy on these and related behaviours such as free-operant CAR responding; and, of course, by the consistent observation of comparable effects after hippocampal lesions. Our drug tests suggest that the pathways that are responsible for the 'disinhibitory' effects of septal lesions in these diverse tests may not synapse in the septal area but do appear to have cholinergic synapses in the hippocampus. Our conclusions about the lack of septal synapses applies, strictly speaking, only to cholinergic synapses, which have been the focus of our investigation, because a large and internally consistent literature (see Carlton & Markiewicz 1971 for review) demonstrates that systemic injections of anticholinergic compounds consistently produce disinhibitory effects on behaviour that are all-but-identical to those seen after septal or hippocampal lesions. No other class of drugs has similarly selective effects on responding suppressed by non-reward, suggesting that the pathways that process the inhibitory effects of extinction may rely on central cholinergic synapses. We have made less-than-systematic efforts to investigate the effects of intraseptal injections of other putative transmitters such as noradrenaline, serotonin, and related compounds, without discovering any indication that these neurohumours affect any of the behaviours we have investigated.

The perhaps most surprising result of our investigation is the overwhelming evidence that the inhibitory effects of punishment do not appear to be processed by the same central pathways that mediate the effects of non-reward. In our pharmacological studies, intraseptal atropine or scopolamine mimicked the effects of lesions in the classic passive avoidance paradigm but not in any of a number of tests where septal lesions 'disinhibited' non-rewarded behaviour. Conversely, intrahippocampal as well as intrahypothalamic injections of anticholinergic compounds significantly increased responding during extinction without 'disinhibiting' punished operant or ingestive behaviour. The results of our lesion studies support the distinction by demonstrating that different regions of the septum are involved in the response to non-reward and punishment, and our transection experiments indicate that amygdaloseptal connections (no directionality intended) rather than septo-hippocampal projections may mediate the influences on punished behaviour. Since entorhinal lesions produce passive avoidance deficits similar to those seen after hippocampal, periamygdaloid, and amygdaloid lesions (Ross *et al.* 1973; Grossman *et al.* 1975), it appears possible that the pathways that control reactivity to punishment may follow this route to or from the hippocampus.

Although lesions along the hippocampal-entorhinal-periamygdaloid-amygdaloid trajectory also modify CAR acquisition (Ross *et al.* 1973; Grossman *et al.* 1975), none of the lesions reproduce the peculiar pattern of facilitated shuttle box acquisition combined with impaired acquisition of one-way CAR tasks that is typical of animals with septal lesions. An interruption of amygdalo-fugal and -petal connections with the septal region and upper brainstem had no effect on active avoidance acquisition in either situation —further evidence that the septal influence on this type of behaviour must rely on a different set of connections. Shuttle box CAR acquisition was facilitated by fornicotomy as well as transection of the ventral connections of the septum but an analysis of the differential effects of these two cuts on the patterning of avoidance responses in our 'optional' free-operant CAR paradigm suggests that fornicotomy may facilitate shuttle box avoidance acquisition mainly or even exclusively because of its general 'disinhibitory' effects on responding, whereas the transection of the ventral connections of the septum may have resulted in more specific effects on CAR acquisition in this situation. The results of our drug studies suggest that this second pathway, but not the former, may have a cholinergic synapse in the septum.

Another quite unexpected result of our investigation was the clear anatomical dissociation of the facilitatory and inhibitory effects of septal lesions on CAR acquisition in different test paradigms. Although the specific influence of situational variables has never been adequately explained, it has generally been assumed that the effects of septal lesions on CAR acquisition in different paradigms reflect a disruption of the same neural functions. The results of our drug studies supported this conclusion by showing that intraseptal atropine injections (which did not duplicate the effects of septal lesions in many other test situations) produced clear evidence of facilitation in the shuttle box and equally unambiguous evidence of disruption or inhibition in one-way avoidance tests. These results suggest that both types of avoidance behaviours may be mediated by pathways which have cholinergic synapses in the septal region. However, the results of our transection studies indicated that different afferent and/or efferent connections must be involved, the MFB and, perhaps, the fornix playing a significant role in shuttle CAR acquisition, the stria medullaris in the one-way avoidance paradigms.

Our investigation of 'aggressive' behaviours indicates that the effects of septal lesions on different behaviours that are commonly classed together as 'aggressive' may also reflect an interruption of different pathways. Septal lesions increased intraspecific shock-induced fighting and facilitated interspecific attack (mouse killing) but resulted in a loss of dominance and aggressive interactions in a food-competition situation. The latter effect was quite specifically reproduced by lesions in the bed nucleus of the stria terminalis, periamygdaloid cortex or cortical nuclei of the amygdala. Lesions in the ventromedial hypothalamus or surgical interruption of its rostral connections reproduced the increase in reflexive fighting as well as the loss of food-competition aggression but had no effect on muricide.

Our investigations of the changes in ingestive behaviour that are seen after septal lesions indicate that the hyperdipsia as well as the sudden and irretrievable weight loss are specifically related to an interruption of the ventral connections of the septum with the hypothalamus and lower brainstem—a conclusion that is in excellent agreement with a substantial literature showing that water intake and body weight are regulated mainly by lower brainstem mechanisms. Finickiness also appeared after transection of the ventral connections of the septum but the association was not unique. Rats with fornicotomies showed similar effects in all of our tests and we could detect no qualitative or quantitative differences between the results of the two experimental procedures. Since we used only highly palatable foods and continuous reinforcement in all of our tests, it is possible that some of these results may reflect 'disinhibited' responding. Further tests using aversive foods or solutions may help to clarify this issue.

The results of our investigation have consistently indicated that the septal lesion syndrome consists of a number of possibly unrelated behavioural dysfunctions that appear to be the result of an interference with different components of the septal area. The results of our recent transection studies indicate that each of the four major classes of behaviours which have been shown to be affected by septal lesions is specifically related to one of the four major fibre systems that terminate, originate, and, to some extent, pass through the septal area. The disinhibitory effects on non-rewarded behaviour are related to an interruption of septo-hippocampal connections; the passive avoidance deficit to the stria terminalis; the impaired avoidance acquisition seen in most CAR paradigms to the stria medullaris; and the facilitated CAR acquisition in the shuttle box, at least some of the effects on 'aggressive' behaviours, and the various changes in ingestive behaviour, to the ventral connections of the septum. Only for the avoidance behaviours (both active and passive) do we, as yet, have evidence for a synapse in the septum.

Do these results imply that the septum is, after all, merely a funnel or perhaps a relay station between subcortical and cortical structures which are more specifically concerned with the organization of the different behaviours that are affected by septal lesions? Our results are compatible with such an interpretation but there is nothing in our data to rule out the possibility that some or all of the information that appears to be funnelled through the septal area may be subjected to important processes of integration so that the sum total of its influence on behaviour may be far greater than a simple addition of the many inputs that appear to impinge on it. To come closer to an answer to this fundamental question, a technological breakthrough is needed which would allow us to selectively affect only the perikarya of the neurons in the septal area, leaving fibres of passage intact and functioning.

ACKNOWLEDGEMENTS

The research reported here has been supported by grant MH 10130 from the U.S. Public Health Service. The preparation of the manuscript was supported, in part, by grant MH 26934 from the U.S. Public Health Service. Special thanks are due to L. Grossman and the many students who have laboured patiently on the research presented here.

References

AHMAD, S. S. & HARVEY, J. A. (1968) Long-term effects of septal lesions and social experience on shock-elicited fighting in rats. *J. Comp. Physiol. Psychol.* 66, 596–602

ANDY, O. J. & STEPHAN, H. (1964) *The Septum of the Cat*, Thomas, Springfield, Ill.

BEATTY, W. W. & SCHWARTZBAUM, J. S. (1967) Enhanced reactivity to quinine and saccharine solutions following septal lesions in the rat. *Psychon. Sci.* 8, 483–484

BEATTY, W. W. & SCHWARTZBAUM, J. S. (1968a) Commonality and specificity of behavioral dysfunction following septal and hippocampal lesions in rats. *J. Comp. Physiol. Psychol.* 66, 60–68

BEATTY, W. W. & SCHWARTZBAUM, J. S. (1968b) Consummatory behavior for sucrose following septal lesions in the rat. *J. Comp. Physiol. Psychol.* 65, 93–102

BESCH, N. F. & VAN DYNE, G. C. (1969) Effects of locus and size of lesions on consummatory behavior in the rat. *Physiol. Behav.* 4, 953–958

BIGNAMI, G., ROSIC, N., MICHALEK, H., MILOSEVIC, M. & GATTI, G. L. (1975) Behavioral toxicity of anticholinesterase agents: methodological, neurochemical, and neuropsychological aspects, in *Behavioral Toxicology* (Weiss, B. & Laties, V. G., eds.), pp. 155–215, Plenum Press, New York

BLANCHARD, R. J. & BLANCHARD, D. C. (1968) Limbic lesions and reflexive fighting. *J. Comp. Physiol. Psychol.* 66, 603–605

BLASS, E. M. & HANSON, D. G. (1970) Primary hyperdipsia in the rat following septal lesions. *J. Comp. Physiol. Psychol.* 70, 87–93

BLASS, E. M., NUSSBAUM, A. I. & HANSON, D. G. (1974) Septal hyperdipsia: specific enhancement of drinking to angiotensin in rats. *J. Comp. Physiol. Psychol.* 87, 422–439

BRADY, J. V. & NAUTA, W. J. H. (1953) Subcortical mechanisms in emotional behavior: affective changes following septal forebrain lesions in the albino rat. *J. Comp. Physiol. Psychol.* 46, 339–346

BRADY, J. V. & NAUTA, W. J. H. (1955) Subcortical mechanisms in emotional behavior: the duration of affective changes following septal and habenular lesions in the albino rat. *J. Comp. Physiol. Psychol.* 48, 412–420

BUNNELL, B. N. & SMITH, M. H. (1966) Septal lesions and aggressiveness in the cotton rat, *Sigmodon hispidus*. *Psychon. Sci.* 6, 443–444

CAPLAN, M. & STAMM, J. (1967) DRL acquisition in rats with septal lesions. *Psychon. Sci.* 8, 5–6

CAREY, R. J. (1967) Contrasting effects of increased thirst and septal ablations on DRL responding in rats. *Physiol. Behav.* 2, 287–290

CAREY, R. J. (1968) A further localization of inhibitory deficits resulting from septal ablation. *Physiol. Behav.* 3, 645–649

CAREY, R. J. (1969) Contrasting effects of anterior and posterior septal injury on thirst motivated behavior. *Physiol. Behav.* 4, 759–764

CARLSON, N. R. (1970) Two-way avoidance behavior of mice with limbic lesions. *J. Comp. Physiol. Psychol.* 70, 73–78

CARLTON, P. L. & MARKIEWICZ, B. (1971) Behavioral effects of atropine and scopolamine, in *Pharmacological and Biophysical Agents and Behavior* (Furchtgott, E., ed.), pp. 346–374, Academic Press, New York

CHIARAVIGLIO, E. (1969) Effect of lesions in the septal area and olfactory bulbs on sodium chloride intake. *Physiol. Behav.* 4, 693–697

CLARK, C. V. H. & ISAACSON, R. L. (1965) Effect of bilateral hippocampal ablation on DRL performance. *J. Comp. Physiol. Psychol.* 59, 137–140

COSCINA, D. V., GRANT, L. D., BALAGURA, S., & GROSSMAN, S. P. (1972) Hyperdipsia following serotonin-depleting midbrain lesions. *Nature New Biol.* 235, 63–64

DALLARD, T. (1970) Response and stimulus perseveration in rats with septal and dorsal hippocampal lesions. *J. Comp. Physiol. Psychol.* 71, 114–118

DONOVICK, P. J. (1968) Effects of localized septal lesions on hippocampal EEG activity and behavior in rats. *J. Comp. Physiol. Psychol.* 66, 569–578

DONOVICK, P. J., BURRIGHT, R. G. & ZUROMSKI, E. (1970) Localization of quinine aversion within the septum, habenula, and interpeduncular nucleus of the rat. *J. Comp. Physiol. Psychol. 71*, 376–383

ELLEN, P. & POWELL, E. W. (1962a) Effects of septal lesions on behavior generated by positive reinforcement. *Exp. Neurol. 6*, 1–11

ELLEN, P. & POWELL, E. W. (1962b) Temporal discrimination in rats with rhinencephalic lesions. *Exp. Neurol. 6*, 538–547

ELLEN, P. & WILSON, A. S. (1963) Perseveration in the rat following hippocampal lesions. *Exp. Neurol. 8*, 310–317

ELLEN, P., WILSON, A. S. & POWELL, E. W. (1964) Septal inhibition and timing behavior in the rat. *Exp. Neurol. 10*, 120–132

FISHER, A. E. & COURY, J. (1962) Cholinergic tracing of a central neural circuit underlying the thirst drive. *Science (Wash. D.C.) 138*, 691–693

FOX, S. S., KIMBLE, D. P. & LICKEY, M. E. (1964) Comparison of caudate nucleus and septal-area lesions on two types of avoidance behavior. *J. Comp. Physiol. Psychol. 58*, 380–386

FRIED, P. A. (1970) Pre- and post-operative approach training and conflict resolution by septal and hippocampal lesioned rats. *Physiol. Behav. 5*, 975

FRIED, P. A. (1972) Septum and behavior: a review. *Psychol. Bull. 78*, 292–310

GITTELSON, P. L. & DONOVICK, P. J. (1968) The effects of septal lesions on the learning and reversal of kinesthetic discrimination. *Psychon. Sci. 13*, 131–138

GROSSMAN, S. P. (1964a) Some neurochemical properties of the central regulation of thirst, in *Thirst in the Regulation of Body Water* (Wayner, M., ed.), pp. 487–510, Pergamon Press, New York

GROSSMAN, S. P. (1964b) Effects of chemical stimulation of the septal area on motivation. *J. Comp. Physiol. Psychol. 58*, 194–200

GROSSMAN, S. P. (1966) The VMH: a center for affective reaction, satiety, or both? *Physiol. Behav. 1*, 1–10

GROSSMAN, S. P. (1967) Neuropharmacology of central mechanisms contributing to control of food and water intake, in *Handbook of Physiology*, Section 6: *The Alimentary Canal*, vol. 1: *Food and Water Intake* (Code, C. F., ed.), pp. 287–362, Williams & Wilkins, Baltimore

GROSSMAN, S. P. (1970a) Avoidance behavior and aggression in rats with transections of the lateral connections of the medial or lateral hypothalamus. *Physiol. Behav. 5*, 1103–1108

GROSSMAN, S. P. (1970b) Modification of emotional behavior by intracranial administration of chemicals, in *Physiological Correlates of Emotion* (Black, P., ed.), pp. 73–93, Academic Press, New York

GROSSMAN, S. P. (1972) Cholinergic synapses in the limbic system and behavioral inhibition. *Res. Publ. Assoc. Res. Nerv. Ment. Dis. 50*, 315–326

GROSSMAN, S. P. (1976) Behavioral functions of the septum: a re-analysis, in *The Septal Nuclei* (DeFrance, J., ed.), pp. 361–422, Plenum Press, New York

GROSSMAN, S. P. & GROSSMAN, L. (1970) Surgical interruption of the anterior or posterior connections of the hypothalamus: effects on aggressive and avoidance behavior. *Physiol. Behav. 5*, 1313–1317

GROSSMAN, S. P. & GROSSMAN, L. (1977) Food and water intake in rats after transections of fibers en passage in the tegmentum. *Physiol. Behav. 18*, 647–658

GROSSMAN, S. P., GROSSMAN, L. & WALSH, L. L. (1975) Functional organization of the rat amygdala with respect to avoidance behavior. *J. Comp. Physiol. Psychol. 88*, 829–850

HAMILTON, L. W. (1969) Active avoidance impairment following septal lesions in cats. *J. Comp. Physiol. Psychol. 69*, 420–431

HAMILTON, L. W. (1970) Behavioral effects of unilateral and bilateral septal lesions in rats. *Physiol. Behav. 5*, 855–859

HAMILTON, L. W. & GROSSMAN, S. P. (1969) Behavioral changes following disruption of central cholinergic pathways. *J. Comp. Physiol. Psychol. 69*, 76–82

HAMILTON, L., MCCLEARY, R. & GROSSMAN, S. P. (1968) Behavioral effects of cholinergic blockade in the cat. *J. Comp. Physiol. Psychol. 66*, 563–568

HAMILTON, L. W., KELSEY, J. E. & GROSSMAN, S. P. (1970) Variations in behavioral inhibition following different septal lesions in rats. *J. Comp. Physiol. Psychol. 70*, 79–86

HARRISON, J. M. & LYON, M. (1957) The role of septal nuclei and components of the fornix in the behavior of the rat. *J. Comp. Neurol. 108*, 120–137

HARVEY, J. A. & HUNT, H. F. (1965) Effects of septal lesions on thirst in the rat as indicated by water consumption and operant responding for water reward. *J. Comp. Physiol. Psychol. 59*, 49–56

HARVEY, J. A. & LINTS, C. E. (1971) Lesions in the medial forebrain bundle: Relationship between pain sensitivity and telencephalic content of serotonin. *J. Comp. Physiol. Psychol. 74*, 28–36

HARVEY, J. A., LINTS, C. E., JACOBSON, L. W. & HUNT, H. F. (1965) Effects of lesions in the septal area on conditioned fear and discriminated instrumental punishment in the albino rat. *J. Comp. Physiol. Psychol. 59*, 37–48

HOTHERSALL, D., JOHNSON, D. A. & COLLEN, A. (1970) Fixed-ratio responding following septal lesions in the rat. *J. Comp. Physiol. Psychol. 73*, 470–476

ISAACSON, R. L., DOUGLAS, R. J. & MOORE, R. Y. (1961) The effect of radical hippocampal ablation on acquisition of avoidance responses. *J. Comp. Physiol. Psychol. 54*, 625–628

JARRARD, L. E., ISAACSON, R. L. & WICKELGREN, W. O. (1964) Effects of hippocampal ablation and intertrial interval on runway acquisition and extinction. *J. Comp. Physiol. Psychol. 57*, 442–444

KAADA, B. R., RASMUSSEN, E. W. & KVEIM, O. (1962) Impaired acquisition of passive avoidance behavior by subcallosal, septal, hypothalamic and insular lesions in the rat. *J. Comp. Physiol. Psychol. 55*, 661–670

KARLI, P., VERGNES, M. & DIDIERGEORGES, F. (1969) Rat-mouse interspecific aggressive behavior and its manipulation by brain ablation and by brain stimulation, in *Aggressive Behaviour* (Garattini, S. & Sigg, E.B., eds.), pp. 47–55, Wiley, New York

KELSEY, J. E. & GROSSMAN, S. P. (1969) Cholinergic blockade and lesions in the ventromedial septum of the rat. *Physiol. Behav. 4*, 837–845

KELSEY, J. E. & GROSSMAN, S. P. (1971) Nonperseverative disruption of behavioral inhibition following septal lesions. *J. Comp. Physiol. Psychol. 75*, 302–311

KELSEY, J. E. & GROSSMAN, S. P. (1975) Influence of central cholinergic pathways on performance on free-operant and DRL schedules. *Pharmacol. Biochem. Behav. 3*, 1043–1050

KENYON, J. & KRIECKHAUS, E. E. (1965) Decrements in one-way avoidance learning following septal lesions in rats. *Psychon. Sci. 3*, 113–114

KIMBLE, D. P. (1968) Hippocampus and internal inhibition. *Psychol. Bull. 70*, 285–295

KIMBLE, D. P. & KIMBLE, R. J. (1965) Hippocampectomy and response perseveration in the rat. *J. Comp. Physiol. Psychol. 3*, 474–476

KIMURA, D. (1958) Effects of selective hippocampal damage on avoidance behavior in the rat. *Can. J. Psychol. 12*, 213–218

KING, B. M., ALHEID, G. F. & GROSSMAN, S. P. (1977) Factors influencing active avoidance behavior in rats with ventromedial hypothalamic lesions. *Physiol. Behav. 18*, 901–913

KING, F. A. (1958) Effects of septal and amygdaloid lesions on emotional behavior and conditioned avoidance responses in the rat. *J. Nerv. Ment. Dis. 126*, 57–63

LASH, L. (1964) Response discriminability and the hippocampus. *J. Comp. Physiol. Psychol. 57*, 251–256

LINTS, C. E. & HARVEY, J. A. (1969) Altered sensitivity to footshock and decreased brain content of serotonin following brain lesions in the rat. *J. Comp. Physiol. Psychol. 67*, 23–31

LISS, P. (1968) Avoidance and freezing behavior following damage to the hippocampus or fornix. *J. Comp. Physiol. Psychol. 66*, 193–197

LORENS, S. A. & KONDO, C. Y. (1969) Effects of septal lesions on food and water intake and operant responding for food. *Physiol. Behav. 4*, 729–732

LORENS, S. A., SORENSEN, J. P. & HARVEY, J. A. (1970) Lesions in the nuclei accumbens septi of the rat: behavioral and neurochemical effects. *J. Comp. Physiol. Psychol. 73*, 284–290

LUBAR, J. F. (1964) Effect of medial cortical lesions on the avoidance behavior of the cat. *J. Comp. Physiol. Psychol. 58*, 38–46

LUBAR, J. F., BOYCE, B. A. & SCHAEFER, C. F. (1968) Etiology of polydipsia and polyuria in rats with septal lesions. *Physiol. Behav. 3*, 289–292

LUBAR, J. F., SCHAEFER, C. F. & WELLS, D. G. (1969) The role of the septal area in the regulation of water intake and associated motivational behavior. *Ann. N.Y. Acad. Sci. 157*, 875–893

MCCLEARY, R. A. (1961) Response specificity in the behavioral effects of limbic system lesions in the cat. *J. Comp. Physiol. Psychol. 54*, 605–613

MCCLEARY, R. A. (1966) Response modulating functions of the limbic system: initiation and suppression, in *Progress in Physiological Psychology*, vol. 1 (Stellar, E. & Sprague, J., eds.), pp. 209–272, Academic Press, New York

MCNEW, J. J. & THOMPSON, R. (1966) Role of the limbic system in active and passive avoidance conditioning in the rat. *J. Comp. Physiol. Psychol. 61*, 173–180

MACDOUGALL, J. M., VAN HOESEN, G. W. & MITCHELL, J. C. (1969) Anatomical organization of septal projections in maintenance of DRL behavior in rats. *J. Comp. Physiol. Psychol. 68*, 568–575

MARCZYNSKI, T. J. (1967) Topical application of drugs to subcortical brain structures and selected aspects of electrical stimulation. *Ergeb. Ges. Physiol. 59*, 86–159

MATALKA, E. S. & BUNNELL, B. N. (1968) Septal ablation and CAR acquisition in the golden hamster. *Psychon. Sci. 12*, 27–28

MICZEK, K. A. & GROSSMAN, S. P. (1972) Effects of septal lesions on inter- and intraspecies aggression in rats. *J. Comp. Physiol. Psychol. 79*, 37–45

MICZEK, K. A., KELSEY, J. E. & GROSSMAN, S. P. (1972) Time course of effects of septal lesions on avoidance, response suppression, and reactivity to shock. *J. Comp. Physiol. Psychol. 79*, 318–327

MICZEK, K. A., BRYKCZYNSKI, T. & GROSSMAN, S. P. (1974) Differential effects of lesions in the amygdala, periamygdaloid cortex or stria terminalis on aggressive behaviors in rats. *J. Comp. Physiol. Psychol. 87*, 760–771

MIDDAUGH, L. D. & LUBAR, J. F. (1970) Interaction of septal lesions and experience on the suppression of punished responses. *Physiol. Behav. 5*, 233–238

MOORE, R. Y. (1964) Effects of some rhinencephalic lesions on retention of conditioned avoidance behavior in cats. *J. Comp. Physiol. Psychol. 57*, 65–71

MORAN, J. S. & BLASS, E. M. (1976) Inhibition of drinking by septal stimulation in rats. *Physiol. Behav. 17*, 23–27

NEGRO-VILAR, A., GENTIL, C. G. & COVIAN, M. R. (1967) Alterations in sodium chloride and water intake after septal lesions in the rat. *Physiol. Behav. 2*, 167–170

NEILL, D. B., ROSS, J. F. & GROSSMAN, S. P. (1974) Comparison of the effects of frontal, striatal and septal lesions in paradigms thought to measure incentive motivation or behavioral inhibition. *Physiol. Behav. 13*, 297–305

NIELSON, H. C., MCIVER, A. H. & BOSWELL, R. S. (1965) Effect of septal lesions on learning, emotionality, activity and exploratory behavior in rats. *Exp. Neurol. 11*, 147–157

OLTON, D. S. & ISAACSON, R. L. (1968) Hippocampal lesions and active avoidance. *Physiol. Behav. 3*, 719–724

PAPEZ, J. W. (1937) A proposed mechanism of emotion. *Arch. Neurol. Psychiatr. 38*, 725–743

PELLEGRINO, L. (1968) Amygdaloid lesions and behavioral inhibition in the rat. *J. Comp. Physiol. Psychol. 65*, 483–491

PERETZ, E. (1965) Extinction of a food-reinforced response in hippocampectomized cats. *J. Comp. Physiol. Psychol. 60*, 182–185

POWELL, E. W. & HINES, G. (1975) Septohippocampal interface, in *The Hippocampus*, vol. 1: *Structure and Development* (Isaacson, R. L. & Pribram, K. H., eds.), pp. 41–60, Plenum Press, New York

RAISMAN, G. (1966) The connexions of the septum. *Brain 89*, 317–348

RECH, R. H. (1968) The relevance of experiments involving injection of drugs into the brain, in

Importance of Fundamental Principles in Drug Evaluation (Tedeschi, D. H. & R. E., eds.), pp. 325–360, Raven Press, New York

Ross, J. F. & Grossman, S. P. (1974) Intrahippocampal application of cholinergic agents and blockers: effects on rats in DRL and Sidman avoidance paradigms. *J. Comp. Physiol. Psychol.* 86, 590–600

Ross, J. F. & Grossman, S. P. (1975) Septal influences on operant responding in the rat. *J. Comp. Physiol. Psychol.* 89, 523–536

Ross, J. F. & Grossman, S. P. (1977) Transections of stria medullaris or stria terminalis in the rat: effects on aversively controlled behavior. *J. Comp. Physiol. Psychol.* 91, 907–917

Ross, J. F., Walsh, L. L. & Grossman, S. P. (1973) Some behavioral effects of entorhinal cortex lesions in the albino rat. *J. Comp. Physiol. Psychol.* 85, 70–81

Ross, J. F., Grossman, L. & Grossman, S. P. (1975a) Some behavioral effects of transection of ventral or dorsal fiber connections of the septum. *J. Comp. Physiol. Psychol.* 89, 5–18

Ross, J. F., McDermott, L. J. & Grossman, S. P. (1975b) Disinhibitory effects of intrahippocampal or intrahypothalamic injections of anticholinergic compounds. *Pharmacol. Biochem. Behav.* 3, 631–640

Schwartzbaum, J. S. & Donovick, P. J. (1968) Discrimination reversal and spatial alternation associated with septal and caudate dysfunction in rats. *J. Comp. Physiol. Psychol.* 65, 83–92

Schwartzbaum, J. S., Kellicutt, M. H., Spieth, T. M. & Thompson, J. B. (1964) Effects of septal lesions in rats on response inhibition associated with food-reinforced behavior. *J. Comp. Physiol. Psychol.* 58, 217–224

Sclafani, A. & Grossman, S. P. (1969) Hyperphagia produced by knife cuts between the medial and lateral hypothalamus in the rat. *Physiol. Behav.* 4, 533–538

Sclafani, A. & Grossman, S. P. (1971) Reactivity of hyperphagic and normal rats to quinine and electric shock. *J. Comp. Physiol. Psychol.* 74, 157–166

Sibole, W., Miller, J. J. & Mogenson, G. J. (1971) Effects of septal stimulation on drinking elicited by electrical stimulation of the lateral hypothalamus. *Exp. Neurol.* 32, 466–477

Siegel, A. & Edinger, H. (1976) Organization of the hippocampal-septal axis, in *The Septal Nuclei* (DeFrance, J. F., ed.), pp. 79–114, Plenum Press, New York

Singer, G. & Montgomery, R. B. (1973) Theoretical review. Specificity of chemical stimulation of the rat brain and other related issues in the interpretation of chemical stimulation data. *Pharmacol. Biochem. Behav.* 1, 211–221

Singh, D. & Meyer, D. R. (1968) Eating and drinking by rats with lesions of the septum and the ventromedial hypothalamus. *J. Comp. Physiol. Psychol.* 65, 163–166

Sodetz, F. J., Matalka, E. S. & Bunnell, B. N. (1967) Septal ablation and affective behavior in the golden hamster. *Psychon. Sci.* 7, 189–190

Sorenson, J. P. & Harvey, J. A. (1971) Decreased brain acetylcholine after septal lesions in rats: correlation with thirst. *Physiol. Behav.* 6, 723–725

Stark, P. & Henderson, J. K. (1966) Increased reactivity in rats caused by septal lesions. *Int. J. Neuropharmacol.* 5, 379–384

Stumpf, C. (1965) Drug action on the electrical activity of the hippocampus. *Int. Rev. Neurobiol.* 8, 77–138

Swanson, L. W. & Cowan, W. M. (1976) Autoradiographic studies of the development and connections of the septal area, in *The Septal Nuclei* (DeFrance, J. F., ed.), pp. 37–64, Plenum Press, New York

Teitelbaum, P. (1955) Sensory control of hypothalamic hyperphagia. *J. Comp. Physiol. Psychol.* 48, 156–163

Teitelbaum, P. & Epstein, A. N. (1962) The lateral hypothalamic syndrome: recovery of feeding and drinking after lateral hypothalamic lesions. *Psychol. Rev.* 69, 74–90

Turner, B. H. (1970) Neural structures involved in the rage syndrome of the rat. *J. Comp. Physiol. Psychol.* 71, 103–113

Ursin, H. (1965) Effect of amygdaloid lesions on avoidance behavior and visual discrimination in cats. *Exp. Neurol.* 11, 298–317

VAN HOESEN, G. W., MACDOUGALL, J. M. & MITCHELL, J. C. (1969) Anatomical specificity of septal projections in active and passive avoidance behavior in rats. *J. Comp. Physiol. Psychol.* 68, 80–89

VINOGRADOVA, O. S. (1975) Functional organization of the limbic system in the process of registration of information: facts and hypotheses, in *The Hippocampus*, vol. 2: *Neurophysiology and Behavior* (Isaacson, R. L. & Pribram, K. H., eds.), pp. 3–70, Plenum Press, New York

VOTAW, C. L. (1960) Study of septal stimulation and ablation in the macaque monkey. *Neurology 10*, 202–209

WILSON, J. R., MITCHELL, J. C. & VAN HOESEN, G. W. (1972) Epithalamic and ventral tegmental contributions to avoidance behavior in rats. *J. Comp. Physiol. Psychol.* 78, 442–449

YUTZEY, D. A., MEYER, P. M. & MEYER, D. A. (1964) Emotionality changes following septal and neocortical ablations in rats. *J. Comp. Physiol. Psychol. 58*, 463–467

ZUCKER, I. (1965) Effect of lesions of the septal-limbic area on the behavior of cats. *J. Comp. Physiol. Psychol. 60*, 344–353

Discussion

Weiskrantz: You suggested that there may be functional labels that can be attached to some of these behavioural components, such as 'disinhibition', for example. One question is therefore how many components there are. (Operationally, one way of attacking that question convincingly is to look for double dissociation, and you had several examples of that.) Secondly, how are these components to be characterized functionally? How does one justify the use of any particular label? There are also a number of detailed questions about whether the lesions are being selected on independent criteria and how they are to be interpreted in the light of anatomical projections.

Grossman: I believe that the results of our studies unequivocally indicate that septal lesions influence at least six different categories of behaviour. When one considers the differential rate of recovery of some of these behaviours, the effects of smaller and more restricted lesions, the consequences of intraseptal (and intrahippocampal) drug injections, and the results of our transection studies, the conclusion seems inescapable that different components of the septum influence each of the six classes of behaviour we have identified. I should hasten to add that the behavioural categories were defined purely on empirical grounds, using mainly the results of experiments done in my own laboratory. There is, as yet, no theoretical basis for this classification and no reason to believe that it could not be expanded or contracted on the basis of the results of future experiments or, for that matter, already available results of other lines of investigation.

Azmitia: I would like clarification on how you made these lesions, and what controls were used, especially for the knife cut. Was the needle lowered in the midline through both the superior and inferior sagittal sinuses?

Grossman: All knife cuts were made with an encephalotome similar to that described earlier (Sclafani & Grossman 1969). With the aid of this instrument, a very thin (30 gauge) stainless steel guide cannula was stereotaxically implanted into the brain. A wire 'knife' of a diameter roughly equal to that of a human hair was then extended from the tip of that guide cannula and the entire assembly raised or lowered (for cuts in the coronal or parasagittal plane) or rotated (for cuts in the horizontal plane). For the fornicotomy, the guide cannula was inserted 1.8 mm lateral to the midline and lowered to the dorsal edge of the corpus callosum. A 2.6 mm long 'knife' was then extended from the tip of the cannula so that it projected medially at an angle of approximately 90° to the guide shaft. The assembly was then lowered 2.4 mm in order to transect all fibres of the fimbria-fornix system by a cut in the coronal plane. The procedure for stria terminalis and stria medullaris transections was identical except that the knife assembly was inserted into the brain slightly more lateral (2.2 and 2.3 mm) to the midline and deeper (3.4 mm below dura). The cut itself (i.e. the length of the knife and the excursion of the encephalotome) was also somewhat shorter and shallower than that used to transect the fornix-fimbria system. In the case of the horizontal undercut of the septum that was used to transect its ventral connections with the hypothalamus and lower brainstem, the cannula guide was inserted 1.6 mm lateral to the midline so that its tip came to rest slightly below the lateral septal nucleus. A 2.6 mm long wire knife was then extended and the assembly rotated 75° in each direction from a line drawn perpendicular to the midline. All procedures were repeated contralaterally. Damage to the sagittal sinuses was not seen in any animal.

Vinogradova: So there was also a lesion of the lateral septum in these cases?

Grossman: In all cases, the 30 gauge needle passes through the lateral septum. Since this needle has an outside diameter of less than 0.2 mm and its tip is smooth and round, there is very little damage to the area—so little, in fact, that we typically cannot find the cannula track when we examine histological materials to determine the extent of the cuts. In earlier studies we examined the behavioural effects of intraseptal implantations of much larger cannulas (used to microinject drugs into the area) and have never found significant effects.

Azmitia: What concerns me about the septal lesions is that a large effect is produced very early on and then declines dramatically between four and eight days. That might suggest that the changes observed were due to a non-specific effect of the lesion itself, maybe upsetting the ventricular system or causing general damage to the brain; rather than a specific effect of the lesion localized to the septum.

Grossman: First, these septal lesions are all-but-identical to those used by nearly all investigators in the field, although most do not report the immediate, post-surgical consequences of their procedures. Second, I cannot agree that the changes that occur in many behaviours during the first days or weeks after a lesion must reflect 'non-specific' effects. It is true that any electrolytic lesion in the brain produces transient effects (including irritation or oedema) on surrounding tissues and that large septal lesions undoubtedly upset the normal flow and pressure of the ventricular fluid. You must not forget, however, that the behavioural changes seen during the initial postoperative period are quantitative rather than qualitative with one notable exception. Only aggressive behaviours recover, apparently completely, within 10–15 days after a septal lesion. Other classes of behaviour (such as passive avoidance) show some degree of recovery but also a permanent deficit. Indeed, the facilitation of avoidance acquisition in the shuttle box that is such a prominent component of the septal lesion syndrome was not observed immediately after the lesion, when your concern about non-specific influences is most justified, and appeared full-blown on Day 5. It is possible that what you have called 'non-specific' effects of the lesion may contribute to some of the changes we saw during the immediate postoperative period but the very different patterns of recovery or development of the different behavioural effects would seem to argue against any interpretation in terms of general factors such as ventricular pressure changes.

Azmitia: How would you explain the rapid decline in some of these behavioural changes if they are due to septal neuronal damage, since it is unlikely that regeneration or sprouting occurs within such a short time?

Grossman: There is plenty of evidence of functional recovery at the neural level 10–12 days after central nervous system lesions, including supersensitivity of denervated postsynaptic receptor sites, and increased presynaptic transmitter production, release and re-uptake. This could certainly account for the apparently complete recovery of aggressive behaviours and may influence other postoperative recovery processes as well. The partial recovery of passive avoidance behaviour that appears to take place primarily during the first few days after the lesion may reflect a lessening of the irritative and disruptive effects of the lesion on functionally related mechanisms elsewhere in the brain (my guess would have to be the stria terminalis system) combined, perhaps, with a gradual reorganization of function in other parts of the brain. This is a particularly appealing interpretation of the partial recovery in this instance because numerous areas of the brain (including the frontal lobe, hippocampus, amygdala and hypothalamus) appear to take part in the organization of behavioural responses to punishment. I have no truly satisfactory explanation for the late development of the septal lesion effect on shuttle box avoidance acquisition. The animals are clearly capable of learning and performing the required behaviour even on the first or second day after surgery and there are no overt signs of debilitation that might account for the lack of facilitatory lesion effects at this time. We (Grant et al. 1973) have observed a similar time course (i.e. no effect immediately after the lesion and near-maximal effects 5–8 days later) for the effects of serotonin-depleting lesions in the raphe nuclei of the brainstem on muricide. In this case, we observed a rather nice correlation between the gradual development of the behavioural effects and the gradual depletion of serotonin from the brain. On the first and second day after these lesions, the turnover of serotonin (i.e. 5-hydroxyindoleacetic acid level) is, in fact, higher than normal and interspecific aggressive behaviour is normal. Serotonin turnover (as well as absolute level) dropped sharply during the next few days and reached near asymptotic low levels by Day 6 (when interspecific aggressive behaviour was markedly affected). Although correlation certainly cannot be equated with causation, the hypothesis occurred to us that muricide might be under the control of neural mechanisms which respond to a tonic serotonergic input. Raphe lesions did not remove the transmitter in the first few days after the lesion and may even have resulted in increased release from degenerating terminals. Although the mechanisms normally controlling the release of serotonin were, of course, not functional, the continued presence of the transmitter may have prevented the highly abnormal muricide response from appearing in the first few days after surgery. Septal lesions are known to deplete serotonin and it has been suggested that this might be responsible for their effects on shock sensitivity (Lints & Harvey 1969). Although our time course studies indicate that the transient lowering of shock-thresholds is probably not responsible for the facilitated acquisition of shuttle box avoidance responses, the gradual depletion of serotonin might provide a mechanism (or, at least a model for a mechanism) that could account for the peculiar time course of the septal lesion effect on CAR facilitation.

I would also remind you that partial or apparently complete recovery of function has been seen, often within days after surgery, after lesions in other parts of the brain. One example is the effect of lateral hypothalamic lesions (or surgical transections of fibres that course through the area) on food and water intake. Rats and cats are aphagic and adipsic after such lesions but recover voluntary ingestive behaviour, often within a few days after the lesion. Teitelbaum & Epstein (1962) have described the recovery process and it is clear that recovery begins within 2–3 days after lateral hypothalamic lesions and continues through a series of 'stages' until voluntary food and water intake, in the absence of complicating influences, is essentially normal. Exactly what happens at the neuronal level in this gradual process of recovery is currently being debated, with a range of hypotheses extending from encephalization of function (Teitelbaum & Cytawa 1965) to functional recovery in catecholamine projections to the striatum (Stricker & Zigmond 1976). The fact that we do not fully understand the mechanisms responsible for behavioural recovery of function need not imply that the original deficit must be related to what you seem to consider uninteresting non-specific effects of CNS lesions. I would argue that *more* attention needs to be paid to the development of lesion effects in the immediate postoperative period if we are to understand the persisting impairments.

Azmitia: Damage of the ventricular system is only an example of something which you could be disrupting which has no direct bearing on the septal system. The possibility of widespread generalized damage makes it difficult to interpret the effects of cuts in the medial forebrain bundle, since this tract itself contains both ascending and descending projections arising from both rostral and caudal structures. If one is unable to be precise about the anatomy, it must surely be even more difficult to be precise about the type of behaviour disrupted?

Grossman: Your comments on the anatomical complexity of the projection systems affected by our cuts, especially the undercut of the ventral septum, are quite correct but not germane to the arguments in my paper. We don't know which of the many diffuse fibre systems that enter or leave the ventral aspects of the septum is responsible for the behavioural effects of the ventral cut. The nature of the deficits (enhanced shuttle box acquisition and a disruption of regulatory functions related to body weight and water intake) suggests an interruption of connections with specific hypothalamic mechanisms (lesions in the ventromedial hypothalamus produce similar effects on avoidance behaviour and lesions in the lateral hypothalamus disrupt regulatory functions related to water intake and body weight). We (or someone else) will have to follow up this promising lead. Much the same could be said about the other cuts which involve complex projection systems that undoubtedly include heterogeneous pathways interconnecting different aspects of the hippocampus, subiculum, entorhinal cortex and amygdala with the septum, thalamus and lower brainstem. Our present methods clearly cannot make the fine distinctions you ask me to make.

I nonetheless believe that what we have done so far represents a significant advance in our understanding of septal functions. In contrast to the widely held belief that the complex behavioural effects of septal lesions reflect a disruption of a single behavioural or psychological function (such as affective reactivity, behavioural inhibition, etc.), our data show that the 'septal lesion syndrome' consists of at least six independent or partly independent components reflecting the disruptive effects of septal lesions on several different neural substrates. That, and only that, is the conclusion one can safely draw from our work so far, but I hope it will provide the basis of further research that will eventually result in the detailed specification of the relationship between neural substrate and behavioural function for which you ask.

Azmitia: Despite the large amount of degeneration found, you made no independent assessment of the systems destroyed. For instance, the serotonin projections to the septum may only be partially destroyed since the fibres into the lateral septum derived from the dorsal raphe nucleus may be spared. The medial septal serotonin fibres might be preferentially destroyed.

Grossman: The ventral connections of the septum are, as you know, enormously diffuse. They also ascend and descend through a large part of the rostral forebrain and it is, indeed, doubtful that one could completely disconnect septum and lower brainstem. Functions relayed exclusively or mainly by pathways that enter or leave the lateral-most aspects of the ventral septum (or, perhaps, its most caudal or rostral portions) could have been affected little or not at all by our cuts, which were designed to sever the majority of the septal connections with the hypothalamus and lower brainstem. This, in fact, might account for the seemingly baffling fact that lesions in the ventromedial hypothalamus as well as knife cuts rostral to that area reproduce many of the effects of septal lesions that are not reproduced by our undercut of the septum itself. I quite agree that we cannot yet assign the very clear-cut behavioural effects of these cuts to any anatomically or pharmacologically specified neural system and must be cautious in interpreting the plethora of 'negative' results observed after ventral cuts. However, specification of pathways was not the primary goal of the initial series of experiments that I described. We were concerned with the more general question of whether anatomical methods could be used to demonstrate the diversity of the behavioural functions of the septum. The interesting anatomical questions you raise have to be deferred until the nature of the behavioural dysfunctions is understood precisely enough for us to be able to look for a specific relationship between anatomical substrate and function.

Weiskrantz: This issue is an extremely important one but it would take a long time to resolve, fibre tract by fibre tract. How much of the histology of these procedures has been published?

Grossman: Most of the data from our transection studies are presented in four papers with J. Ross (quoted in my paper here). They all include extensive discussions of methodology and the anatomical consequences of our knife cuts. One paper (Ross *et al.* 1975) includes brief descriptions of the degeneration pattern seen in Fink-Heimer silver stained material after septal lesions, or transections of the dorsal or ventral connections of the septum. I did not discuss this material here because it only confirms what is obvious on the basis of our routine histology—that we did transect the fibre systems we intended to cut. There are many excellent descriptions by experts of the degeneration patterns seen after damage to each of these fibre systems or their sites of origin. Our work adds nothing to this literature.

Rawlins: Did you make recordings of the effects of your lesions on spontaneous hippocampal activity, which might be a way of differentiating between some of them?

Grossman: No.

Rawlins: In the paper with Hamilton and Kelsey (Hamilton *et al.* 1970) you at times found no effects of total septal lesions on tasks which were affected by smaller septal lesions; the fibre cuts you reported here also seem to affect some tasks which your total septal lesions did not. Does this mean that there are mutual inhibitory processes within the septum, or is there some other explanation?

Grossman: We have on several occasions observed larger behavioural effects of small lesions in portions of the septum than after larger lesions that involved the same areas in addition to other aspects of the septal area. A particularly nice example of this is our observation (Hamilton *et al.* 1970) that dorsolateral lesions in the septal area produced a massive disinhibitory effect on punished ingestive behaviour whereas larger lesions that involved the medial as well as lateral nuclei did not. It seems intuitively obvious that an area that exerts pronounced influences on behavioural responses to punishment must have a variety of relevant inputs, including some which facilitate and some that inhibit the final output of the system. Even though the effect of the septal area on punished behaviour appears to be inhibitory, this influence needs to be modulated. One would thus expect differential effects of various small lesions in the septal area, unless the neural substrate of this function is very diffusely represented or our behavioural tests are not sufficiently sensitive to pick up differential lesion effects.

Gray: The aspect of your data that I find most puzzling concerns punishment. You say that a dorsolateral septal lesion disinhibits punished responding; that is something that Nick Rawlins and Joram Feldon have found in my laboratory (Gray *et al.* 1978). I don't see how that fits with the effect of the stria terminalis cut, because there is no stria terminalis projection to or from the dorsolateral septal area and you shouldn't invade the stria terminalis with the dorsolateral lesion.

Ursin: These could be two different phenomena. I found that lesions of the stria terminalis or medial amygdala (Ursin 1965) gave a passive avoidance deficit in the cat.

Gray: That is all right, provided it is not being suggested that the effect of the septal lesion on punishment is due to inadvertent damage to the stria terminalis.

Ursin: A large septal lesion will also do this.

Gray: The point is that a small lateral septal lesion does it as well.

Weiskrantz: Dr Grossman has been concentrating quite properly on subsections of the fornix and the septo-hippocampal axis, but some of the effects of these cuts can be obtained by lesions quite outside the system. Orbitofrontal lesions, for example, produce disinhibitory effects. Amygdalar lesions affect avoidance behaviour. While we want to talk about specific effects of lesions in this particular system on particular kinds of behaviour, it cannot be assumed that these behaviours are affected only by lesions in this system or some part of it.

Grossman: Numerous other areas of the brain have certainly been shown to influence the behaviours we (as well as many others) have found to be affected by septal lesions or transections of its major connections. Indeed, one point that the results of our transection studies make clear is the fact that none of the classic components of the septal lesion syndrome reflects a dysfunction of septal origin plain and simple. Our data clearly indicate that, in all cases, the effects of septal lesions are due to a disruption of functions which involve specific connections of the area with other limbic system or brainstem structures (or fibres of passage that may not synapse in the septal area). That, of course, applies to the influence of septal area components on punished behaviour. My own principal problem with the pattern of effects that emerged from our transection studies is why one should obtain such pronounced effects of hippocampal as well as medial hypothalamic lesions on punished behaviour but none after transection of the septal area's connections with these regions.

Gray: I am still not clear whether you are suggesting that the stria terminalis cut and the lateral septal cut are related to each other anatomically and that both affect punishment for this reason, or whether the stria terminalis is just another system that affects punishment.

Grossman: You are asking a very difficult question. Lesions in the hippocampus (Kimura 1958), amygdala (Grossman *et al.* 1975; Ursin 1965), habenula (Van Hoesen *et al.* 1969) and ventromedial hypothalamus (Kaada *et al.* 1962; King *et al.* 1978) have all been reported to result in disinhibitory effects on punished behaviour that are apparently similar to those seen after septal lesions. We fully expected, therefore, to see passive avoidance deficits after each of our cuts and were surprised to find that only transection of the stria terminalis reproduced the effects of septal lesions. It seems unlikely that these results are due to general test insensitivity since we obtained rather dramatic effects of septal lesions and stria terminalis transections.

In the case of our ventral cut, it is possible that it might miss some important caudal projections of the septum, particularly since we tried to keep the cut from entering the ventricles. In the case of the stria medullaris cut, we are fairly certain that we cut that bundle completely and, in fact, also transected some additional antero-medial projections of the thalamus. I nonetheless consider this aspect of our data base a bit soft because the experimental animals did return to the electrified drinking spout somewhat more often than the controls and did so against a very high baseline. The effect of the lesion was far from statistically reliable in this test but I would not entirely rule out the stria medullaris until we have repeated this experiment in more optimal conditions. I am quite confident of our observation that fornicotomy did not affect punished behaviour, particularly since it is in excellent agreement with earlier reports of the effects (or lack thereof) of lesions in the fornix (e.g. Van Hoesen *et al.* 1969).

We have seen passive avoidance deficits after entorhinal cortex lesions (Ross *et al.* 1973) as well as more anterior periamygdaloid cortex damage (Grossman *et al.* 1975) and it seems plausible that a punishment-related projection might involve a route between the hippocampus and the septum via the entorhinal area, periamygdaloid cortex, amygdala, and stria terminalis. It is not likely to just stop there and I would guess that some projection to the hypothalamus and, perhaps, the lower brainstem provides the most probable route. The lateral septal nuclei have direct inputs to the medial forebrain bundle as well as indirect access to the hypothalamus via relays in the medial septum–diagonal band complex.

Black: We have analysed the published data on the effects of lesions of the fornix and hippocampal lesions on punishment and have done some experiments of our own on this topic (Black *et al.* 1977). I agree that lesions of the fornix don't have an effect on punishment in the Skinner box where the animal is punished for drinking or eating. In addition, lesions of the fornix don't interfere with most step-down and step-through situations. For example, suppose that an animal is placed on a small platform which is surrounded by an electrified grid, and is punished if it steps down from the platform. In this case, there seems to be no effect of fornical lesions. The fornical and hippocampal lesions seem to have a consistent effect in only two types of punishment situations. In one, the animal is placed on an electrified grid, and then has to run to a safe area and stay there. We think that this effect is a result of the inability of animals with hippocampal lesions to identify safe places. The second effect of punishment occurs in runway situations. Here, the effect of punishment on animals with fornical lesions is selective. There is a deficit in the start box, and to some extent in the runway, but very little deficit in the goal box. Again, we think that this selective effect of fornix lesions on punishment can be attributed to difficulty in learning about places. It seems to me, therefore, that it is incorrect to say that hippocampal lesions produce a deficit in punishment situations. Hippocampal and fornical damage do not have an effect in many such situations. One has to explain the selective effects of fornical and hippocampal lesions on punishment in order to provide an adequate account of the data.

Grossman: The literature on the effects of hippocampal lesions on passive avoidance behaviour is, indeed, confusing. As you know, Kimble *et al.* (1966) suggested some time ago that only recently acquired behaviours showed disinhibitory effects; Isaacson (1974) has maintained that the strength of the approach tendency may be the critical variable that determines whether or not one sees passive avoidance deficits after hippocampal damage; Wishart & Mogenson (1970) have proposed that preoperative experience with punishment contingencies abolishes the effects of hippocampal lesions; and a number of investigators (e.g. Snyder & Isaacson 1965) have published data in support of your suggestion that the nature of the task must be considered. It is my own feeling that tasks that are sensitive to what I have called the general disinhibitory effects of septal and hippocampal lesions (i.e. disinhibitory effects on non-rewarded rather than punished behaviour) may be the ones that are sensitive to hippocampal (as well as fornix) damage. A really satisfactory reply to your comment would require data from hippocampectomized animals obtained in the passive avoidance test used to demonstrate the effects of septal lesions (and the lack of effect of fornicotomy) in our own experiments. I unfortunately do not have these data.

Weiskrantz: Your own very nice analysis, Dr Grossman, of a DRL task with alternation between two responses would appear to distinguish between response perseveration *per se* and, if you like, act disinhibition.

Winocur: You described disinhibitory effects of hippocampal lesions in DRL: that is the case, providing there had been prior training on a different schedule.

Grossman: In the traditional single-lever DRL paradigm, that is always the case. Our results were obtained in an apparatus consisting of an alley with goal boxes at either end. In each of the goal boxes there was a lever and the rat learned *(a)* to alternate responses to the two levers and *(b)* to wait at least 30 seconds between successive responses (the DRL contingency). We devised this apparatus because we wanted to distinguish 'perseverative' from 'anticipatory' errors but found that it also permitted DRL training without prior experience with a continuous reinforcement schedule. Because the animal is initially unsure of the alternating requirements, a fair amount of time elapses most of the time between successive responses to the two levers. The DRL contingency is in effect, but probably affects behaviour very little during the initial stages of acquisition. Once the animal has learned to alternate, the DRL contingencies begin to exercise control over the behaviour. In this situation we have seen pronounced effects of septal lesions, fornicotomy, and intrahippocampal drug injections. We have not tested animals with hippocampal lesions but the results of our fornicotomies and intrahippocampal drug injections lead me to predict that they too would show disinhibitory effects in this situation.

O'Keefe: To follow up the general point that Abe Black was making, how many of the various tasks were done in the same piece of apparatus, and how many in different apparatuses?

Grossman: The disinhibitory effects of our experimental surgery on operant responding were observed either in a standard Skinner box (FI and DRL contingencies) or in the alley apparatus just described, which requires alternation of responses to two levers placed at opposite ends of an alley in addition to the spacing of successive responses in accordance with DRL contingencies. The effects of punishment were observed in two situations, one involving the classic passive avoidance paradigm where the rat receives a painful shock whenever it drinks a highly palatable liquid diet from a metal spout, the other involving signalled or unsignalled punishment of food-rewarded lever pressing in a Skinner box. Aggressive interactions are observed in the so-called shock-induced fighting paradigm where two rats are placed into a small box with a grid floor which is electrified at unpredictable intervals; a food-competition situation where the rat is barred from a tasty morsel of food by another rat; and muricide. Our shuttle box avoidance apparatus is all-but-identical to that used by other investigators although we use a somewhat unusual shock source. One-way avoidance acquisition is examined in a totally automated apparatus where the rat learns to respond to a signal by opening a door to an adjoining 'safe' compartment. It is a 'one-way' avoidance task because the animal is not required to return to the compartment where he was most recently shocked or threatened as in the shuttle box but, instead, continues to pursue a forward course away from that compartment.

O'Keefe: That is different from the standard one-way avoidance apparatus.

Grossman: I'm not sure that there is a 'standard' one-way apparatus. Our situation is different from the two-compartment apparatus (essentially a shuttle box that is used only in one direction) that many investigators have used, but we consistently obtain similar effects of septal (and other brain) lesions which, in the case of septal lesions and stria medullaris transection, are opposite in direction to the classic facilitatory effects of septal lesions in the shuttle box. I would, in fact, argue that it is the shuttle box one needs to look at more carefully because it is the only avoidance apparatus, in my experience, which allows one to demonstrate this facilitatory effect. All other tests we have used, including lever pressing and shelf-jumping, give rise to opposite, inhibitory effects.

O'Keefe: You could in fact have drawn up a large matrix of variables on which all these tasks vary, including the spatial one—the question of whether the animal has to go back to the same place or not, the size of the box, the response actually measured, and so on. Is it possible to devise one piece of apparatus where *all* these behaviours could be tested, to try to hold as many of the other variables constant, so that one could begin to see if it is solely the difference between emitting or withholding a response which is the crucial variable between your tasks?

Grossman: This is an interesting point. It is true that we have not examined all the behavioural effects of septal lesions or related knife cuts in the same apparatus. I doubt, in fact, that it could be done without changing possibly important aspects of the test environment from one experiment to another. One should not forget, however, that lever pressing has been a common feature in many of our experimental situations, including tests of general disinhibitory effects of septal lesions, tests of decreased responsiveness to punishment, tests of 'finickiness', and tests of active avoidance behaviour.

The issue you raise has another side. In our continued study of the effects of septal lesions and related knife cuts, we have been greatly impressed by the fact that their behavioural consequences are not situation-specific. Thus, shuttle box avoidance acquisition is consistently facilitated regardless of the specific configuration of the apparatus (we have used shuttle boxes with hurdles, doors, small partitions, and even no partition at all; lights as well as buzzers, punished as well as unpunished inter-trial contingencies, etc.). In our experience all other types of avoidance tests, including our automated 'one-way' apparatus, lever-pressing, shelf-jumping and pole climbing, always result in the demonstration of inhibitory rather than facilitatory effects of septal lesions (although the magnitude of the effect varies from test to test) and this generalization seems to hold true for cats as well as rats. As I indicated, septal lesions reduce the effectiveness of punishment in various tests of 'passive' avoidance (where the response of eating or drinking itself is punished) as well as in tests where instrumental responses (usually lever presses but alley-running has also been used) result in punishment (as well as positive reward). It doesn't seem to make any difference whether the punishment is predictable (i.e. signalled) or not. Nor does it seem important what positive (or negative) reinforcers are used to maintain the punished behaviour (we have observed appropriate effects of septal lesions in experiments where avoidance responses rather than food-rewarded lever presses were punished) or what contingencies are used to programme either the punishment or the reinforcer that maintains the behaviour. General disinhibitory effects of operant responding have been observed in standard Skinner boxes, straight runways, alleys that require alternation, and alleys that complicate the alternation with DRL contingencies. Once again, changes in the reward contingencies may enhance or decrease the inhibitory effects of septal lesions but do not abolish them. We have not yet duplicated all these tests in animals with selective transections of the principal septal connections but I am confident that the pattern established in our initial tests will hold when we expand our repertoire of tests in each of the categories. In the unlikely event that the pattern established in animals with septal lesions holds only partially in rats with a particular knife cut, we would, in fact, have gained a very important clue to the nature of the dysfunction.

Weiskrantz: There are two possible strategies: one is to devise an apparatus which would hold all the features constant, and one would just vary one contingency. I would think it would be virtually impossible to do that here. The other approach is to vary the situation widely, so that you could study punishment, for example, in a variety of situations, and see whether the effects hang together across situations. The latter approach at least gives one a chance of arriving at strong generalizations.

Grossman: That is what we are trying to do. We have four measures of aggressive behaviours, four tests of active avoidance acquisition (although only two have, so far, been used with animals that sustained selective cuts of septal connections), three different measures of the 'disinhibitory' effects of septal lesions on non-rewarded behaviour, and at least three (we are now experimenting with several additional ones) of their effects on punished behaviour. The generality of the effects of our cuts requires further study, but we must view this task in proper perspective. Even the admittedly incomplete story described here is the result of many years of intensive labour by a number of industrious graduate students! We shall undoubtedly 'persevere' on the course you indicate, but the complete story will take a great deal more time and patience. But I doubt that the *general* pattern of effects is likely to change drastically in these extensions of the earlier part of the experiments.

References

BLACK, A. H., NADEL, L. & O'KEEFE, J. (1977) Hippocampal function in avoidance learning and punishment. *Psychol. Bull. 84*, 1107–1129

GRANT, L. D., COSCINA, D. V., GROSSMAN, S. P. & FREEDMAN, D. X. (1973) Muricide after serotonin-depleting lesions of midbrain raphe nuclei. *Pharmacol. Biochem. Behav. 1*, 77–80

GRAY, J. A., RAWLINS, J. N. P. & FELDON, J. (1978) Brain mechanisms in the inhibition of behavior, in *Mechanisms of Learning and Motivation: A Memorial Volume for Jerzy Konorski* (Dickinson, A. & Boakes, R. A., eds.), Erlbaum, Hillsdale, N.J.

GROSSMAN, S. P., GROSSMAN, L. & WALSH, L. L. (1975) Functional organization of the rat amygdala with respect to avoidance behavior. *J. Comp. Physiol. Psychol. 88*, 829–850

HAMILTON, L. W., KELSEY, J. E. & GROSSMAN, S. P. (1970) Variations in behavioural inhibition following different septal lesions in rats. *J. Comp. Physiol. Psychol. 70*, 79–86

ISAACSON, R. L. (1974) *The Limbic System*, Plenum Press, New York

KAADA, B. R., RASMUSSEN, E. W. & KVEIM, O. (1962) Impaired acquisition of passive avoidance behavior by subcallosal, septal, hypothalamic and insular lesions in the rat. *J. Comp. Physiol. Psychol. 55*, 661–670

KIMBLE, D. P., KIRKBY, R. J. & STEIN, D. G. (1966) Response perseveration interpretation of passive avoidance deficits in hippocampectomized rats. *J. Comp. Physiol. Psychol. 61*, 141–143

KIMURA, D. (1958) Effects of selective hippocampal damage on avoidance behavior in the rat. *Can. J. Psychol. 12*, 213–218

KING, B. M., CARRINGTON, C. D. & GROSSMAN, S. P. (1978) Passive avoidance behavior in lean and obese rats with ventromedial hypothalamic lesions. *Physiol. Behav. 20*, 57–66

LINTS, C. E. & HARVEY, J. A. (1969) Altered sensitivity to footshock and decreased brain content of serotonin following brain lesions in the rat. *J. Comp. Physiol. Psychol. 67*, 23–31

ROSS, J. F., WALSH, L. L. & GROSSMAN, S. P. (1973) Some behavioral effects of entorhinal cortex lesions in the albino rat. *J. Comp. Physiol. Psychol. 85*, 70–81

ROSS, J. F., GROSSMAN, L. & GROSSMAN, S. P. (1975) Some behavioral effects of transection of ventral or dorsal fiber connections of the septum. *J. Comp. Physiol. Psychol. 89*, 5–18

SCLAFANI, A. & GROSSMAN, S. P. (1969) Hyperphagia produced by knife cuts between the medial and lateral hypothalamus in the rat. *Physiol. Behav. 4*, 533–538

SNYDER, D. R. & ISAACSON, R. L. (1965) Effects of large and small bilateral hippocampal lesions on two types of passive avoidance responses. *Psychol. Rep. 16*, 1277–1290

STRICKER, E. M. & ZIGMOND, M. J. (1976) Recovery of function damage to central catecholamine-containing neurons: a neurochemical model for the lateral hypothalamic syndrome, in *Progress in Psychobiology and Physiological Psychology*, vol. 6 (Sprague, J. M. & Epstein, A. N., eds.), pp. 121–188, Academic Press, New York

TEITELBAUM, P. & CYTAWA, J. (1965) Spreading depression and recovery from lateral hypothalamic damage. *Science (Wash. D.C.) 147*, 61–63

TEITELBAUM, P. & EPSTEIN, A. N. (1962) The lateral hypothalamic syndrome: recovery of feeding and drinking after lateral hypothalamic lesions. *Psychol. Rev. 69*, 74–90

URSIN, H. (1965) Effect of amygdaloid lesions on avoidance behavior and visual discrimination in cats. *Exp. Neurol. 11*, 298–317

VAN HOESEN, G. W., MACDOUGALL, J. M. & MITCHELL, J. C. (1969) Anatomical specificity of septal projections in active and passive avoidance behavior in rats. *J. Comp. Physiol. Psychol. 68*, 80–89

WISHART, T. B. & MOGENSON, G. J. (1970) Effects of lesions of the hippocampus and septum before and after passive avoidance training. *Physiol. Behav. 5*, 31–34

The role of the septo-hippocampal system and its noradrenergic afferents in behavioural responses to non-reward

JEFFREY A. GRAY, J. FELDON, J. N. P. RAWLINS, S. OWEN and N. McNAUGHTON

Department of Experimental Psychology, University of Oxford

Abstract Our experiments were designed with two purposes: (*i*) to examine the effects on one behaviour of differing interventions in the septo-hippocampal system; (*ii*) to compare these effects with those of minor tranquillizers. The behaviour studied (in rats) is extinction in the alley after continuous (CRF) or partial (PRF) reinforcement. Minor tranquillizers and large septal lesions produce three effects: (1) resistance to extinction is increased after CRF; (2) resistance to extinction is decreased after PRF; (3) the partial reinforcement extinction effect (PREE) is abolished. Small septal lesions fractionate this syndrome: either effect (1) or an actual increase in the size of the PREE is produced by medial septal lesions abolishing hippocampal theta; effects (2) and (3), but not (1), are produced by lateral septal lesions sparing theta. Dorso-medial fornix section, abolishing theta, reproduces the effects of medial septal lesions. Fimbrial section, sparing theta, reproduces some of the effects of lateral septal lesions. Minor tranquillizers produce a rise in the threshold for septal driving of hippocampal theta specifically at 7.7 Hz. This effect is reproduced by blockade of noradrenergic transmission or destruction of the dorsal noradrenergic bundle with 6-hydroxydopamine. This lesion reproduces all three behavioural changes listed above. These results suggest a model for the role of the septo-hippocampal system and its noradrenergic inputs in the PREE. This model is compared with other approaches to the septo-hippocampal system.

'The hippocampus—an organ of hesitation and doubt'
(P. V. Simonov 1974)

Our experiments were designed with two purposes: *(i)* to examine the effects on one form of behaviour of differing interventions in the septo-hippocampal system, and thus to attempt to construct a functional 'wiring diagram' for this behaviour; *(ii)* to compare the behavioural effects observed after these interventions with those produced by minor tranquillizers such as barbiturates, benzodiazepines and alcohol. Before these experiments are described a brief introduction is necessary to establish the background which led us to choose the particular behaviour on which we have concentrated; and also to explain the connection between our interests in the septo-hippocampal system and in the minor tranquillizers.

MINOR TRANQUILLIZERS AND THE SEPTO-HIPPOCAMPAL SYSTEM

If one surveys the literature on the behavioural effects of minor tranquillizers, on the one hand, and either large septal or large hippocampal lesions on the other, a remarkable congruity emerges. First, in the great majority of behavioural paradigms for which data are available on both lesions, their effects are highly similar (J. A. Gray & N. McNaughton, in preparation). Given the close physiological interrelations between the septal area and the hippocampus, this is hardly surprising. But, second, whenever the effects of the two lesions do resemble each other, and corresponding drug data are available (Gray 1977), the direction of the behavioural change (or the lack of any change) produced by these substances is the same as that produced by the lesions. This pattern of similarity prompted the hypothesis (Gray 1970) that the minor tranquillizers influence behaviour by way of an action on the septo-hippocampal system.

Now, the minor tranquillizers are used clinically to control anxiety. Thus, if they reduce anxiety by impairing the normal functioning of the septo-hippocampal system, one might conclude that this system is the neural substrate of anxiety. Given the very different—and very diverse—views of the behavioural functions of the septo-hippocampal system held by the participants in this symposium, this statement seems provocative, but needlessly so. For, if we first make it clear what we mean by 'anxiety', it will become apparent that it is possible to make the same statement in a way which reveals the similarities between our view of the functions of this system and at least some of its rivals.

The theory of anxiety which one of us has attempted to develop (Gray 1972a, 1976, 1978) is, in fact, based very closely on the behavioural effects of minor tranquillizers in animals (Gray 1977), together with a general view of emotional learning (Gray 1975). This theory holds that anxiety consists of activity in a hypothetical 'behavioural inhibition system', defined in functional terms but supposed to correspond to real structures in the brain. The adequate inputs to which this system responds are of three kinds: conditioned stimuli which have been associated with punishment; conditioned stimuli similarly associated with frustrative non-reward; and novel stimuli. The behavioural outputs produced by the system are also of three kinds (but all outputs are elicited by any of the inputs): inhibition of ongoing instrumental behaviour and of Pavlovian conditioned reflexes; an increment in the level of arousal (Gray 1964); and increased attention to the environment, especially to any novel features in it. The function discharged by the behavioural inhibition system is thought to be the suppression of old but now maladaptive patterns of behaviour, while the animal scans its environment for cues as to possible new alternatives. But, under conditions where there is no better alternative than to continue with the old behaviour, the behavioural inhibition system is also responsible for developing the necessary added persistence in performing this behaviour under adverse circumstances. *Ex hypothesi*, activity in the behavioural inhibition system constitutes 'anxiety' and is antagonized by the minor tranquillizers.

For a number of years the guiding hypothesis which has governed the selection and design of experiments in our laboratory has been this: that there is some correspondence between the behavioural inhibition system, as just defined, and the septo-hippocampal system. Let us be clear: we are not making a take-over bid for all the functions of the septo-hippocampal system; nor (and this will be clearer later) do we necessarily suppose that the behavioural inhibition system makes use of no structures outside the septo-hippocampal system. We believe, rather, that: *(a)* the septo-hippocampal system, among its functions, includes participation in the behavioural inhibition system; and *(b)* the behavioural inhibition system, among its neural structures, includes the septo-hippocampal system.

We shall first describe some of the experiments we have done to test these hypotheses (and in this way, we hope, make them more concrete); and then, in the final section of this paper, we shall consider how our views of the functions of the septo-hippocampal system relate to those held by other workers.

THE PARTIAL REINFORCEMENT EXTINCTION EFFECT

Many of our experiments have made use of a behavioural phenomenon termed the 'partial reinforcement extinction effect' (PREE). If rats are run in a straight alley for a food or water reward, this reward may be presented on every trial (a continuous reinforcement, CRF, schedule) or on a random selection of trials (a partial reinforcement, PRF, schedule). Resistance to extinction of the behaviour is subsequently much greater in the PRF-trained animals. This is the PREE, a well-known and highly robust effect. As argued by Amsel (1962) the PREE may be considered a consequence of the development in PRF animals of tolerance for the aversive effects of non-reward (a phenomenon known as counter-conditioning). (There is plenty of evidence that non-reward, and conditioned stimuli associated with it, *are* aversive: Gray 1975.) But this tolerance extends beyond the aversive effects of non-reward itself. It was shown by Brown & Wagner (1964) that there is a cross-tolerance between non-reward and electric shock: rats trained on a PRF schedule are more resistant to the response-suppressant effect of shock than are rats trained on CRF. The converse is also true: behavioural tolerance for shock is accompanied by greater resistance to extinction (Brown & Wagner 1964). There is other evidence indicating a similar cross-tolerance between cold stress and shock (Weiss et al. 1975); and Amsel et al. (1973) have shown that resistance to extinction may be increased by exposure during acquisition to an intense auditory stimulus. Thus it is probable that the PREE is one aspect of a general tolerance for stress which results from repeated exposure to any kind of aversive event. A similar view has been proposed by Amsel (1972) in his general theory of persistence.

Work in our laboratory has shown that the PREE may be completely abolished if animals are treated with sodium amylobarbitone or chlordiazepoxide during acquisition; if either of these drugs is given during extinction, the PREE is intact, but there is a general increase in resistance to extinction. Our recent results (J. Feldon & J. A. Gray, in preparation) using chlordiazepoxide are presented in Fig. 1. This figure shows in addition, that, if the drug is present during both acquisition and extinction, there is a compromise between the acquisition and the extinction effects of the drug. The PREE is absent and resistance to extinction in the PRF group is reduced (both consequences of the presence of the drug during acquisition); in addition, resistance to extinction in the CRF group is increased (due to the presence of the drug during extinction). This pattern of results is consistent with the hypothesis, supported also by many other findings (Gray 1977), that the minor tranquillizers reduce or even eliminate the behavioural effects of stimuli (in this case, those encountered in the startbox and stem of the alley) which signal non-reward. Thus, during extinction, the drugged animal is slower to abandon the non-rewarded response; and during acquisition on a PRF schedule it is less able to learn about non-reward.

SEPTAL LESIONS AND PREE

Given these effects of the minor tranquillizers on the PREE, and the hypothesis that these drugs act on behaviour by impairing the normal functioning of the septo-hippocampal system, it follows that damage to this system might be expected to produce the same kind of change in the PREE.

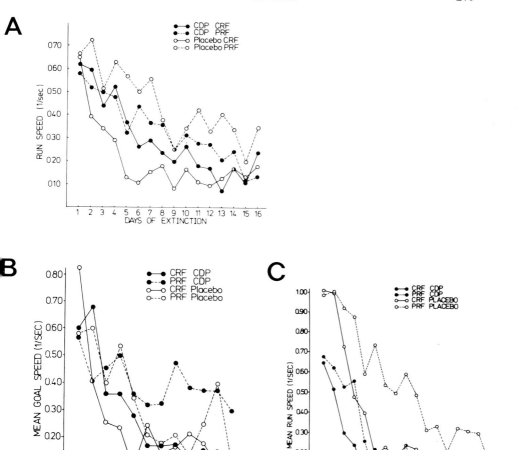

FIG. 1. The effects of chlordiazepoxide (CDP) 5 mg/kg on resistance to extinction in the straight alley after continuous (CRF) or partial (PRF) reinforcement in acquisition.
A. Drug administration during acquisition and extinction.
B. During acquisition only.
C. During extinction only.

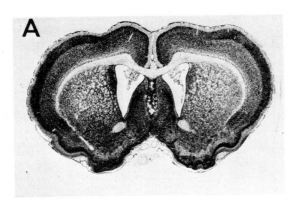
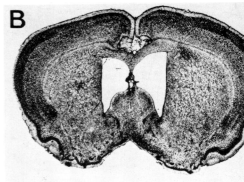
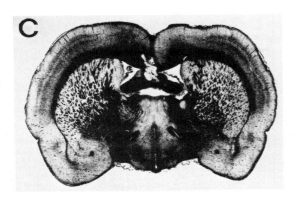
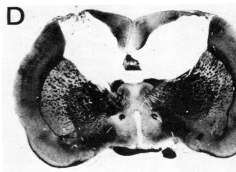
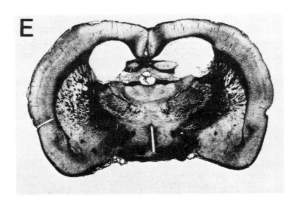

In confirmation of this prediction Gray et al. (1972b) showed that electrolytic lesions of the septal area attenuated the PREE by increasing resistance to extinction in CRF-trained rats (a finding which had been reported several times before), but also by decreasing resistance to extinction in PRF-trained animals. Gray et al. (1972b) used 72 acquisition trials at an inter-trial interval (ITI) of 4–5 minutes, and, while the effects they reported were statistically significant, they were not large. Henke (1974) repeated their experiments under two conditions: 48 or 96 acquisition trials. With 48 trials he obtained a complete abolition of the PREE, with 96 only an increase in resistance to extinction in CRF animals, the PREE remaining fully present. Thus the intermediate result obtained by Gray et al. with an intermediate number of trials falls neatly into line with Henke's results.

Gray et al.'s (1972b) lesions were intended to be confined to the medial septal area. However, our more recent experiments make it clear that those findings were due to partial lateral, in addition to medial, septal damage. In these experiments we have mainly used an inter-trial interval of 24 hours, rather than one of 4–5 minutes. For we have since shown that this interval is optimum for bringing out the effects of the minor tranquillizers on the PREE (J. Feldon & J. A. Gray, in preparation; Gray 1972a; see Fig. 1). Our lesion parameters were first developed in pilot experiments, using as criteria their effects on the hippocampal theta rhythm (J. N. P. Rawlins, J. Feldon, L. E. Jarrard & J. A. Gray, in preparation). We defined a successful medial septal lesion as the smallest damage to ventral midline structures which caused virtually total loss of theta; and a successful lateral septal lesion as the largest damage to dorsal lateral structures which caused little or no loss of theta (see Figs. 2 and 3). Theta was recorded in these experiments from a bipolar electrode in the dorso-medial subiculum (Gray 1972b; James et al. 1977).

We have completed several experiments on the effects of these two lesions on the PREE (J. Feldon & J. A. Gray, in preparation). At a 24-hour inter-trial interval these experiments demonstrate the following points. (1) A general increase in resistance to extinction (similar to that produced by the minor tranquillizers injected during extinction only) is produced by medial, and not

---

FIG. 2. Frontal sections of rat brain illustrating
A. A medial septal lesion (note ventricles are enlarged).
B. A dorsolateral septal lesion (note ventricles are enlarged).
C. A dorsomedial fornix lesion.
D. A fimbrial lesion.
E. The same fimbrial lesion in a more posterior section.

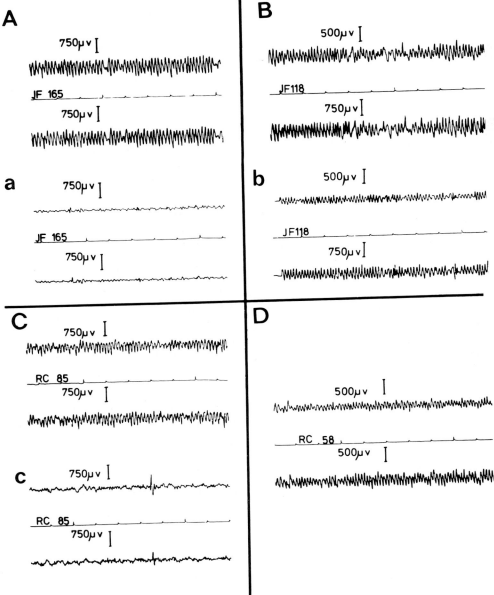

FIG. 3. Recordings via electrodes chronically implanted in the left and right subiculum (upper and lower tracings, respectively, in each pair) from the same subjects as illustrated in Fig. 2.
A. Before and *a*: after a medial septal lesion.   C. Before and *c*: after a fornix lesion.
B. Before and *b*: after a lateral septal lesion.   D. After a fimbrial lesion.
Amplification is indicated for each recording; time marker is in seconds.

by lateral lesions. This increase is sometimes substantial (Fig. 4), but in other experiments we fail to see it; we do not yet know the reason for this variability. (2) Reduction or abolition of the PREE is produced by lateral, and not by medial lesions. This reduction, like that produced by minor tranquillizers injected during acquisition only, is due to a fall in resistance to extinction in PRF animals (Fig. 5). Medial lesions sometimes produce the opposite effect: an actual increase in the size of the PREE (Fig. 4).

These conclusions apply with only slight modification also to the multi-trial case. Only the lateral lesion reduces the PREE. In confirmation of Henke's (1974) findings with large septal lesions, the PREE was abolished by lateral lesions when 48 acquisition trials were run but left intact with

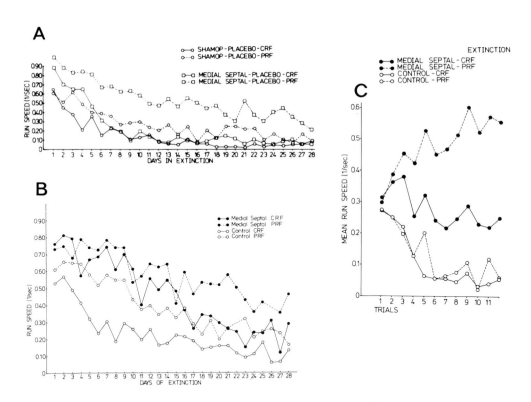

FIG. 4. The effects of medial septal lesions on resistance to extinction after continuous (CRF) or partial (PRF) reinforcement in acquisition in the alley.
A. An increased partial reinforcement extinction effect (PREE) at a 24-hour inter-trial interval (ITI).
B. Increased resistance to extinction with no change in the PREE at a 24-hour ITI.
C. PREE present in lesioned subjects only at a two-minute ITI.

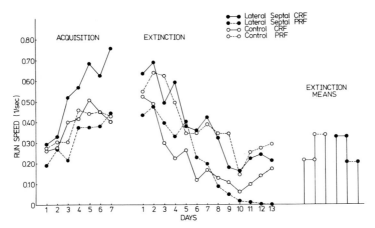

FIG. 5. The effects of lateral septal lesions on resistance to extinction after continuous (CRF) or partial (PRF) reinforcement in the alley at a 24-hour inter-trial interval.

96 trials. Medial lesions produced a remarkable increase in the PREE, as well as a substantial increase in resistance to extinction, when we used only six acquisition trials (Fig. 4). At 48 and 96 trials, however, we observed only a slight and non-significant increase in resistance to extinction after medial lesions; and (for the only time) we also observed an increase in resistance to extinction after CRF training in animals with lateral lesions.

Taken overall, our findings constitute a clear double dissociation between the behavioural effects of medial and lateral septal lesions. The PREE is reduced or abolished by lateral septal lesions, owing to a fall in resistance to extinction in PRF animals; the medial lesion never has this effect, and often has the opposite one, increasing resistance to extinction and the size of the PREE.

The clarity of this double dissociation suggests that the electrophysiological criteria we used to establish our lesions reflect functional changes which are behaviourally significant. If so, it should be possible to produce similar behavioural effects using other interventions in the septo-hippocampal system which have equivalent electrophysiological consequences.

LESIONS TO THE FORNIX AND FIMBRIA

Myhrer (1975) reported that sectioning a restricted portion of the dorso-medial fornix abolishes theta. We copied his method and obtained the same result (J. N. P. Rawlins, J. Feldon, L. E. Jarrard & J. A. Gray, in preparation).

The lesion we made is illustrated in Fig. 2, and its effect on hippocampal theta in Fig. 3. Myhrer's recording site was in dorsal CA1; ours in the dorso-medial subiculum. Per Andersen (personal communication 1977) has also seen theta disappear in the dorsal hippocampus after similar lesions. If, instead of lesioning the fornix, one destroys the fimbria bilaterally, as in the case illustrated in Fig. 2, theta is of normal amplitude and abundance in our subicular recording site (Fig. 3) (Rawlins et al., in preparation).

These results have several implications.

First, they apparently indicate that the medial septal efferents which control hippocampal theta course exclusively in the fornix. To establish this with certainty it would be necessary also to record from the ventral hippocampus after fornix and fimbria lesions, for there is evidence that the fornical projection innervates only the dorsal hippocampus (Lewis & Shute 1967; Meibach & Siegel 1977).*

Second, our results cast doubt on the accepted view that the medial septal efferents which control theta are cholinergic. They may be. But there is some evidence that the cholinergic projection from the medial septal area to the hippocampus travels wholly by the fimbrial route (Dudar 1975). Thus, if the 'theta' fibres take exclusively the fornical route, it is at present not known what neurotransmitter they use.

Third, our results imply that the massive projection from the medial septal area which takes the fimbrial route to the hippocampus, and which is cholinergic (Dudar 1975; Lewis & Shute 1967), has nothing to do with the control of theta.

Fourth, our results rule out the hypothesis proposed by McLennan & Miller (1974, 1976), according to which the theta rhythm is controlled by a hippocampal input to the lateral septal area travelling in the fimbria. For neither lateral septal lesions nor destruction of the fimbria caused a loss of theta. Vinogradova & Brazhnik (this volume) come to a similar conclusion.

A MODEL FOR COUNTER-CONDITIONING

These findings were the basis for a model linking the anatomical and electrophysiological organization of the septo-hippocampal system to our behavioural observations with septally lesioned rats in the alley. According to this hypothesis (Fig. 6) the medial septal area is the recipient of information,

---

*Note added in proof. We have now completed these experiments. It is clear that ventral hippocampal theta is controlled by fibres travelling in the fimbria, not in the fornix; and there is good correspondence between loss of theta and loss of staining for cholinesterase, suggesting that the 'theta' fibres are indeed cholinergic (Rawlins et al., in preparation).

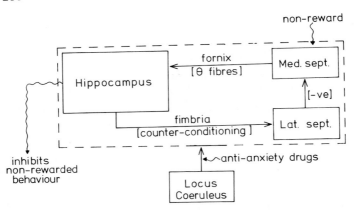

Fig. 6. A model for counter-conditioning. For explanation, see text.

conveyed (via an as yet unspecified route) by conditioned frustrative stimuli, concerning the imminence of non-reward. This information is conveyed to the hippocampus by way of the theta-producing fibres which travel in the dorsal fornix. The hippocampus has the job of inhibiting the non-rewarded behaviour (by an unknown route) while determining the best behavioural strategy in the changed circumstances. (This period of behavioural inhibition and uncertainty is subjectively experienced as 'anxiety'.) Under conditions in which the best strategy is in fact to continue with the original behaviour (as on a PRF schedule), the hippocampus sends a message (via the fimbria) to the lateral septal area which, in turn, via septal interneurons (DeFrance 1976), inhibits or otherwise alters the medial septal input to the hippocampus. The operation of this hippocampo-septal pathway underlies the phenomenon of counter-conditioning. This hypothesis may easily be generalized to stimuli which signal imminent punishment and to novel stimuli (the two other classes of stimuli which activate the behavioural inhibition system); although it should be noted that all of the data we report in this paper are from experiments using non-reward.

This model predicts that sectioning the dorso-medial fornix (thereby abolishing theta) should mimic the behavioural effects of a medial septal lesion. We have completed only a pilot experiment testing this prediction. The number of rats used was small and the extent of theta loss produced by the lesion was unsatisfactory (55% on average). Nonetheless, the results (Fig. 7) were encouraging. The PREE at one trial a day was significantly enhanced in the lesioned group, owing to an increase in resistance to extinction in PRF animals. This effect, as mentioned above, is also obtained with medial septal lesions.

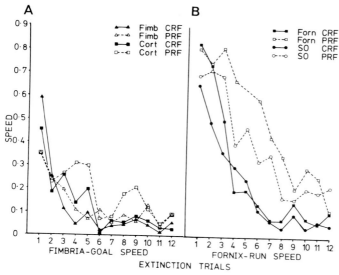

FIG. 7. The effects of A. Fimbrial and cortical control lesions, and B. Dorsomedial fornix lesions on resistance to extinction after continuous (CRF) or partial (PRF) reinforcement at a 24-hour inter-trial interval.

If the effects of the medial septal lesion on resistance to extinction are exclusively due to loss of hippocampal theta, a fimbrial lesion should *not* produce the same effects. To test this point we ran rats in the one-trial-a-day PREE experiment after the fimbrial lesion shown in Fig. 3. The results (Fig. 7) showed a *faster* rate of extinction in both CRF- and PRF-trained rats. Thus, in spite of the fact that the great bulk of medial septal efferents travel in the fimbria, and only a relatively small number in the fornix (Meibach & Siegel 1977), the effects of a medial septal lesion on resistance to extinction are reproduced by section of the latter, but not the former pathway.

The effects we obtained with fimbrial section did, however, resemble those obtained with lateral septal lesions, in that resistance to extinction was significantly reduced in PRF-trained rats (Fig. 9). But, unlike animals with lateral septal lesions, the rats with fimbrial lesions also showed a reduction in resistance to extinction after CRF training and continued to display a PREE. This discrepancy is something of a mystery, since it means that, while the PREE may be attenuated or abolished (depending on the experimental parameters) by either large septal lesions or small dorso-lateral septal lesions, this effect cannot be reproduced by sectioning either of the two pathways connecting the septum to the hippocampus.

It is possible, of course, that the lateral septal involvement with the PREE is unrelated to its connections with the hippocampus. This is unlikely, however, since three of the four experiments which have examined the PREE or analogous phenomena after hippocampal damage have found similar effects to those we have seen after lateral septal damage—namely, a smaller difference in resistance to extinction between CRF- and PRF-trained groups than among controls (Franchina & Brown 1970; Amsel et al. 1973; Brunner et al. 1974). A further possibility is that the effects of sectioning the hippocampal afferents to the lateral septal area are masked by the simultaneous section of the medial septal efferents which take the fimbrial route to the hippocampus. Unfortunately, we know of no method of selectively destroying one or other of these projections. Perhaps it is in any case naive to expect that damaging some of the afferents to a structure will mimic all of the effects of damage to the structure itself; this, after all, implies that the structure concerned acts simply as a relay station. Applying this caveat to the present case, however, would mean that the complex task of differentiating a PRF from a CRF schedule is performed by the lateral septum alone; and we find this conclusion improbable. We are at present, therefore, re-examining the effects of fimbrial damage on the PREE.

THE SIGNIFICANCE OF THETA FREQUENCY

The results of our experiments using septal lesions make it difficult to construct a parsimonious account of the behavioural effects of the minor tranquillizers if we suppose that these drugs act in the septal area itself. One might suggest that they act sequentially, first on the lateral septal area (during acquisition), then on the medial (during extinction). This proposal would account for our results; but it is implausible. A more attractive possibility is that they act on the hippocampus itself. A third possibility is that they act on the input which initiates activity in the model illustrated in Fig. 6. Recently we have completed some experiments which support this last suggestion.

These experiments have their origin in some observations reported by Gray & Ball (1970). The theta rhythm was recorded from rats in our standard alley situation, and it was noticed that the frequency of theta varied predictably depending on the rat's behaviour and on what was happening to it. When the animal was in the process of consuming the reward (water) in the goalbox, theta frequency was about 6–7 Hz; when it was running down the alley towards the goalbox, theta frequency was about 9–10 Hz; and an intermediate frequency, with a mean value of 7.7 Hz, was recorded in the goalbox on non-rewarded trials. This same intermediate frequency was also observed when the rat explored the alley for the first time. Values close to 7.7 Hz have since been reported by Kimsey et al. (1974) on non-rewarded trials in the alley, by P. Soubrié (personal communication 1977) on non-rewarded trials at a drinking spout, and by Kurtz (1975) when sexual behaviour was not rewarded.

Since the minor tranquillizers attenuate the behavioural effects of non-reward and novelty, but not those of reward, these data imply that any alteration produced by these drugs in septal control of hippocampal theta (which might underlie their effects on behaviour) is restricted to frequencies lying close to 7.7 Hz. Gray & Ball (1970) therefore proposed such a 'frequency-specific' hypothesis of the action of these drugs. According to this hypothesis, theta consists of three functionally distinct frequency bands: a low frequency band (less than about 7 Hz in the rat) is related to fixed action patterns, including consummatory behaviour; a middle frequency band (centred on 7.7 Hz) is related to the activity of the behavioural inhibition system as described in this paper; and a high frequency band (above about 8.5 Hz) is related to the performance of goal-directed behaviour (rewarded or active avoidance). The minor tranquillizers, on this hypothesis, alter septal control of theta specifically in the middle frequency band.

This hypothesis has guided our research over the last few years. Had we known at the time we proposed it (Gray & Ball 1970) of the extensive work of Vanderwolf and his associates (e.g., 1975) showing a close correlation between theta frequency and the intensity of motor behaviour, we would probably have disregarded our own findings as being merely a consequence of the degree to which the animal was moving under our different conditions of observation. And, indeed, R. G. M. Morris & A. H. Black (in preparation) have recently shown that it is possible to predict theta frequency under conditions of mixed reward and non-reward by knowledge of the exact motor behaviour engaged in by the rat when it encounters either of these conditions of reinforcement. Fortunately, we did not know Vanderwolf's work at that time and instead conceived the idea that there is something special about theta frequencies close to 7.7 Hz; in particular, that such frequencies are related to the behaviour patterns which anti-anxiety drugs impair. Morris and Black's data leave us unrepentant about this idea, since we are neither so mentalistic as to suppose that states of 'frustration' or 'anxiety' can occur without visible effect on behaviour; nor so behaviouristic as to suppose that physiological psychology stops when behavioural observations have been correlated with physiological ones.

Testing the frequency-specific hypothesis has led us to a number of new findings (which fit rather well with it); and has produced a number of new ideas about the action of the minor tranquillizers and the organization of the septo-hippocampal system. The methods used in these tests have included observation of the behavioural effects of experimentally inducing theta at 7.7 Hz (Gray 1972b; Glazer 1974), and of experimentally blocking spontaneous occurrences of 7.7 Hz theta (Gray et al. 1972a). The former technique produced effects opposite in kind to those produced by minor tranquillizers, the latter effects similar in kind; both types of change were in accordance with our predictions. However, the method which has taken us furthest has been to explore the effect of such drugs on the elicitation of theta by septal electrical stimulation.

As first reported by Stumpf (1965) and his colleagues in the anaesthetized rabbit, and as we have replicated in the free-moving rat, short pulses (0.5 ms in our experiments) delivered to the septal area at a frequency lying within the naturally occurring theta range are able artificially to drive the hippocampal theta rhythm, in the sense that for each pulse there appears a wave filling most of the inter-pulse interval. If one plots the current threshold for driving theta in this way as a function of stimulation frequency in the free-moving male rat, one obtains a characteristic 'theta-driving curve' with a minimum threshold situated exactly at 7.7 Hz (Gray & Ball 1970; James et al. 1977). From the frequency-specific hypothesis of the action of minor tranquillizers one would predict that these drugs would have a maximal effect at this same frequency of 7.7 Hz. This indeed is the case, since all the tranquillizing drugs so far tested (sodium amylobarbitone, chlordiazepoxide, diazepam, nitrazepam, alcohol) eliminate the 7.7 Hz minimum in the theta-driving curve by selectively raising the threshold at this frequency (Gray & Ball 1970; McNaughton et al. 1977; M. Nettleton, personal communication) (Fig. 8). The 7.7 Hz minimum, incidentally, is also eliminated by fimbrial section (J. N. P. Rawlins & J. A. Gray, unpublished observations). It is possible, therefore, that McLennan & Miller's (1974, 1976) observations of rhythmicity in the fimbrial projection to the lateral septal area relate, not to the production of theta as such (as they proposed), but to the relative ease with which different frequencies of theta circulate round the septo-hippocampal loop.

THE DORSAL ASCENDING NORADRENERGIC BUNDLE

The effects of the minor tranquillizers on the theta-driving curve offer strong support for the frequency-specific hypothesis. We therefore tried to establish their neuropharmacological basis. To this end we attempted to mimic their characteristic effect on the theta-driving curve by using other agents with better-understood effects on putative neurotransmitters. We

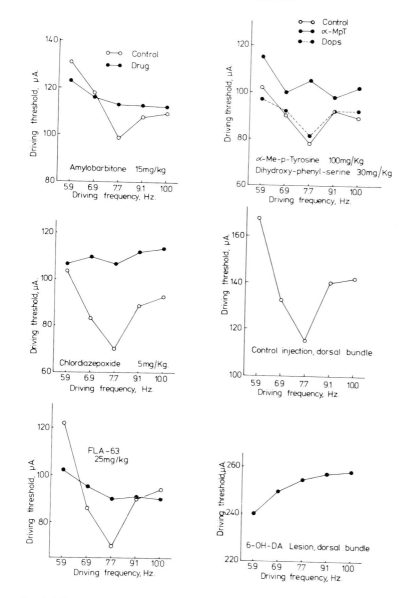

Fig. 8. Effects of various treatments (as indicated in each panel) on septal driving of hippocampal theta as a function of stimulation frequency. FLA-63 inhibits noradrenaline synthesis. α-Methyl-p-tyrosine inhibits both dopamine and noradrenaline synthesis; subsequent injection of dihydroxy-phenylserine restores only noradrenaline synthesis. The 6-hydroxydopamine (6-OH-DA) injection in the dorsal ascending noradrenergic bundle lowered hippocampal noradrenaline levels to less than 5% of normal values.

found that we were unable to mimic the effect of minor tranquillizers by altering cholinergic, serotonergic or dopaminergic transmission; selective blockade of noradrenergic function, however, gave an effect (Fig. 8) which was highly similar to that of the minor tranquillizers (Gray et al. 1975; McNaughton et al. 1977).

These results prompted us to look for the neural substrate of the effect we had obtained by pharmacological blockade of noradrenaline. The natural candidate is the dorsal ascending noradrenergic bundle (DANB), which originates in the locus coeruleus and innervates the septal area and the hippocampus (Livett 1973). To ascertain whether it was involved in the effects shown in Fig. 8, we used a local injection into the DANB of 6-hydroxydopamine, a poison specific to noradrenergic and dopaminergic neurons. Regional assay of catecholamines in the brains of the experimental animals showed levels of noradrenaline in the hippocampus which were well below 10% of control values, with no evidence of damage to dopaminergic fibres or the ventral noradrenergic bundle. Animals which had sustained this selective and virtually total destruction of the DANB did not display a minimum of 7.7 Hz in the theta-driving curve (Gray et al. 1975) (Fig. 8).

In the light of these findings we proposed (Gray et al. 1975) that the DANB is an important site of action for the behavioural effects of the minor tranquillizers. Independent evidence for this view had been obtained earlier by Corrodi et al. (1971; Lidbrink et al. 1972), who showed that stress increases turnover of noradrenaline in the forebrain and that this increase is antagonized by a range of minor tranquillizers. It follows from this hypothesis that destruction of the DANB by injection of 6-hydroxydopamine ought to reproduce the behavioural effects of the minor tranquillizers. To test this prediction we had recourse to our standard PREE experiment, using 50 trials with a 4–5 minute inter-trial interval, parameters which show up well the effects of total (Henke 1974) or lateral (J. Feldon & J. A. Gray, in preparation) septal lesions. Our results (Fig. 9) were extremely clearcut. We observed the full pattern of action of the minor tranquillizers given both during acquisition and extinction: increased resistance to extinction in CRF-trained animals, decreased resistance to extinction in PRF-trained animals, and a complete absence of the PREE (S. Owen, M. Boarder, J. A. Gray & M. Fillenz, in preparation). Increased resistance to extinction in CRF-trained animals has also been reported by Mason & Iversen (1975) after destruction of the DANB. Like these authors, we found no deficit in acquisition of the rewarded response in the lesioned animals, a finding which is difficult to reconcile with the hypothesis advanced by Anlezark et al. (1973) that the DANB mediates the behavioural effects of reward.

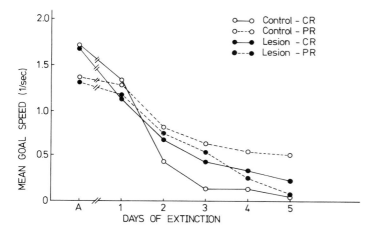

FIG. 9. The effects of a 6-hydroxydopamine lesion of the dorsal ascending noradrenergic bundle on resistance to extinction after continuous (CR) or partial (PR) reinforcement during acquisition in the alley at a 5–8 minute inter-trial interval. The point 'A' on the abscissa is the last day of acquisition.

These findings provide important support for the inclusion in our model for counter-conditioning (Fig. 6) of: *(a)* the hypothesis that the DANB carries a critical input which initiates or facilitates activity in the septo-hippocampal system; and *(b)* the hypothesis that the minor tranquillizers act on the DANB to dampen or eliminate this input. They also strongly support that part of the frequency-specific hypothesis of hippocampal theta which attributes a role to middle frequency theta (about 7.7 Hz) in the operation of the behavioural inhibition system. The special role of middle frequency theta is also supported by other evidence (e.g. Drewett *et al.* 1977). Nonetheless, most workers in the field have accepted the view advanced by Vanderwolf (1971) that theta is related to voluntary movement. It is necessary, therefore, to examine a little more closely the differences between this view and our own.

In fact, the discrepancy between the two views is smaller now than it was when the two hypotheses were each advanced (Gray 1970; Vanderwolf 1971). For since that time the Canadian group (Vanderwolf *et al.* 1975) has suggested that there are two kinds of theta: one with frequencies below about 7 Hz, unrelated to movement, and cholinergic; the other with higher frequencies, related to movement and non-cholinergic. Thus they now accept that, in the rabbit and cat often, and in the rat occasionally, theta may occur without movement (Vanderwolf *et al.* 1975; Kemp & Kaada 1975). Nor is there any real difference in our respective views of the behavioural correlates of high frequency theta ('voluntary movement' meaning much the same thing as 'behaviour performed for reward or active avoidance') or low frequency theta (fixed action patterns). We part company from them at two points, however. First, our pharmacological studies of septal driving of theta provide no evidence that low frequencies are affected by cholinomimetic or cholinolytic drugs any more than high frequencies (McNaughton et al. 1977). Second, to their two kinds of theta, we add a third, allocating to it the bottom 1.5 Hz or so of their high frequency band and supposing it to be associated with the activity of the behavioural inhibition system. We return to the relation between theta frequency and behaviour in the final section of this paper.

FUNCTIONS OF THE SEPTO-HIPPOCAMPAL SYSTEM

In this final section we try to show how our view of the functions of the septo-hippocampal system relates to those held by other workers.

Views which are obviously close to ours are the early theories attributing inhibitory functions to both the septal area (Kaada et al. 1962; McCleary 1966) and the hippocampus (Douglas 1967; Kimble 1969). Indeed, ours is a rather old-fashioned view of the septo-hippocampal system. But it is a mistake to take our phrase, 'behavioural *inhibition* system', as indicating that all this system does is to 'slam on the brakes', to use Nadel et al.'s (1975) dismissive metaphor. We would rather state the functions of the behavioural inhibition system in terms of the stimuli to which it responds (conditioned punishing stimuli, conditioned frustrative stimuli and novelty). Clearly, not all responses to these stimuli are inhibitory in any obvious manner. For example, one of the things an animal does when faced with a novel environment is to freeze; but another is to explore it. Similarly, there is good evidence that one of the effects of conditioned frustrative stimuli is to increase the diversity of information that an animal acquires from its environment (Sutherland 1966).

This aspect of our theory brings us very close to those early views of the theta rhythm which related it to the orienting response (Grastyán 1959); and, above all, close to the detailed model of septo-hippocampal circuitry proposed by Vinogradova (1975; Vinogradova & Brazhnik, this volume). Just as she sees the CA3–lateral septal link as underlying habituation of reactions to novel stimuli, so we see it as underlying the habituation of reactions to signals of non-reward ('counter-conditioning'). As outlined at the outset of this paper, our theory of anxiety treats both kinds of stimuli as adequate to activate the behavioural inhibition system. Thus, putting together our work on the PREE and Vinogradova's experiments on the habituation of reactions to novelty, one may readily extend the counter-conditioning model (Fig. 6) to cover those cases in which the initiating input to the septo-hippocampal system is a novel stimulus. This extension is also supported by Mason & Fibiger's (1977) report that habituation is blocked by DANB lesions.

Another historically early view is that the septo-hippocampal system is related to memory (Milner 1959). This view cannot easily be reconciled to the one developed here. However, recent work on human amnesia (from which the memory hypothesis was originally derived) has caused a radical re-evaluation of the amnesic syndrome. As Weiskrantz & Warrington (1975) put it: 'an amnesic subject is amnesic when prior learning is allowed to interfere with subsequent learning, and scarcely amnesic at all when false positive responses are not allowed to intrude' (p. 418). Thus the human amnesic's problem lies in inhibiting responses which under earlier conditions were correct but no longer are so. Under this description, the data which were formerly seen as supporting the memory hypothesis are easily compatible with the general approach to the septo-hippocampal system adopted here; though much further theoretical and experimental work will certainly be needed before we know in what way the inhibition of an incorrect 'association' in human learning corresponds to the inhibition of a non-rewarded response in animals.

The views of septo-hippocampal function which are furthest from ours are those which emphasize a role in voluntary movement (Vanderwolf 1971) or spatial analysis (O'Keefe & Nadel 1978).

The major evidence supporting the former hypothesis is the close correlation between the theta rhythm and motor behaviour, on which we have already commented. But the evidence from manipulative experiments completely rules out the possibility that the septo-hippocampal system is concerned in the production or control of movement. There is no impairment in 'voluntary movement' (however one wishes to define this) after massive damage to the hippocampus, after destruction of the septal area eliminating theta, or after transection of the fornix also eliminating theta (Myhrer 1975; and our own observations). Conversely, the production of theta by septal stimulation does not produce movement, and may even reduce the speed of movement (Gray 1972b); though it is true that movement is normally seen when theta is produced by hypothalamic or reticular stimulation (Bland & Vanderwolf 1972; James et al. 1977), and that blocking theta by stimulation of the septum (Gray et al. 1972a) or the hippocampus (Bland & Vanderwolf 1972) causes behavioural arrest. It is clear, therefore, that the correlation between theta and movement requires some other explanation than the attribution of movement-producing functions to the septo-hippocampal system.

The spatial hypothesis (O'Keefe & Nadel 1978), according to which the septo-hippocampal system forms and uses spatial 'maps', is one way of providing such an alternative explanation. We accept the evidence offered by O'Keefe and his collaborators that single units in the septo-hippocampal system sometimes fire in correlation with particular places in the animal's environment; and that some tasks involving complex spatial analysis are severely disrupted after damage to this system. But the existing data again rule out the possibility that it is concerned *only* with the analysis and use of spatial information. There is clear evidence of impairment after both hippocampal and septal lesions in tasks which cannot be regarded as predominantly, or even in some cases marginally, spatial in character. For example, extinction of bar-pressing is clearly retarded after both kinds of lesion; and reversal learning is impaired in a number of non-spatial modalities after hippocampal lesions (J. A. Gray & N. McNaughton, review in preparation). Thus what is needed is a theory of septo-hippocampal function which incorporates the data gathered by Vanderwolf and O'Keefe and their respective collaborators, while not discarding the explanatory power of earlier theories for earlier data.

As a first move in this direction it would seem wise to perform on O'Keefe & Nadel's (1978) somewhat abstract notion of space the same kind of operation that Weiskrantz & Warrington (1975) performed on the notion of memory loss. The tasks which O'Keefe & Nadel have analysed as requiring spatial maps would, on this view, simply be tasks which are rich in sources of potential interference. It is a critical feature of spatial performance that the way to get to a place depends on where you start. In (again old-fashioned) S-R terms, this requires that the animal stores lists of *different* responses to the *same* stimuli (depending on where it is going); and different *sequences* of such lists (depending both on where it is going and where it is coming from). It is just such lists as these which maximize the human amnesic deficit. Thus we suggest that the susceptibility of tasks involving complex spatial analysis to hippocampal damage (O'Keefe & Nadel 1978) is merely a special case of a general deficit in the suppression of interference from responses which are correct under slightly different circumstances; or which have been correct in the past under the same circumstances but no longer are so.

How, on this view of the significance of space, can one account for Vanderwolf's correlations between theta frequency and movement?* A plausible argument can be derived from Vinogradova's (1975) hypothesis

---

*One of us (Gray 1977) has offered a previous account of these correlations; but we now find it disturbingly *ad hoc*.

that theta quantizes the activities of the septo-hippocampal system. On this hypothesis, theta frequency might reflect the rate of flow of information through the system. If this information is used to check actual against expected events (Vinogradova 1975), one might expect its rate of flow to rise as the animal's speed of locomotion increased, giving rise to the correlation between theta frequency and locomotor vigour at high theta frequencies (Vanderwolf et al. 1975). It is also plausible that the execution of fixed action patterns requires a relatively small rate of flow of information, so that only low theta frequencies would be observed during these.

If we apply this hypothesis to the behavioural inhibition system, we must suppose that the theta frequencies observed during its operation reflect the rate of information intake when an on-going chain of behaviour has been disrupted by a mismatch between expected and actual events, causing the animal to re-sample its environment in a search for new cues to action. The motor behaviour observed during such re-sampling would depend on the species. Cats and rabbits are more likely to remain immobile and use visual search; rats (especially albino varieties) are more likely to use vibrissal and olfactory cues, and consequently to engage in locomotor exploration. This species difference would account for the fact that middle theta frequencies have more often been seen in immobile rabbits and cats than in immobile rats (Vanderwolf et al. 1975), though they have also been seen in the latter species (Gray 1971). If this view is correct, then what we have called middle frequency theta would differ from both high and low theta frequencies in that it reflects a deliberate search for information under conditions of uncertainty, as distinct from checking off that a routine behaviour chain (whether learned or innate) is going according to plan.

Given the complexity of the functions we have attributed to the behavioural inhibition system, and the corresponding complexity of the anatomical and electrophysiological substrate to which we have tried to match this system, it would be difficult to sum up the view presented here with a simple label. The theory we have adumbrated bears important similarities to other approaches which have stressed variously the inhibitory role of the septo-hippocampal system, and its role in detecting novelty, in maintaining behavioural flexibility and in selecting the appropriate response chain amongst competing rivals. One stresses, in this way, the similarities between different views at the risk of vagueness; but one ignores them at the risk of attributing too narrow and too simple a function to an obviously complex system. If we were forced to characterize our own position in a phrase, we would opt for Simonov's (1974) felicitous description of the hippocampus as 'an organ of hesitation and doubt'.

ACKNOWLEDGEMENTS

Our research has been supported by the U.K. Medical Research Council. We are grateful to Dr L. E. Jarrard for his help in developing the fimbrial lesion; and to him, Dr M. Fillenz and Dr M. Boarder for allowing us to refer to our as yet unpublished collaborative experiments. J. F. held a Kenneth Lindsay Scholarship from the Anglo-Israel Association.

References

AMSEL, A. (1962) Frustrative nonreward in partial reinforcement and discrimination learning: some recent history and a theoretical extension. *Psychol. Rev.* 69, 306–328

AMSEL, A. (1972) Behavioural habituation, counterconditioning, and a general theory of persistence, in *Classical Conditioning II: Current Research and Theory* (Black, A. H. & Prokasy, W. F., eds.), pp. 409–426, Appleton-Century-Crofts, New York

AMSEL, A., GLAZER, H., LAKEY, J. R., MCCULLER, T. & WONG, P. T. P. (1973) Introduction of acoustic stimulation during acquisition and resistance to extinction in the normal and hippocampally damaged rat. *J. Comp. Physiol. Psychol.* 84, 176–186

ANLEZARK, G. M., CROW, T. J. & GREENWAY, H. P. (1973) Impaired learning and decreased cortical norepinephrine after bilateral locus coeruleus lesions. *Science (Wash. D.C.)* 181, 682–684

BLAND, B. H. & VANDERWOLF, C. H. (1972) Electrical stimulation of the hippocampal formation: behavioural and bio-electric effects. *Brain Res.* 43, 89–106

BROWN, R. T. & WAGNER, A. R. (1964) Resistance to punishment and extinction following training with shock or nonreinforcement. *J. Exp. Psychol.* 68, 503–507

BRUNNER, R. L., HAGGBLOOM, S. J. & GAZZARA, R. A. (1974) Effects of hippocampal X-irradiation-produced granule-cell agenesis on instrumental runway performance in rats. *Physiol. Behav.* 13, 485–494

CORRODI, H. FUXE, K., LIDBRINK, P. & OLSON, L. (1971) Minor tranquilizers, stress and central catecholamine neurons. *Brain Res.* 29, 1–16

DEFRANCE, J. F. (1976) A functional analysis of the septal nuclei, in *The Septal Nuclei* (DeFrance, J. F., ed.), pp. 185–227, Plenum Press, New York

DOUGLAS, R. J. (1967) The hippocampus and behaviour. *Psychol. Bull.* 67, 416–442

DREWETT, R. F., GRAY, J. A., JAMES, D. T. D., MCNAUGHTON, N., VALERO, I. & DUDDERIDGE, H. J. (1977) Sex and strain differences in septal driving of the hippocampal theta rhythm as a function of frequency: effects of gonadectomy and gonadal hormones. *Neuroscience* 2, 1033–1041

DUDAR, J. D. (1975) The effect of septal nuclei stimulation on the release of acetylcholine from the rabbit hippocampus. *Brain Res.* 83, 123–133

FRANCHINA, J. J. & BROWN, T. S. (1970) Response patterning and extinction in rats with hippocampal lesions. *J. Comp. Physiol. Psychol.* 70, 66–72

GLAZER, H. I. (1974) Instrumental conditioning of hippocampal theta and subsequent response persistence. *J. Comp. Physiol. Psychol.* 86, 267–273

GRASTYÁN, E. (1959) The hippocampus and higher nervous activity, in *The Central Nervous System and Behavior* (Brazier, M. A. B., ed.) *(Transactions of the 2nd Conference)*, p. 34i, Josiah Macy, Jr Foundation, New York

GRAY, J. A. (1964) Strength of the nervous system and levels of arousal: a reinterpretation, in *Pavlov's Typology* (Gray, J. A., ed.), pp. 289–366, Pergamon Press, Oxford

GRAY, J. A. (1970) Sodium amobarbital, the hippocampal theta rhythm and the partial reinforcement extinction effect. *Psychol. Rev.* 77, 465–480

GRAY, J. A. (1971) Medial septal lesions, hippocampal theta rhythm and the control of vibrissal movement in the freely moving rat. *Electroencephalogr. Clin. Neurophysiol.* 30, 189–197

GRAY, J. A. (1972a) The structure of the emotions and the limbic system, in *Physiology, Emotion and Psychosomatic Illness* (*Ciba Found. Symp.* no. 8), pp. 87–120, Associated Scientific Publishers, Amsterdam

GRAY, J. A. (1972b) Effects of septal driving of the hippocampal theta rhythm on resistance to extinction. *Physiol. Behav. 8*, 481–490

GRAY, J. A. (1975) *Elements of a Two-Process Theory of Learning*, Academic Press, London

GRAY, J. A. (1976) The behavioural inhibition system: a possible substrate for anxiety, in *Theoretical and Experimental Bases of the Behaviour Therapies* (Feldman, M. P. & Broadhurst, A., eds.), pp. 3–41, Wiley, London

GRAY, J. A. (1977) Drug effects on fear and frustration: possible limbic site of action of minor tranquilizers, in *Handbook of Psychopharmacology*, vol. 8: *Drugs, Neurotransmitters and Behavior* (Iversen, L. L., Iversen, S. D. & Snyder, S. H., eds.), pp. 433–529, Plenum Press, New York & London

GRAY, J. A. (1978) A neuropsychological theory of anxiety, in *Emotions and Psychopathology* (Izard, C. E., ed.), Plenum Press, New York, in press

GRAY, J. A. & BALL, G. G. (1970) Frequency-specific relation between hippocampal theta rhythm, behaviour and amobarbital action. *Science (Wash. D.C.) 168*, 1246–1248

GRAY, J. A., ARAUJO-SILVA, M. T. & QUINTAO, L. (1972a) Resistance to extinction after partial reinforcement training with blocking of the hippocampal theta rhythm by septal stimulation. *Physiol. Behav. 8*, 497–502

GRAY, J. A., QUINTAO, L. & ARAUJO-SILVA, M. T. (1972b) The partial reinforcement extinction effect in rats with medial septal lesions. *Physiol. Behav. 8*, 491–496

GRAY, J. A., MCNAUGHTON, N., JAMES, D. T. D. & KELLY, P. H. (1975) Effect of minor tranquillizers on hippocampal theta rhythm mimicked by depletion of forebrain noradrenaline. *Nature (Lond.) 258*, 424–425

HENKE, P. G. (1974) Persistence of runway performance after septal lesions in rats. *J. Comp. Physiol. Psychol. 86*, 760–767

JAMES, D. T. D., MCNAUGHTON, N., RAWLINS, J. N. P., FELDON, J. & GRAY, J. A. (1977) Septal driving of hippocampal theta rhythm as a function of frequency in the free-moving male rat. *Neuroscience 2*, 1007–1017

KAADA, B. R., RASMUSSEN, E. W. & KVEIM, O. (1962) Impaired acquisition of passive avoidance behaviour by subcallosal, septal, hypothalamic and insular lesions in rats. *J. Comp. Physiol. Psychol. 55*, 661–670

KEMP, I. R. & KAADA, B. R. (1975) The relation of hippocampal theta activity to arousal, attentive behaviour and somato-motor movements in unrestrained cats. *Brain Res. 95*, 323–342

KIMBLE, D. P. (1969) Possible inhibitory functions of the hippocampus. *Neuropsychologia 7*, 235–244

KIMSEY, R. A., DYER, R. S. & PETRI, H. L. (1974) Relationship between hippocampal EEG, novelty, and frustration in the rat. *Behav. Biol. 11*, 561–568

KURTZ, R. G. (1975) Hippocampal and cortical activity during sexual behavior in the female rat. *J. Comp. Physiol. Psychol. 89*, 158–169

LEWIS, P. R. & SHUTE, C. C. D. (1967) The cholinergic limbic system: projections to hippocampal formation, medial cortex, nuclei of the ascending cholinergic reticular formation, and the subfornical organ and supra-optic crest. *Brain 90*, 521–540

LIDBRINK, P., CORRODI, H., FUXE, K. & OLSON, L. (1972) Barbiturates and meprobamate: decreases in catecholamine turnover of central dopamine and noradrenaline neuronal systems and the influence of immobilization stress. *Brain Res. 45*, 507–524

LIVETT, B. G. (1973) Histochemical visualization of peripheral and central adrenergic neurones. *Br. Med. Bull. 29*, 93–99

MCCLEARY, R. A. (1966) Response-modulating functions of the limbic system: initiation and suppression, in *Progress in Physiological Psychology*, vol. 1 (Stellar, E. & Sprague, J. M., eds.), pp. 210–272, Academic Press, New York

MCLENNAN, H. & MILLER, J. J. (1974) The hippocampal control of neuronal discharges in the septum of the rat. *J. Physiol. (Lond.) 237*, 607–624

MCLENNAN, H. & MILLER, J. J. (1976) Frequency-related inhibitory mechanisms controlling rhythmical activity in the septal area. *J. Physiol. (Lond.) 254*, 827–841

McNaughton, N., James, D. T. D., Stewart, J., Gray, J. A., Valero, I. & Drewnowski, A. (1977) Septal driving of the hippocampal theta rhythm as a function of frequency in the male rat: drug effects. *Neuroscience 2*, 1019–1027

Mason, S. T. & Iversen, S. D. (1975) Learning in the absence of forebrain noradrenaline. *Nature (Lond.) 258*, 422–424

Mason, S. T. & Fibiger, H. C. (1977) Altered exploratory behaviour after 6-OHDA lesion to the dorsal noradrenergic bundle. *Nature (Lond.) 269*, 704–705

Meibach, R. C. & Siegel, A. (1977) Efferent connections of the septal area in the rat: an analysis utilizing retrograde and anterograde transport methods. *Brain Res. 119*, 1–20

Milner, B. (1959) The memory defect in bilateral hippocampal lesions. *Psychiatric Research Reports 11*, 43–52

Myhrer, T. (1975) Normal jump avoidance performance in rats with the hippocampal theta rhythm selectively disrupted. *Behav. Biol. 14*, 489–498

Nadel, L., O'Keefe, J. & Black, A. H. (1975) Slam on the brakes: a critique of Altman, Brunner and Bayer's response-inhibition model of hippocampal function. *Behav. Biol. 14*, 151–162

O'Keefe, J. & Nadel, L. (1978) *The Hippocampus as a Cognitive Map*, Oxford University Press, Oxford, in press

Simonov, P. V. (1974) On the role of the hippocampus in the integrative activity of the brain. *Acta Neurobiol. Exp. 34*, 33–41

Stumpf, C. (1965) Drug action on the electrical activity of the hippocampus. *Int. Rev. Neurobiol. 8*, 77–138

Sutherland, N. S. (1966) Partial reinforcement and breadth of learning. *Q. J. Exp. Psychol. 18*, 289–301

Vanderwolf, C. H. (1971) Limbic diencephalic mechanisms of voluntary movement. *Psychol. Rev. 78*, 83–113

Vanderwolf, C. H., Kramis, R., Bland, B. H. & Gillespie, L. A. (1975) Hippocampal rhythmical slow activity and neocortical low voltage fast activity: relations to behavior, in *The Hippocampus*, vol. 2: *Neurophysiology and Behavior* (Isaacson, R. L. & Pribram, K. H., eds.), pp. 101–128, Plenum Press, New York

Vinogradova, O. S. (1975) Functional organization of the limbic system in the process of registration of information: facts and hypotheses, in *The Hippocampus*, vol. 2: *Neurophysiology and Behavior* (Isaacson, R. L. & Pribram, K. H., eds.), pp. 1–70, Plenum Press, New York

Vinogradova, O. S. & Brazhnik, E. S. (1978) Neuronal aspects of septo-hippocampal relations, this volume, pp. 145–177

Weiskrantz, L. & Warrington, E. K. (1975) The problem of the amnestic syndrome in man and animals, in *The Hippocampus*, vol. 2: *Neurophysiology and Behavior* (Isaacson, R. L. & Pribram, K. H., eds.), pp. 411–428, Plenum Press, New York

Weiss, J. M., Glazer, H. I., Pohorecky, L. A., Brick, J. & Miller, N. E. (1975) Effects of chronic exposure to stressors on avoidance-escape behaviour and on brain norepinephrine. *Psychosom. Med. 37*, 522–534

Discussion

Black: Your model (Fig. 6) predicts that one would obtain the same effects from medial septal lesions and from lesions of the fornix. It also predicts, I think, that the partial reinforcement extinction effect (PREE) would be abolished after such lesions. Isn't it disconcerting that the PREE effect is not abolished after these lesions?

Gray: I agree with you; this deduction from our model is correct. However, there is an escape clause. It is exceedingly unlikely that we would have got our knife or electrode on all, and only, the fibres or cells that matter. Suppose there is a group of cells in the medial septal area sending out a bundle of fibres, and that our electrolytic lesion destroys a large part but not all of them, and our knife cut destroys a large part but not all of the fibres. Remember also that if our lesion is too big it extends into the lateral septal area and into the fimbria, and we then begin to bring in opposing effects from the other parts of the system. (This, by the way, is why I think that total fornix lesions may fail to affect punishment; it would be interesting to know what happens to punishment with partial fornix or fimbria lesions.) If that is so, we don't have a total loss of the system producing extinction when we make a small medial septal lesion or dorso-medial fornix cut; we have a very damaged system, and this would be particularly likely to succumb to inhibition from the lateral septal input. So you would get a big partial reinforcement extinction effect because the medial septum was working, but not too well.

Black: Brunner *et al.* (1974) also found a partial reinforcement extinction effect after X-irradiation-produced dentate gyrus granule-cell agenesis. How does your explanation account for these data?

Gray: Their data actually are as follows: the partially reinforced animals with lesions of the dentate granule cells (produced by neonatal X-irradiation) ran more slowly during extinction than partially reinforced controls, but they also had different acquisition asymptotes. The continuously reinforced lesioned animals took longer to extinguish than continuously reinforced controls. Thus the partial reinforcement extinction effect was significantly reduced by X-irradiation of the granule cells.

Black: It was still there, however. Also, it's important to note that X-irradiation results in other deficits that are typical of hippocampal lesions. So, one can't argue that X-irradiation generally does away with the deficits that are produced by damage to the hippocampal formation.

Andersen: Ten per cent of the granule cells were also still there.

Gray: The effects were in the direction we would predict.

Black: You suggest that the hippocampus inhibits ongoing behaviour. I am a little suspicious about terms like that—particularly since we have been videotaping the behaviour of rats with fornical lesions in a variety of situations (Osborne & Black 1978). If you observe the behaviour of control animals and animals with lesions of the fornix after the switch from continuous reinforcement to extinction, you see a wide variety of changes in behaviour of normal animals. They press the lever less often, as one might expect, but they also increase the duration of bouts of lever pressing, as they do the duration of most activities in which they indulge. For example, they increase the amount of time spent biting the lever, the duration of bouts of grooming, the duration of trips away from the area of the level and food cup, and so on. Extinction seems then to make normal animals more perseverative in this sense. One might predict that animals with hippocampal and fornical lesions would persevere even more than normals if they cannot inhibit ongoing behaviour, and display even longer bout lengths of responding. But that doesn't happen. Rats with fornical lesions don't perseverate during extinction as do normals. They change behaviours very rapidly.

Gray: I started with some 'neurophilosophy' precisely because I don't think it is possible to say 'I have a feeling that inhibition is involved', or that 'attention' is involved. I have drawn up a system in which I specify the kinds of information that must act as inputs, the kind of responses you would expect, and the way the information is dealt with. This doesn't mean to say that perseveration will always be increased.

In the situation you are describing, I would ask *you* whether the normal animal's behaviour change is a response to a signal of non-reward, or a signal of punishment, or a novel stimulus. I am not saying that after hippocampal lesions every kind of behaviour lasts longer.

Black: But that is what happens in normal animals in response to non-reward. After non-reward, they do perseverate longer.

Weiskrantz: I want to inject one possible answer for Dr Gray which would be relevant to Dr Vinogradova's studies, namely that a prediction of Dr Gray's model is that exactly the same sort of alterations ought to occur in a habituation situation as occur with PREE and so forth. Dr Vinogradova produces habituation effects measured electrophysiologically; do we have any behavioural measure of habituation which shows similar characteristics and alterations?

Srebro: We have some results, partly published by Dr C. Köhler (1976), which may answer your question. We have studied behavioural habituation of the orienting response to the auditory stimulus in a familiar environment. The orienting response to a sound in the rat may be divided into two components, an early component that consists of a brief response to a stimulus, and a late component which is a post-stimulus search/exploration of the nearest surroundings. Selective lesions in the lateral septum delayed habituation of the orienting response, mainly because of the longer post-stimulus search. Lesions placed in the medial septum tended to speed up the habituation by shortening the post-stimulus exploration. Differential effects of the lateral and medial septal lesions were also observed in a test of exploration in which there was free access to a new environment from the home cage. In such test the lateral lesions enhanced while the medial ones attenuated the rats' exploration of new surroundings. Similar results were seen when the reaction to a new object was studied. Thus, in general, medial septal lesions tend to obliterate the rat's response to novelty, while lateral lesions may enhance such a response in certain experimental conditions. Finally, I would like to stress that many of the behavioural effects observed after selective lesions in the septum are critically dependent on the experimental conditions; moreover, many of the tests do not distinguish the effects of the two types of lesions.

Gray: On the face of it, your data are not the way round I would have expected. We are now setting up experiments in which we shall look at analogues of the partial reinforcement extinction effect with punishment substituted for non-reward (i.e. the partial punishment effect in an alley); and with novelty substituted for non-reward, getting the rat to learn to tolerate distraction in an alley. We shall try to do all these experiments in the alley because we have already found that when we look at punishment in a different situation (a Skinner box) (Gray *et al*. 1978) we don't get anything that quite fits with the model I have reported here, so this model may be situation-specific to some extent.

Vanderwolf: There is a point of view that the nervous system of an animal like the rat is built for doing things like burrowing, avoiding owls, finding food, and so on, in places where rats live. I wonder if we are missing something in focusing exclusively on what rats do in runways and Skinner boxes?

Gray: Runways are exactly the kinds of place in which rats live. The runway was originally devised by psychologists because it is like the tunnels a rat runs around in his burrow.

Vanderwolf: Rats do very well in runways, certainly, but there hasn't been much serious attempt to find out what effect septal or hippocampal lesions have on behaviour in a general sense. The rat, after all, has hundreds of different behaviours that it can exhibit on different occasions and most of them are not looked at in these experimental situations. In fact, what you are counting is *scores*, like runway times; there isn't much real examination of behaviour at all.

Weiskrantz: This is a general point about the strategy of research and there are two schools of thought about it. One is that you look at a whole spectrum of natural behaviour, ethologically; I think Dr Black was pointing in this direction as well. The other way is that one makes specific predictions and tests them in an artificial controlled situation. Surely both are useful at different stages of development of a research problem and they can complement each other.

Gaffan: To take up this point of the generality of the model, I want to question how Dr Gray can maintain that the hippocampus inhibits non-rewarded or punished behaviour as a general statement. In the first place, total fornix transection (the section of the hippocampal–septal axis) does not impair extinction in all situations. It impairs it in big runways but not in a very small runway (Gaffan 1972). Dr G. Winocur (personal communication) has shown the same thing with hippocampal lesions. The extinction deficit is related to the length of the runway.

Gray: It also occurs in Skinner boxes, where there is no runway at all.

Gaffan: There is a small extinction deficit in a Skinner box, but there is more space for exploration in a Skinner box than there was in the smallest runway I could devise, and in that runway there was no extinction deficit at all.

Secondly, there is other evidence that fornix-transected animals can show normal sensitivity to non-reward. Rats with that lesion showed behavioural contrast when a signal of non-reward (S^-) was introduced into a variable-interval schedule in a Skinner box (Gaffan 1973). The same animals also showed inhibitory generalization gradients around S^-; that is one of the clearest indications of a specific inhibitory process. Further, my monkeys with fornix transection (Gaffan 1974) were required to remember when an object was not rewarded on its first presentation, and avoid that object on its second presentation. They performed that task at the same level as control animals. So in some situations there is evidence of unimpaired inhibitory processes in response to stimuli signalling non-reward. The same is also true of stimuli signalling punishment; animals with damage to the hippocampal system are unimpaired in many tests of passive avoidance (Nadel *et al.* 1975).

Thirdly, in those tasks where perseverative deficits do occur in the face of non-reward or punishment, there may be similar deficits when the change is in the opposite direction, from non-reward to reward. When a well-established S^- was converted into an S^+ in a lever-pressing task, the response rate of fornix-lesioned rats rose more slowly than that of controls (Gaffan 1973).

My own view is that perseverative deficits arise because one of the main natural functions of the hippocampal system as a novelty discriminator is to ensure that the animal from time to time explores relatively unfamiliar parts of an environment or tries out new responses, so as not to fall into a stereotyped pattern of behaviour. But how do you reconcile the results I have mentioned with your own model?

Grossman: There are many other behavioural test situations where the effects of non-reward are reduced, at least in the rat and cat, after septal lesions or fornicotomy. Any intermittent reinforcement schedule, particularly a predictable one, or a plain extinction paradigm, clearly shows the effect of hippocampal and septal lesions as well as fornicotomy, not only in runways but in Skinner boxes and other apparatus also.

General discussion II

HIPPOCAMPAL UNITS AND THETA ACTIVITY

Ranck: We have recorded from single neurons in the hippocampus, septal nuclei, and retrohippocampal areas of freely moving rats (Best & Ranck 1975; Feder & Ranck 1973; Fox & Ranck 1975, 1977; Mitchell & Ranck 1977; Ranck 1973a, b, 1975, 1976). There are two categories of neurons, into which almost all neurons can be unequivocally placed. One type, *theta cells*, are defined as neurons which increase their rates of firing if and only if there is a slow wave theta rhythm in the hippocampus. The other type, *complex spike cells*, are defined as neurons which, at one time or another, fire complex spikes. A complex spike is a series of spikes with inter-spike intervals of 2–7 ms duration, in which the amplitude of the extracellularly recorded spike decreases. Complex spike cells usually fire only single spikes which are the same as the first spike of the complex spike. In many cells a complex spike is a rare event, in some only occurring in slow wave sleep. Theta cells never fire complex spikes. There are many other differences between these two types of cells. Complex spike cells rarely fire faster than 10/s, and theta cells almost always fire faster than 10/s. There is a difference in the duration of the extracellularly recorded spike, in the pattern of firing, in the localization (Fox & Ranck 1975), and in their responses to stimulation of hippocampal inputs (Fox & Ranck 1977). We have studied the electrophysiological characteristics of these cells in a preparation which has three independently moveable microelectrodes, one in CA1, one in CA3, and one in dentate, all in the same lamella, and three stimulating electrodes, one in perforant path, one in contralateral ventral hippocampal commissure, and one in the vertical nucleus of the diagonal band. This is all in a freely moving rat. The only cells which can be antidromically driven by hard (i.e. collision) or soft (i.e. less than 200 μs jitter in latency, no change in latency with increasing stimulus strength) criteria are complex spike cells. Complex spike cells show a long and profound inhibition after being excited; theta cells show facilitation and/or mild inhibition. Complex spike cells fire single action potentials on the orthodromic or antidromic population spikes. Theta cells fire multiple action potentials on what is presumed to be the extracellular sign of the hyperpolarizing phase of a pyramidal cell IPSP. Theta cells are less than 6% of the total cells in Ammon's horn. From these data we have argued that theta cells are included in the class of interneurons, and that complex spike cells are pyramidal cells.

Both types of cells fire with a phase relation to a slow wave theta rhythm. The behavioural correlates of theta cells are the same as the behavioural correlates of the slow wave theta rhythm, and hence all theta cells have the same behavioural correlates as those described by Vanderwolf *et al.* (see this volume). Each complex spike cell has a different behavioural correlate. Among the major categories are *(a)* cells which fire during a particular consummatory behaviour and part of the appetitive behaviour before it (an 'approach–consummate cell'); *(b)* cells which fire like *(a)* but also and often most rapidly when a consummatory behaviour which can usually be performed is not able to be performed (an 'approach–consummate–mismatch cell'); and cells which do not fire during any consummatory behaviours except sleep, and only fire during some appetitive behaviours ('appetitive cells'). Some cells also fired when the rat was at a particular place. Recently I have been trying to refine these descriptions, but there are some problems. First, it has been suspiciously easy for all of us who do these kind of studies to find relations of behaviour to cell firing. We are looking at only a tiny fragment of what a rat can do or where he can be. Segal & Olds (1972), Berger *et al.* (1976) and Lynch (see this volume) have evidence suggesting that the circumstances in which the cell fires can change. If this is so and if we try to refine the description, perhaps all we shall be describing will be the firing of the cell in one particular circumstance. To get around this we are looking for the behavioural correlates of the same neuron in three different situations: pup retrieval in a box with a nest, running an eight-arm maze, and performing a DRL-16 with schedule-induced polydipsia (DRL: differential reinforcement for low rate of responding). All three behaviours are disrupted by hippocampal ablation. Preliminary results show that one can find different behavioural correlates of the same cell in these three situations. In pup retrieval some cells fire when the mother is at a particular place, and the cell will fire if she is at that place independent of whether she is retrieving a pup, getting food, getting water, or exploring. These 'place' units usually also require that the rat have a particular orientation or be moving in a particular direction.

As Dr O'Keefe has said here (O'Keefe & Black, this volume), all these 'place cells' increase their rate of firing when the rat goes into slow wave sleep, regardless of where she sleeps.

One of the points of all this is that we must be very cautious about studying these neurons. Mays & Best (1975) attempted to repeat some of the findings of Dr Vinogradova's rabbit studies in rats. They were able to repeat many of her findings, but when they recorded EEG they found that many of the responses of the cells were associated with waking the rat up. There was habituation of the awakening by a tone, and hence habituation of the response of a neuron to the tone.

Vinogradova: Is it possible to awaken an animal once in five seconds?

Ranck: In rats, yes, it is common to be aroused for only a few seconds. I have no idea if this occurs in rabbits. In rats a change from slow wave sleep to arousal can occur with no overt movements, so one would not be aware of it by just looking at the rat. One must record EEG.

Vinogradova: I cannot agree with your classification of hippocampal cells into two types, Dr Ranck, because I have never seen such differences in recording from hippocampal cells in rabbits. We observed complex spike cells in one structure only, the subiculum, but never in the hippocampus, except in two situations: (1) injured cells, and (2) when epileptic foci were present in the hippocampus. These complex spikes resemble synchronized discharges observed in penicillin foci in the hippocampus. Another similarity between complex spikes and epileptic discharges is that hippocampal seizures can be suppressed with a strong sensory stimulus; likewise, the complex spikes disintegrate immediately after a sensory stimulus and are transformed into single random spikes. We normally discard such animals, regarding them as pathological cases.

I also cannot find such fine differences between the other parameters described. There is a difference in the mean frequency of discharge of CA1 and CA3 cells, but not so much for 'theta' and 'non-theta' cells from one field. Moreover, in the hippocampus there are rapidly discharging cells without any traces of theta rhythm, some of them even close to a pacemaker-like regular type of activity. There are also slowly discharging cells, but with regular 4-5/s single spikes which are exactly in the range of theta frequencies for the rabbit.

From our data on perforant path lesions I also cannot see how it is possible to regard 'theta cells' as interneurons. Almost all our cells are converted into 'theta cells' after the perforant path lesion. Do they become interneurons?

Andersen: I want to ask a few specific questions. Does theta activity mean idling, or activity? If you stimulate the dorsal hypothalamus the vast majority of cells in the dentate region increase their activity immensely. So the general conclusion seems to be that theta waves mean activity.

A second question concerns the immediate cause of theta activity. Apart from the work by Fujita & Sato (1964), who showed that the oscillations were due to membrane potential oscillations, some work by P. Schwarzkroin and myself (unpublished observations) can throw some light on it. We made intracellular recordings from dentate granule cells. These are difficult experiments because one has to have a very 'light' animal and the cells are very small. When we stimulated the brainstem or hypothalamus we got regular membrane potential oscillations which were perfectly in phase with the extracellular theta activity. We can conclude that the immediate cause of dentate theta waves is synchronous synaptic potentials. The hyperpolarizing potentials are clearly IPSPs. They reverse with Cl^- injection and change amplitude with changes in the membrane potential. The depolarizing potentials, however, are odd. They are not simple EPSPs, because of their size and duration.

It appears that the rhythm is caused by regularly occurring inhibition, cutting into the ongoing steady background activity. The system is different from the thalamic situation where the cells show post-inhibitory hyperactivity. This does not seem to be the case with granule cells.

If this interpretation is true, the regularity and frequency of theta is a function of the depolarizing pressure on the cells. The more the cell is depolarized the higher the frequency and the greater the regularity. In fact the theta activity can perhaps be regarded as a measure of the excitation the cell is subject to; perhaps it is not too important where it comes from.

Next, are the theta cells a separate class of cells or are these cells just any hippocampal cells in a particular state? If we stimulate the hypothalamus, and record from a granule cell, we produce theta and the cell increases its discharge frequency. When the theta activity was poor, the cell gave random discharges, which I call 'random singles'. With increasing amplitude and regularity of the theta activity, the cell discharged more regularly; I call them 'regular singles'. Then come 'irregular doubles' and 'regular doubles', and triplets and quadruplets. At higher degrees of theta activity the discharges become continuous. So, there is a slot within which you can see oscillations, or rhythmic activity. If you depolarize even more you drive the cells asynchronously; if you drive too little, the cell is quiet. On the basis of these observations I propose that there are *no* particular theta cells, but that there are theta states, in which nearly all hippocampal cells can be.

Finally, we have also recorded from what Dr Ranck would call complex cells, and there is no question that they take part in theta activity.

Ranck: Our suggestion has been that the complex spike cells are pyramidal cells.

Andersen: I agree with you; they are CA1 pyramidal cells.

Ranck: The theta cells, we think, are interneurons. We are not suggesting that all interneurons are theta cells, and we are not suggesting that all theta cells are identical. Both theta cells and complex spike cells have phase relations to the slow wave theta rhythm. We have done quantitative studies on this phase relation and there isn't much difference in the degree of phase relation in the two types of cells. Some theta cells have striking phase relations; others have poor phase relations. Some complex spike cells have striking phase relations and others poor relations. Fig. 1, from a study by S. E. Fox, shows the firing of a theta cell to stimulation of the contralateral ventral hippocampal commissure. The main graph shows that the latency is not constant at a given stimulus strength, and decreases with increasing stimulus strength. Record *(b)* shows the theta cell firing on the positive field potential which Dr Andersen has shown to be due to IPSPs in pyramidal cells

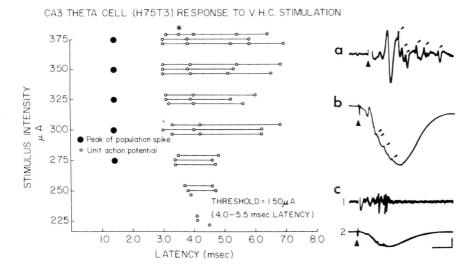

FIG. 1. (Ranck). Graph of the firing of a theta cell to increasing intensity of stimulation of the ventral hippocampal commissure (V.H.C.) in a behaving rat. (Three trials per intensity.) The rat was in slow wave sleep. The pulse was 200 µs in duration. Asterisk (*) marks an action potential at a curiously short interspike interval which was seen several times at higher stimulus intensities (also see *a*). Record *(a)* shows the response to 425 µA stimulus with narrow band filtering. Theta cell action potentials are indicated by arrowheads. The large negative/positive spike is the population spike. Record *(b)* is the same as *(a)* but 475 µA stimulus and wide band filtering (note change in time calibration). Record *(c)* shows two sweeps of narrow band (trace 1) and wide band (trace 2) response to 250 µA stimulus.

Time calibration: *(a)*, 2 ms; *(b)* and *(c)*, 4 ms. Amplitude calibration: *(a)* and *(c1)*, 0.2 mV; *(b)*, 0.75 mV; *(c2)*, 0.5 mV (negative up).

(Andersen *et al.* 1964a). Fox has found that theta cells have all the electrophysiological characteristics which Dr Andersen has shown for presumed basket cells in hippocampus (Andersen *et al.* 1964b, 1969).

We have not proved that theta cells are interneurons. To prove it we would have to inject dye, which could probably only be done in anaesthetized animals or a hippocampal slice. However, theta cells are defined by their behavioural correlates, so we need some other way to identify them in anaesthetized animals or slices. These electrophysiological characteristics are necessary (but not sufficient) conditions for their being interneurons, but these data also give someone a way to identify these cells in preparations in which dye might be injected.

Andersen: We agree completely that our basket cells do take part in theta, but how well? What you have to do is to make phase or cyclic histograms to show *how* modulated they are. In our experience, the dentate cells are extremely well modulated; the CA1 pyramidal cells are also well modulated; the basket cells are poorly modulated. They take part in theta activity but not very well.

Vinogradova: I agree completely with Professor Andersen that there are no theta cells but there is just a theta *state* of cells in the hippocampus. But I cannot agree that theta is merely activation of hippocampal cells, because we observed many cells in CA3 and CA1 and in both septal nuclei which go from random activity into what you call regular one-spike activity in the theta range, having high spontaneous activity before the stimulus. So it can be a traffic in both directions: it can be activation, relative to previous spontaneous activity; it can be inhibition, also relative to previous spontaneous activity. And it can be neither, but just a reorganization from randomized activity into regular theta-burst activity, which is easy to see with computer analysis.

Ranck: These cells are not, in spite of their name, the cause of the slow wave theta rhythm. We find many theta cells in CA3 and there is no generator of theta in CA3. This is also true in the medial septal nucleus.

Andersen: I don't in fact think there is so much disagreement between us as there appears to be. You agree, Dr Ranck, that these complex spike cells also take part in theta activity. I myself think they are all theta cells, *in that sense.*

O'Keefe: The name 'theta' for these cells is unfortunate, since it may be misinterpreted to mean that only they and not the complex spike cells show a phase relationship to the EEG theta. We in fact call them displace cells (O'Keefe 1976) to emphasize their relationship to a class of movements. The primary difference between the theta and complex spike cell in its relation to the EEG theta is that the theta cell always participates in theta activity whenever it occurs while a complex spike cell will only show a bursting firing which is phase-related to the EEG theta if the theta occurs while the rat is in that unit's place field.

Andersen: The relation can perhaps be explained as follows. At a high level of membrane potential a synaptic potential will produce a single spike; when the membrane potential is reduced the same cell can give single, double and complex spikes to the same input. When the cell shows the complex spike behaviour it is just under high depolarizing pressure.

Gray: Is there any clue yet as to why there are theta cells in the septum and CA3, with no theta waves, but in CA1 and in dentate there is theta?

Andersen: The experiment hasn't been done! The slow wave experiments have been done in anaesthetized animals where there is no modulation of CA3 histograms. If the experiment were done in awake animals, it is possible one would find slow wave activity.

Gray: But we know there aren't slow waves in the septum.

Vinogradova: I think Bland & Whishaw (1976) worked in unanaesthetized and unrestrained animals, and they have seen the same thing: absence of theta waves in CA3, though obviously there are theta-modulated cells in this field.

Gray: An important question is whether the division into two cell types is important for the rest of us working in the field, or not.

Ursin: There is a real need to explain the situation. If Jim Ranck said that the theta cell is the basket cell or some interneuron, and the complex cell is the pyramidal cell, Dr Vinogradova may end up with a rabbit with only basket cells!

Segal: I think the distinction exists; I don't think there is just one type of cell. I find in experiments with iontophoresis as well as in experiments in awake rats two types of cells which respond differently to various drugs related to catecholamines and acetylcholine. It doesn't depend on the cells being in one state or another. The distinction is important if you want to build a model on the basis of the types of cells encountered in the hippocampus and the responses that you get; if you pre-select one type of cell you will end up with one type of model.

Weiskrantz: Can we broaden the discussion to include the variety of triggers for these hippocampal cells?

Ranck: We have been able to find a relation between the firing of a hippocampal neuron and the inputs or outputs of our rats. We have called this a behavioural correlate, and it includes relations to the place where the rat is. It is surprising that we have been able almost always to find these relations, because there is certainly a lot going on in the brain which does not have much to do with overt behaviour.

Weiskrantz: It is important to get some agreement about this, because one point that will influence the behaviourists, the anatomists and indeed the neurochemists is the nature of the correlation between single-unit activity and behavioural and environmental effects, not only in the free-moving animal but in the animal subjected to controlled stimulation. If, for example, it turned out that all hippocampal units were spatial, or all comparator type cells, we could all go home. What hints do we get, and what agreement is there about, the kinds of units in the hippocampus, either correlated with behaviour, or triggered by external stimuli? Is there any environmental event that *won't* trigger hippocampal cells?

Lynch: Some cells in the hippocampus are driven by environmental events after conditioning that were not so driven before conditioning. John O'Keefe is studying cells that apparently have nothing to do with prior conditioning, on the other hand. So there are already two kinds of hippocampal cells on this basis.

Andersen: This distinction could be based on a threshold of depolarization.

Weiskrantz: It could also be due to something else that we haven't yet considered. We have concentrated on hippocampal units. There are units in other parts of the brain, as Bureš (1965) and Yoshii & Ogura (1960), for example, found, which change their properties after conditioning. Does the hippocampus have a special affinity towards changes associated with conditioning, or can such changes occur widely throughout the brain? If so, what are the similarities and dissimilarities? Another point is that we have not had a good description of natural environmental features associated with hippocampal single-unit activity, independent of conditioning.

Gray: Aren't you presupposing that there ought to be units that respond to natural, regular environmental stimuli?

Weiskrantz: No; if there aren't any, negative answers needn't worry us, but a positive answer would give us something hard to go on. We have heard about spatial units; what other kinds of units are there in the hippocampus? Perhaps Dr Ranck could tell us about the work on the human hippocampus.

Ranck: Eric Halgren, Tom Babb and Paul Crandall at the University of California at Los Angeles are also attempting to identify the behavioural correlates of units in hippocampus, hippocampal gyrus, and amygdala. However, their recordings are obtained from people, from fine wires implanted during the course of a diagnostic procedure in temporal lobe epileptics.

Simple, non-habituating visual responses are present, especially in posterior hippocampal gyrus. No units respond to odour but many neurons in all structures exhibit an acute sensitivity to hypoxia which may appear during sniffing. Unit activity averaged over a delayed matched-to-sample task reveals, in addition to visually responsive units in all structures, hippocampal gyrus neurons that respond only during situations requiring a choice. During a wide-ranging interview, some units, again confined to hippocampal gyrus, fire at high rates only during the recall of specific recent memories: of word pairs, or of events and locations on the hospital ward. In contrast, some hippocampal units are active during movements requiring effort, because they either require skill or induce pain. Other hippocampal neurons fire during the transition between tasks, or when a task is interrupted.

Lynch: This is the same result as you get by recording from the rat dentate, where sound by itself doesn't activate the granule cell until the rat has to recognize the sound; then the cells fire.

Weiskrantz: There is a technical difference in the use of the word 'recognize' there. In the conditioning situation the stimulus acquires associative significance.

Dr Ranck in his recent studies has stressed spatial units, but is it fair to characterize, say, a large minority of units in the hippocampus as being spatial, or is there any predominance at all?

Olton: We tested rats using a strategy different from that of Dr Ranck and Dr O'Keefe. The testing situation was arranged so that spatial cues were maximized and the animal was unlikely to build up expectations about any other characteristics of the environment (Olton *et al.* 1978). Out of 31 units in the hippocampus, 28 had strong spatial correlations. Thus there are probably two answers to the question. First, there may be a strong preponderance of cells with spatial characteristics. Second, there may be a substantial influence of the testing environment on the behavioural correlates of unit activity. This latter point has been emphasized by Dr Phil Best in his work, and is particularly important when the evidence for non-spatial correlates in other testing procedures is considered (Thompson *et al.* 1976).

Lynch: Thompson and Berger have done a series of experiments in curarized animals in which they place recording microelectrodes next to a cell in the pyramidal layer simply on the basis of spontaneous activity. They find that nearly all these cells respond to a conditioned stimulus after conditioning. Most of the cells found are antidromically activated (see Berger & Thompson 1977). The cells projecting into the septum universally increase their firing rate to the conditioning stimulus. However, they periodically find cells that can only be driven orthodromically from the septum. Of those cells, the great majority are *inhibited* by the conditioned stimulus—by the tone. So we have an animal not doing anything (except giving a nictitating membrane response), and most cells can be conditioned to a tone, and their responses can be differentiated on whether or not they are receiving or projecting into the septum.

Black: I would like to suggest a possible way of relating Dr Vinogradova's work to that of Drs Ranck and O'Keefe in behavioural terms. The latter distinguish between theta cells and complex spike cells. They suggested that theta cells fired whenever theta slow wave activity occurred and that complex spike cells fired when an animal was in a particular place. The cells that Dr Vinogradova described seemed to be quite different. In CA3, cells fired when a stimulus was presented and continued firing for a fairly long time; furthermore, those cells did not fire specifically to one stimulus. The difference, however, may not be really so great. Remember that Dr Vinogradova's rabbits were in a small, enclosed cage, and were relatively immobile. Therefore, one is unlikely to see place cell activity, and also unlikely to see theta cell activity that is correlated with movement. But atropine-sensitive theta occurs in the rabbit in an immobile animal in response to sensory stimulation, according to R. Kramis (unpublished). It may be, therefore, that Dr Vinogradova's CA3 cells are theta cells that are firing during atropine-sensitive theta that is produced by sensory stimulation.

Andersen: There may be a species difference here.

Gray: We should take seriously something that Olga Vinogradova said after Dr O'Keefe's paper (p. 197). She said that experimenters tend to find what they are looking for. She may have meant this in a general way, that experimenters in any science tend to find what they look for; or she might have meant it in a way which is particularly appropriate to this field.

Let us suppose that one of the things the hippocampus does is to register regularities in the animal's environment. This is a point of view common to many of us. We may diverge about what it does with those regularities. If it does register regularities in the environment, and it registers them temporarily, then of course one will find in the hippocampus the regularities that one builds into one's particular environment. People interested in spatial units will find spatial units because they provide the animal with a spatially regular environment and, as Olga Vinogradova said, people looking for temporal regularities will find units that respond according to time. In experiments with conditioning in curarized rats, you will find something else again. This is surely vital for a proper understanding of what is going on in these experiments.

Weiskrantz: That is a provocative summary of a position with which people may or may not agree; whether the hippocampus is a regularity register.

It is a hypothesis that would unite a certain amount of material.

Ursin: Surely we can reach the answer to this apparently straightforward problem: does Dr Andersen agree with and accept what Dr Ranck said, that there is an anatomical distinction between cells responsible for complex spikes and cells responsible for other spikes, or does he not? If Dr Ranck believes that, and Dr Andersen does not, then there is a difference between them, it seems.

Ranck: Our evidence is not perfect, but at least in the rat there is a great deal of evidence that there is this anatomical difference. This kind of evidence has not been looked for in other species.

Andersen: All the CA1 and dentate cells that we have recorded from show theta pacing. Dr Ranck thinks that only some do; I think that all do. I think there is a spectrum, of which he sees two parts only. But I would like Dr Ranck to identify the cells, because this can be done. Then we shall have the answer.

Vinogradova: I do think it is necessary that we try to exchange approaches; *we* should try behavioural situations, and you, Dr Ranck and Dr O'Keefe, should test sensory stimuli more rigorously, because we can surely have a double approach to the brain, as a region which controls behaviour, and also as an information-processing system. So far, your approach has been more traditionally behaviouristic (the response side of the stimulus-response sequence) and ours has been more on the stimulus side and concerned with information processing.

MECHANISM OF THETA ACTIVITY

Azmitia: The understanding of the mechanism of theta requires that we begin to isolate those anatomical regions which combine to generate this electrical phenomenon. First, one should consider that theta can be produced from stimulation as far away as the midbrain, and the hypothalamus. Second, theta is most effectively driven by stimulation of the medial septal nucleus, and unit discharge in the system occurs synchronously with hippocampal theta activity. Finally, the septal driving of theta is completely disrupted by lesions in the dorsal part of the fimbria-fornix (as shown by Gray *et al.*, this volume; Rawlins 1977). This may encompass the anatomical circuit by which midbrain, medial septum, and theta may be harmonized (Fig. 1). Two possibilities are the dorsal fornix (fornix superior) and the dorsal part of the fimbria fornicis. It is interesting that the dorsal fornix (fornix superior) pathway was shown by Valenstein & Nauta (1959) to carry cingular and subicular short fibres to the medial septal nucleus in all specimens examined

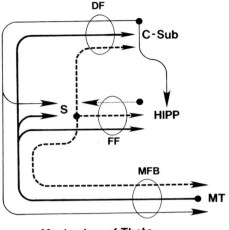

Mechanism of Theta

FIG. 1. (Azmitia). Schematic diagram of some of the main connections thought to be associated with the generation of theta rhythm in the hippocampus. C-Sub, cingulate and subicular cortex; DF, dorsal fornix (fornix superior); FF, fimbria fornicis; MT, mesencephalic tegmentum; MFB, medial forebrain bundle; S, septum.

and long fibres to the midbrain tegmentum, most prominent in the guinea pig and cat (see also Nauta 1958). Likewise both the midbrain tegmentum (i.e. serotonin fibres) and the medial septal nucleus project through the dorsal fornix to reach the subiculum and cingulate cortex. The projections from these cortical areas to the granule cells of the dentate gyrus have been previously described (Domesick 1969) and have recently been demonstrated autoradiographically in my laboratory. Thus a bidirectional route from the midbrain to the dentate gyrus exists which may be disrupted by a small localized lesion of the dorsal fornix.

Gray: It is correct that the lesion which, in our laboratory, eliminates theta and destroys an area in the dorso-medial part of the fornix (see Fig. 2, p. 280) encroaches on the fornix superior (König & Klippel 1963), thus disrupting the fibre systems of which Dr Azmitia speaks. However, we have two reasons for believing that the critical damage as regards theta is to fibres which originate in the medial septal area and travel towards the hippocampus. First, there is the general evidence (Stumpf 1965; James *et al.* 1977; Gray 1971; Ball & Gray 1971) that the medial septal area contains the pacemaker cells for theta—that is, the final common pathway which provides the hippocampus with the rhythmic input necessary for the normal production of this waveform. Second, we find (Rawlins 1977) theta to be still present in rats which sustain a more posterior lesion to the dorso-medial fornix than that illustrated in

Fig. 2 of our paper. This lesion damages the fornix superior, and thus causes essentially the same disruption of long fibre systems as the lesion which is effective in eliminating theta. We suppose that it nonetheless spares theta because septal afferents are able to reach the hippocampus anterior to the lesion site.

Azmitia: I agree that the medial septum is the major final cellular group to project to the hippocampal theta cells, and may serve as a 'pacemaker'. However, a mechanism is, by definition, a system of mutually adapted parts working together. Thus electrical flow may have several synaptic interruptions while still constituting a single mechanism.

Andersen: It was not the dorsal fornix, in fact, in Nauta's study. The fibres destroyed were different ones, just lateral to the dorsal fornix; between the dorsal fornix fibres and the fimbria is a narrow gate through which the theta-inducing fibres go. As soon as they enter the hippocampal formation they fan out and distribute to wide areas.

Azmitia: It seems the problem may be one of terminology. In a recent study by Meibach & Siegel (1977) the dorsal fornix is designated as those fibres which constitute the dorsolateral component of the fimbria fornicis. However, the dorsal fornix (fornix superior) was clearly differentiated from the fimbria fornicis by Valenstein & Nauta (1959). They equated the dorsal fornix tract to the fornix longus of Forel described by Cajal (Ramon y Cajal 1911), as those fibres which travel inferior to the corpus callosum and superior to the dorsal psalterium. This tract, which joins with the fimbria fornicis in the septum to form the fornix column, contained all the long fibres in the guinea pig which projected monosynaptically to the midbrain tegmentum. In certain species, most notably the cat and monkey, these long fibres were absent, and the connection to the midbrain tegmentum was at least disynaptically organized—the important point being that the fornix superior fibres are anatomically linked with the medial septum and the midbrain tegmentum as well as with the cingulum, subiculum and hippocampus.

Thus, it seems that the theta-driving pathway is not the fornix superior as described by Nauta but the dorsal fornix as described by Meibach & Siegel.

FUNCTIONAL SIGNIFICANCE OF THETA ACTIVITY

Weiskrantz: An important further question about theta activity concerns its functional significance and the question of causation or correlation—the fact that in the absence of theta there are many of the same behaviours that in the intact animal are associated with theta. What do we think theta means and how necessary is it for the integrity of different kinds of behaviour?

Vanderwolf: This is a question that we can't begin to answer without thinking about what the function of the whole forebrain is. It isn't just a question of what the hippocampus does. A rat without a forebrain (Woods 1964) walks perfectly well, face-washes, bites and so on and can perform most actions, whether they are normally accompanied by theta or not. This is true also, as I have seen myself (Vanderwolf *et al.* 1978), in rats from which the entire neocortex and hippocampus and a lot of the piriform cortex have been removed. The only way to understand this is by adopting a principle from Hughlings Jackson (Taylor 1958) and thinking in terms of different levels of function. There are probably output circuits which involve brainstem, cerebellum and spinal cord and which coordinate the activity of spinal motor neurons to produce stepping and keep the animals upright. These output circuits must be acted on by other circuits involving the neocortex or the hippocampus, and so on, as well as by the sensory inputs that influence behaviour. The forebrain then has to activate or suppress these lower-level output circuits. The specific role of the hippocampus in that, as well as the specific role of the neocortex, is something I do not pretend to understand.

The relations of slow wave activity to behaviour that I discussed in my paper occur throughout the hippocampus, but also in the neocortex. Our data show that by the use of atropine, but other drugs do this too. Schwartzbaum (Pond & Schwartzbaum 1972; Schwartzbaum & Kreinick 1973) showed that neocortical evoked potentials to a sensory stimulus vary in relation to behaviour, and Racine *et al.* (1975) showed the same thing for the trans-callosal, monosynaptic evoked potential. These types of potentials, electrically or sensorily induced, are modulated in relation to behaviour in the same way as theta waves occur in relation to behaviour. When the rat walks, the evoked potential is smaller than it is when the rat is standing still or washing its face.

There are also data on units in the reticular formation which have this kind of relation to behaviour (Malmo & Malmo 1976; Schwartzbaum 1975; Siegel & McGinty 1977; Vertes 1977). At least one component of the reticular formation is active in a phasic way in relation to overt behaviours like walking, and not in relation to behaviours like standing still or face-washing. This influence seems to be felt by the neocortex, the hippocampus and also by the caudate nucleus. It is a widespread influence. What it does, I don't know, but it seems significant that a lot of the ascending reticular activity influences the forebrain in a way which has a definite relation to on-going behaviour.

Weiskrantz: There is still the question of causal mechanisms; for what is theta necessary and of what, at the other extreme, might it be merely an epiphenomenon?

Vinogradova: From my point of view there is no mystery about the theta rhythm; just like other brain rhythms, such as alpha, beta and delta rhythms, theta activity simply reflects a certain level of arousal in the brain—a certain level of brain activity. In this way, as was stated already in initial studies of the functional significance of theta (e.g. Green & Arduini 1954) it may be regarded as an indication of arousal—that is, of a rather high optimal level of active brain state. This is a source of controversy over the functional significance of the theta rhythm, because a certain optimal level of activation is necessary for voluntary movements, for the motivational state, for attention, for learning, for detecting novelty, and so on. That is why it is a common factor which enters into all these types of activity. There *is* a certain mystery about the theta rhythm when we come to the hippocampus, because this kind of activity seems to be specially arranged for the hippocampus; it is specifically sent to the hippocampus from the medial septal nucleus, and so it should have a certain teleological significance, in Granit's sense of the word (Granit 1972).

Here I will propose an idea of my own. If we regard the hippocampus as a comparing device, which works on the basis of interaction between the two inputs coming in from opposite sides, scanning all this structure on the way, is it not possible to regard the septum as a synchronizing device for the comparator? It is known that septal influences are widespread; they influence the whole dendritic system of the pyramidal cell, as well as (possibly) basket cells. This means that a synchronized septal burst can influence the state of a pyramidal cell as a whole, so that any other kind of signal, from the perforant path or any other source, can interact, can influence the cell, can be compared, only if it comes in a very strict phase or time relation to the septal burst. Otherwise it will just miss the possibility of influencing a cell. Such a device is used in the comparator systems of computers. One can conceive that such a synchronizing device would considerably increase the precision of action of a match-mismatch system and allow it to compare only related events in both inputs.

Gray: Are you saying that the synchronization is so that two inputs onto one neuron can be compared, or that synchronization brings together the activities of different neurons or different sub-fields of the hippocampus?

Vinogradova: I am trying to reason in regard to events in a single neuron, although, of course, the process is synchronous for big neuronal populations, and can also be important for interaction between hippocampal fields, dentate, subiculum, neocortex, and so on.

Koella: When we talk about the functional significance of theta waves, can we try to put the idea of 'synchronization' into somewhat more physiological terms? Synchronization of EEG waves actually suggests highly synchronized discharge of cells in the area picked up by our macro electrodes. This means that we have a good deal of loss of gradients between the units in this area; this again means loss of information content as well as information transfer. So how can we imagine, particularly in states where we would like the hippocampus to do a great deal of information processing, that at least some of the hippocampal cells are in a state where they contain and transfer very little information? Do we have to conclude from this that possibly theta waves (and I agree that theta waves may be a manifestation of interneurons) act to set a state, perhaps somewhat similar to what Olga Vinogradova says, for the remaining neurons, so that at one particular time they are in a similar state of excitability, so that they can handle incoming information in an equalized way?

Black: I doubt that theta waves are related to information processing—at least in any simple way. Some years ago we recorded the theta waves of dogs in an experiment in which they were required to process information while moving and also required to process information while holding still (Black & Young 1972). In the presence of one discriminative stimulus (SD), a given dog had to press a lever rapidly for a period of time in order to avoid shock; in the presence of a second SD the same dog had to stay immobile in order to avoid shock. Clearly, the dog had to distinguish between these two stimuli, process information about them, and select the appropriate response. If theta was related to information processing, one would expect it to occur in the presence of both SD's, but it did not. It only occurred in the presence of the SD for which the appropriate response was movement. Data such as these make it difficult for me to think of theta as a correlate of information processing.

Vanderwolf: Or as a correlate of arousal states?

Black: The dogs were aroused during both SD's; yet theta occurred only in the presence of one. Therefore, it seems unlikely that theta is a simple correlate of arousal.

References

ANDERSEN, P., ECCLES, J. C. & LOYNING, Y. (1964*a*) Location of postsynaptic inhibitory synapses on hippocampal pyramids. *J. Neurophysiol.* 27, 592–607

ANDERSEN, P., ECCLES, J. C. & LOYNING, Y. (1964*b*) Pathway of postsynaptic inhibition in the hippocampus. *J. Neurophysiol.* 27, 608–619

ANDERSEN, P., GROSS, G. N., LØMO, T. & SVEEN, O. (1969) Participation of inhibitory and excitatory interneurones in the control of hippocampal cortical output, in *The Interneuron* (Brazier, M. A. B., ed.), pp. 415–465, University of California Press, Berkeley & Los Angeles

BALL, G. G. & GRAY, J. A. (1971) Septal self-stimulation and hippocampal activity. *Physiol. Behav.* 6, 547–549

BERGER, T. W. & THOMPSON, R. F. (1977) Limbic system interrelations: functional divisions among hippocampal-septal connections. *Science (Wash. D.C.) 197*, 587–589

BERGER, T. W., ALGER, B. & THOMPSON, R. F. (1976) Neuronal substrate of classical conditioning in the hippocampus. *Science (Wash. D.C.) 192*, 483–485

BEST, P. J. & RANCK, J. B., JR (1975) Reliability of the relationship between hippocampal unit activity and behavior in the rat. *Abstr. Soc. Neurosci.* (abstr. 837)

BLACK, A. H. & YOUNG, G. A. (1972) Electrical activity of the hippocampus and cortex in dogs operantly trained to move and to hold still. *J. Comp. Physiol. Psychol. 79*, 128–141

BLAND, B. H. & WHISHAW, I. Q. (1976) Generators and topography of hippocampal theta in the anaesthetized and freely moving rat. *Brain Res. 118*, 259–280

BUREŠ, J. (1965) in *Anatomy of Memory* (Kimble, D. P., ed.), pp. 49–52, Science and Behavior Books, Palo Alto, California

DOMESICK, V. B. (1969) Projections from the cingulate cortex in the rat. *Brain Res. 12*, 296–320

FEDER, R. & RANCK, J. B., JR (1973) Studies on single neurons in dorsal hippocampal formation and septum in unrestrained rats. II. Hippocampal slow waves and theta cell firing during bar pressing and other behaviors. *Exp. Neurol. 41*, 532–555

FOX, S. E. & RANCK, J. B., JR (1975) Localization and anatomical identification of theta cells and complex spike cells in dorsal hippocampal formation of rats. *Exp. Neurol. 49*, 299–313

FOX, S. E. & RANCK, J. B., JR (1977) Hippocampal complex-spike and theta cell activity evoked by stimulation of limbic structures in unrestrained rats. *Abstr. Soc. Neurosci.* (abstr. 613)

FUJITA, Y. & SATO, T. (1964) Intracellular records from hippocampal pyramidal cells in rabbit during theta rhythm activity. *J. Neurophysiol. 27*, 1011–1025

GRANIT, R. (1972) In defence of teleology, in *Brain and Human Behavior* (Karczmar, A. G. & Eccles, J. C., eds.), pp. 400–408, Springer-Verlag, Berlin

GRAY, J. A. (1971) Medial septal lesions, hippocampal theta rhythm and the control of vibrissal movement in the freely moving rat. *Electroencephalogr. Clin. Neurophysiol. 30*, 189–197

GREEN, J. D. & ARDUINI, A. A. (1954) Hippocampal electrical activity in arousal. *J. Neurophysiol. 17*, 532–546

JAMES, D. T. D., MCNAUGHTON, N., RAWLINS, J. N. P., FELDON, J. & GRAY, J. A. (1977) Septal driving of the hippocampal theta rhythm as a function of frequency in the free-moving male rat. *Neuroscience 2*, 1007–1017

KÖNIG, J. F. R. & KLIPPEL, R. A. (1963) *The Rat Brain*, Williams & Wilkins, Baltimore

MALMO, H. P. & MALMO, R. B. (1976) Movement-related forebrain and midbrain multiple unit activity in rats. *Electroencephalogr. Clin. Neurophysiol. 42*, 501–509

MAYS, L. E. & BEST, P. J. (1975) Hippocampal unit activity to tonal stimuli during arousal from sleep and in awake rats. *Exp. Neurol. 47*, 268–279

MEIBACH, R. C. & SIEGEL, A. (1977) Efferent connections of the hippocampal formation in the rat. *Brain Res. 124*, 197–224

MITCHELL, S. J. & RANCK, J. B., JR (1977) Firing patterns and behavioral correlates of neurons in entorhinal cortex of freely moving rats. *Abstr. Soc. Neurosci.* (abstr. 631)

NAUTA, W. J. H. (1958) Hippocampal projections and related neural pathways to the midbrain in the cat. *Brain 81*, 319–340

O'KEEFE, J. (1976) Place units in the hippocampus of the freely moving rat. *Exp. Neurol. 51*, 78–109

OLTON, D. S., BRANCH, M. & BEST, P. (1978) Spatial correlates of hippocampal unit activity. *Exp. Neurol. 58*, 387–409

POND, F. J. & SCHWARTZBAUM, J. S. (1972) Interrelationships of hippocampal EEG and visual evoked responses during appetitive behavior in rats. *Brain Res. 43*, 119–137

RACINE, R., TUFF, L. & ZAIDE, J. (1975) Kindling, unit discharge patterns and neural plasticity. *Can. J. Neurol. Sci. 2*, 395–405

RAMON Y CAJAL, S. (1911) *Histologie du Système Nerveux de l'Homme et des Vertèbres*, vol. II, p. 183, Maloine, Paris

RANCK, J. B., JR (1973a) A movable electrode for recording from single neurons in unrestrained rats, in *Brain Unit Activity During Behavior* (Phillips, M. I., ed.), pp. 76-79, University of Iowa Press, Iowa City

RANCK, J. B., JR (1973b) Studies on single neurons in dorsal hippocampal formation and septum in unrestrained rats. I. Behavioral correlates and firing repertoires. *Exp. Neurol.* 41, 461-531

RANCK, J. B., JR (1975) Behavioral correlates and firing repertoires of neurons in dorsal hippocampal formation and septum of unrestrained rats, in *The Hippocampus*, vol. 2: *Neurophysiology and Behavior* (Isaacson, R. L. & Pribram, K. H., eds.), pp. 207-244, Plenum Press, New York

RANCK, J. B., JR (1976) Behavioral correlates and firing repertoires of neurons in the septal nuclei in unrestrained rats, in *The Septal Nuclei* (DeFrance, J., ed.), pp. 423-462, Plenum Press, New York

RAWLINS, J. N. P. (1977) *Behavioural and Physiological Correlates of Limbic System Activity*. Unpublished D. Phil. Thesis, Oxford University

SCHWARTZBAUM, J. S. (1975) Interrelationship among multiunit activity of the midbrain reticular formation and lateral geniculate nucleus, thalamocortical arousal, and behavior in rats. *J. Comp. Physiol. Psychol.* 89, 131-157

SCHWARTZBAUM, J. S. & KREINICK, C. J. (1973) Interrelationships of hippocampal electroencephalogram, visually evoked response, and behavioral reactivity to photic stimuli in rats. *J. Comp. Physiol. Psychol.* 85, 479-490

SEGAL, M. & OLDS, J. (1972) Behavior of units in hippocampal circuit of the rat during learning. *J. Neurophysiol.* 35, 680-690

SIEGEL, J. M. & MCGINTY, D. J. (1977) Pontine reticular formation neurons: relationship of discharge to motor activity. *Science (Wash. D.C.)* 196, 678-680

STUMPF, C. (1965) Drug action on the electrical activity of the hippocampus. *Int. Rev. Neurobiol.* 8, 77-138

TAYLOR, J. (ed.) (1958) *Selected Writings of John Hughlings Jackson*, London, Staples Press

THOMPSON, R. F., BERGER, T. W., CEGAVSKE, C. F., PATTERSON, M. M., ROEMER, R. A., TEYLER, T. J. & YOUNG, R. A. (1976) The search for the engram. *Am. Psychol.* 31, 209-227

VALENSTEIN, E. S. & NAUTA, W. J. H. (1959) A comparison of the distribution of the fornix system in the rat, guinea pig, cat, and monkey. *J. Comp. Neurol.* 113, 337-363

VANDERWOLF, C. H., KOLB, B. & COOLEY, R. K. (1978) The behavior of the rat following removal of the neocortex and hippocampal formation. *J. Comp. Physiol. Psychol.*, in press

VERTES, R. P. (1977) Selective firing of rat pontine gigantocellular neurons during movement and REM sleep. *Brain Res.* 128, 146-152

WOODS, J. W. (1964) Behavior of chronic decerebrate rats. *J. Neurophysiol.* 27, 635-644

YOSHII, N. & OGURA, H. (1960) Studies on the unit discharge of brainstem reticular formation in the cat. I. Changes of reticular unit discharge following conditioning procedure. *Med. J. Osaka Univ.* 11, 1-17

The function of septo-hippocampal connections in spatially organized behaviour

DAVID S. OLTON

Department of Psychology, The Johns Hopkins University, Baltimore

Abstract The role of septo-hippocampal connections in spatial behaviour is examined in lesion and stimulation experiments in rats. Destruction of septo-hippocampal connections produces a severe and enduring deficit in the ability to perform a spatial memory task. Furthermore, crossed unilateral lesions of the entorhinal area and fimbria–fornix produce the same deficit as bilateral lesions in either the entorhinal area or the fimbria–fornix, as predicted from a disconnection analysis of hippocampal lesions. Evidence also suggests that septo-hippocampal damage produces a greater deficit in spatial behaviours when a flexible response is required than when a consistent response is required, although this difference is a relative one rather than an absolute one. Finally, disruptive stimulation of the hippocampus while rats are performing a spatial memory task produces retrograde amnesia but not proactive interference. Taken together, these data indicate an important role of the septo-hippocampal system in spatial behaviour, and in the maintenance of spatial memories.

The experiments I describe here all test rats in a radial arm maze. This testing procedure has several important characteristics. First, it tests memory for a list of items. List-learning experiments in humans have generated many of our concepts about memory and the development of an animal analogue provides the opportunity to characterize animal memory in terms of these same concepts. Second, the task is learned rapidly and performed well. Our experiments were designed to examine the brain mechanisms underlying long-term stable performance (rather than acquisition) and we wanted a test in which acquisition was relatively simple and performance was consistently good for extended periods of time. Third, the task is flexible and with small modifications can address a number of different issues. The fewer the variables that differ between tasks, the easier is the identification of the particular components responsible for alterations in behaviour. In the radial arm maze, we have brought together different physiological techniques as well as different psychological tests to examine the functional role of the septo-hippocampal system in spatial behaviour. Finally, the radial arm maze test evokes behaviour that appears to be relatively natural and other experiments in our laboratory demonstrate that the choice behaviour of rats in this task is very similar to their food-searching behaviour in semi-natural environments.

Our initial experiments demonstrated that rats learned rapidly and performed well in this task. After approximately ten tests (one test per day), they consistently made at least seven correct responses in the first eight choices and with continued training performed almost perfectly (Olton & Samuelson 1976). Subsequent experiments examined the possible role of response sequences (such as choosing adjacent arms). Rats were confined to the centre platform between choices (Olton *et al.* 1977) and were forced to choose arms in a particular order (Olton & Schlosberg 1978). Neither of these procedures had any influence on choice accuracy. Another set of experiments examined the possible role of intra-maze cues such as the odour or sight of food, odour 'trails' left by the rats as they moved down the arm, or cues characteristic of the arms themselves. Here, the maze was rotated between choices so that unchosen arms were placed in chosen spatial locations (and chosen arms in unchosen spatial locations). Rats rewarded for choosing particular spatial locations were generally unaffected by this manipulation (Olton & Samuelson 1976; Olton *et al.* 1977), while rats rewarded for choosing particular arms were unable to perform above chance (D. S. Olton & C. Collison, unpublished observations). Taken together, these results demonstrate that rats do not use response sequences or intra-maze cues to solve the radial arm maze task. Rather, the rats identify each arm on the basis of extra-maze spatial stimuli and remember which arms have been chosen by developing a list of the relevant spatial locations. The memory of this list of spatial locations is characterized by limited capacity, proactive interference, little or no decay, no serial order effect, and resetting (Olton 1978; Olton *et al.* 1977; Olton & Samuelson 1976).

An important point about the radial arm maze procedure is that it is a test of spatial memory, and not of cognitive maps in the sense that O'Keefe and Black (this volume) use the term. The distinction between the use of a particular stimulus as an isolated cue on the one hand, and as a component of a map on the other, is of primary importance both in the design of experiments and in the interpretation of results (O'Keefe & Nadel 1978). Rats may develop a 'map' of the environment when they solve the radial arm maze task, but they do not have to do so, and we have not done the relevant experiments which would answer this question. Consequently, the radial arm maze may be considered as a test of spatial memory but at least for the present may not be considered as a test of cognitive maps.

BEHAVIOURAL EFFECTS OF BILATERAL LESIONS

A series of experiments have demonstrated the importance of the septo-hippocampal system for normal performance in the radial arm maze task (Olton *et al.* 1978*a*). Rats were first trained preoperatively to a criterion of at least seven correct responses in the first eight choices for five consecutive days. This preoperative training–postoperative retention procedure has several advantages. First, it reduces variability because all animals are brought to the same level of performance before surgery. Second, it allows a within-subject comparison of performance with each animal being used as his own control. Third, it ensures that any deficit which follows the lesion must be a substantial one because the animals learn all the information necessary to solve the task before the operation. Surgery was then carried out, with lesions being placed in a variety of septo-hippocampal structures. Two periods of postoperative testing took place. One began within five days of surgery to determine the acute effects of the lesion. The other began fifty days later to determine the extent of behavioural recovery associated with the extensive neuroanatomical changes that follow lesions in the septo-hippocampal system. After this testing, the rats were killed and their brains stained for cell bodies and myelinated fibres at the site of the lesion, and for acetylcholinesterase in the hippocampus.

The results of this experiment are presented in Fig. 1. The ordinate is the probability of a correct response, corrected for chance and presented as a fraction of maximum performance. A score of 0 indicates chance performance. The more positive the score, the greater choice accuracy, with a score of 1.00 reflecting perfect performance. A negative score indicates performance worse than chance; that is, the rat selectively chose incorrect rather than correct arms. The abscissa presents data from choices two through eight. Data for the first choice are not presented because all arms contained food on that choice and any choice had to be correct; data after the eighth choice are not presented because control rats often performed perfectly and there were no data available for analysis.

As can be seen from this figure, rats with control operations and posterior neocortical operations generally performed well after surgery and exhibited the slight decline in choice accuracy across choices that is characteristic of performance in this task (Olton 1978). In contrast, rats with lesions in the

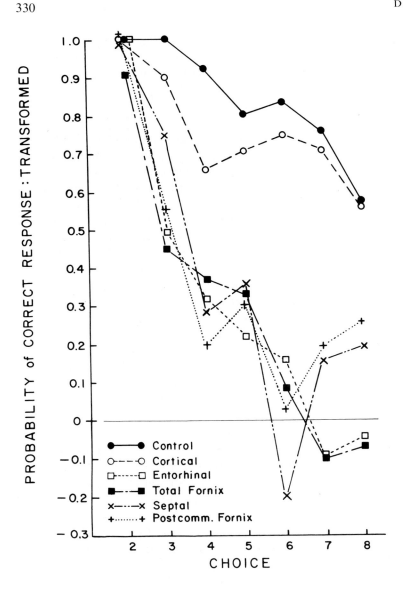

FIG. 1. Postoperative performance in the radial arm maze. The ordinate presents the probability of a correct response, transformed to a fraction of the maximum performance score. A score of 0 is chance. A score greater than 0 indicates choice accuracy above chance with a score of 1.00 reflecting perfect performance. A score of less than 0 indicates performance worse than chance; the rats were actively choosing incorrect rather than correct arms.

septo-hippocampal system performed poorly. After the third choice, accuracy decreased to chance levels and remained there throughout the test. There was virtually no difference in the performance of rats with lesions in the septum, fimbria-fornix, postcommissural fornix, or entorhinal area. (Histological details are presented in Olton *et al.* 1978*a*.) Finally, there was no evidence of recovery of function. The performance of rats tested 50 days after surgery was virtually the same as the performance of rats tested five days after surgery.

This lack of recovery in a preoperative acquisition–postoperative retention paradigm is important because it demonstrates that an intact septo-hippocampal system is necessary for normal performance. I would like to suggest a distinction between two types of behavioural changes following septo-hippocampal damage. One I would call 'relative', indicating that the performance of animals with lesions is different from that of normals at the beginning of training but becomes indistinguishable from that of normals at the end of training. The changes in avoidance behaviour after septo-hippocampal lesions (Grossman, this volume; Olton 1973) are an example of this type of short-term change. Another type of change after lesions I would call 'absolute', indicating that the performance of animals with lesions is permanently altered in spite of substantial postoperative recovery time and testing experience. The deficits in radial arm maze performance and spontaneous alternation (Johnson *et al.* 1977) are examples of this type of long-term change. The distinction between short-term and long-term changes in performance is important because these results have different implications for brain function. Relative changes after lesions imply that the brain area destroyed is normally used in the behaviour but that other brain systems can compensate for its absence. Absolute changes imply that the brain area destroyed is a prerequisite for normal behaviour and that the rest of the system cannot compensate for its absence. The absolute changes after lesions also seem to make a stronger statement about the functional importance of the brain structures in question and should probably be given more weight in any attempt to provide a coherent summary of the role of the septo-hippocampal system in behaviour.

Rats with septo-hippocampal damage exhibited a number of changes in general behaviour on the radial arm maze. First, they ran faster than normals. Second, they chose arms relatively far apart and gave the impression that their speed of running led to ballistic tendencies which made sharp turns difficult for them. Third, in a free-choice situation they had a tendency to fall into perseverative patterns of responding, repeating the same sequence of choices (such as arms 1, 5 and 7) many times (Olton *et al.* 1978*a*). Fourth, when confined to the centre of the apparatus between choices (by guillotine doors which block access to the arms), they generally ran into the closest arm when the guillotine doors were raised. Normal rats usually paused at the entrance to an arm when the doors were raised and entered only if the arm was correct (Olton *et al.* 1977). In contrast, rats with septo-hippocampal lesions rarely paused and the probability of entering the arm by which they were standing was very high and not different for correct and incorrect arms. Response changes similar to these have been observed by others in both mazes and open field tasks, and any theory of septo-hippocampal function must explain the appearance of these marked and consistent patterns.

One other point should be made about the effect of bilateral lesions on radial arm maze performance. There is a strong dissociation between the effects of lesions in the septo-hippocampal system and lesions in other brain areas. We are now testing rats with bilateral damage to the caudate nucleus, posterolateral neocortex, dorsomedial thalamus, dorsomedial frontal cortex, and sulcal cortex. None of these lesions have any marked effect on performance. Grossman (this volume) and Ursin (this volume) have both discussed the importance of dissociation within the septo-hippocampal system, but an equally important dissociation is between this system and other brain areas. Our results demonstrate that rats with damage to a variety of other brain areas can do the radial arm maze task quite well, demonstrating a selective involvement of the septo-hippocampal system in this behaviour.

BEHAVIOURAL EFFECTS OF CROSSED LESIONS

Lesions of the fimbria–fornix and the entorhinal area both produced a substantial and enduring deficit in the radial arm maze task, suggesting that these septo-hippocampal components might function as an integrated system which is required to support normal choice behaviour. An explicit way of testing this notion is to use the crossed-lesion design of Mishkin (1958, 1966). Here, unilateral lesions are placed in different structures on opposite sides of the brain and the commissures which interconnect the relevant structures are cut. If the different structures are both involved in an integrated system, then the crossed-unilateral lesions in different structures ought to produce the same behavioural effect as a bilateral lesion in either structure.

The application of the crossed-lesion approach to limbic function has been discussed by Geschwind (1965), and this approach has been applied by Horel & Keating (1969, 1972). These authors argue that the limbic system forms a major interconnecting path between cortical and subcortical structures and that all sets of lesions which produce a cortical-subcortical disconnection ought to have the same behavioural effect.

Can the functional organization of the septo-hippocampal system in rats be conceptualized in the same manner as Horel and Keating have suggested for the temporal lobe system in monkeys? Anatomically and electrophysiologically, the hippocampus forms a major link interconnecting cortical areas through its entorhinal connections and subcortical areas through its septal connections. Behaviourally, the question can be addressed by a series of lesions in the entorhinal area and the fimbria–fornix. The general design of the experiment is outlined in Fig. 2. Each diagram in the figure is a schematic of the septo-hippocampal–entorhinal system as viewed from above the brain. The septum is the round structure in dotted lines at the top of the diagram, the hippocampus is the curved structure in solid lines in the middle, and the entorhinal area is the dotted structure at the bottom. For this experiment, lesions were placed either in the fimbria–fornix just posterior to the septum, or in the entorhinal area, just posterior to the hippocampus. A fimbria–fornix lesion disrupted both the subcortical projections of the hippocampal system through the septum and the commissural projections from the hippocampal system on one side of the brain to the hippocampal system on the other side that traverse the ventral psalterium. An entorhinal lesion disrupted the

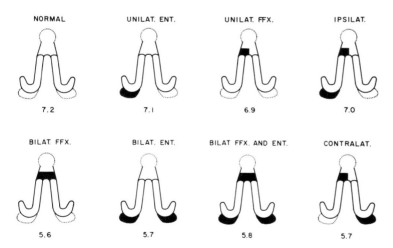

FIG. 2. A summary of the design and results of the disconnection analysis. Each diagram is a schematic of the septo-hippocampal system as seen from above the brain. The septum is represented by the round structure outlined in dotted lines at the top, the hippocampus by the elongated structure outlined by solid lines in the middle, and the entorhinal area by the small structure outlined by dotted lines at the bottom. A lesion is represented by a blackened area. The mean number of correct responses in the first eight choices is presented under each diagram for the animals in that group. Further explanation is given in the text.

cortical projections of the hippocampal system, cutting both the perforant path and the angular bundle.

We made the following predictions about the behavioural results to be expected after different sets of these lesions. First, any set of unilateral lesions on the same side of the brain should have no behavioural effect. A unilateral lesion of the entorhinal area (top row, second diagram), of the fimbria–fornix (top row, third diagram), or of both fimbria–fornix and the entorhinal area (top row, fourth diagram) should result in normal behaviour because the contralateral septo-hippocampal system is still able to make its cortical-subcortical connections in the normal fashion. Second, any set of bilateral lesions should produce an equivalent deficit in performance. Bilateral lesions of the fimbria–fornix (bottom row, first diagram), entorhinal area (bottom row, second diagram), or both the fimbria–fornix and entorhinal area (bottom row, third diagram) all produce the same functional disconnection of cortical and subcortical areas and should have the same behavioural effects. Third, unilateral lesions of the fimbria–fornix and the entorhinal area, when they are contralateral to each other (bottom row, fourth diagram), should have the same behavioural effect as bilateral lesions; cortical information is unable to get into the hippocampal system on the side with the entorhinal lesion, and subcortical information is unable to get into the hippocampal system on the side with the fimbria–fornix lesion. Since the fimbria–fornix lesion also cuts the commissural fibres in the ventral psalterium, there is no pathway for the hippocampal system which has the cortical information (on the side with an intact entorhinal area) to communicate with the hippocampal system that has the subcortical information (on the side with an intact fimbria–fornix). This set of unilateral lesions produces a functional disconnection that is equivalent to that following a bilateral lesion.

Rats were trained preoperatively to a criterion of seven correct responses in the first eight choices. After criterion performance, each rat was given a lesion of either the fimbria–fornix or the entorhinal area. After a five-day recovery period, each rat received ten more tests. A second lesion was made in the structure not destroyed the first time (fimbria–fornix in animals which had received an entorhinal lesion previously, entorhinal area in animals which had received a fimbria–fornix lesion previously). After a five-day recovery period, each rat was given ten more tests. The animals were killed and the brain was cut in half so that coronal sections could be taken throughout the area of the fimbria–fornix lesion and anterodorsal hippocampus and horizontal sections throughout the area of the entorhinal lesion and postero-ventral hippocampus. Adjacent sections were stained with luxol fast blue and cresyl violet for myelinated fibre tracts and cell bodies, respectively, and for acetylcholinesterase. The luxol fast blue–cresyl violet stain allowed assessment of the lesion at the site of damage, while the acetylcholinesterase stain allowed assessment of the lesion at the hippocampus.

Histological data indicated that the lesions were successful. The fimbria–fornix lesions destroyed all of the fimbria–fornix on the side of the lesion except the most lateral tips and when the lesion was unilateral it spared the opposite side. Acetylcholinesterase in the hippocampus on the side of the lesion was eliminated, while that on the intact side was normal. The entorhinal lesions destroyed the medial and lateral entorhinal area and some of the adjacent subiculum; the dentate gyrus and the hippocampus proper were generally intact. Acetylcholinesterase in the hippocampus was intensified in a band superficial to the dentate granule cells. Thus both lesions were effective in disrupting the intended connections. (Examples of similar lesions may be found in Olton et al. 1978a.)

Fig. 2 summarizes the behavioural data for each group of animals under each diagram in terms of the mean number of correct responses in the first eight choices. All rats had attained criterion performance preoperatively and control animals (top row, first diagram) were unaffected by surgery, so a figure of 7.0 or better can be considered as normal performance. Let us consider the three predictions made by the disconnection analysis. First, all rats with a single unilateral lesion or with both unilateral lesions on the same side of the brain ought to perform normally. The relevant groups to test this prediction are the three operated ones on the top row, and as can be seen from Fig. 2 these rats performed essentially the same as controls. Second, all rats with a bilateral lesion ought to have an equivalent deficit. The relevant groups are the first three on the bottom row, all of which have a substantial impairment which is of approximately the same size. Third, rats with unilateral lesions which are contralateral to each other ought to perform the same as rats with bilateral lesions. The relevant group is the last one in the bottom row, which exhibits a deficit of the same magnitude as those of the groups with bilateral lesions.

What can we conclude from these data? First, we can eliminate two possible hypotheses. The first says that the septo-hippocampal system is organized so that normal behaviour can be supported by either an intact set of cortical connections or an intact set of subcortical connections. Since both the fimbria–fornix and the entorhinal area carry reciprocal connections to the hippocampal system, the hippocampus has a source of both input and output with either set of connections intact and could conceivably function to direct behaviour. But, at least in this task, there is no dissociation between the effects of bilateral fimbria–fornix and bilateral entorhinal lesions, so that this idea can be rejected. A second hypothesis says that the behavioural effects observed after either a bilateral fimbria–fornix lesion or a bilateral entorhinal lesion are due not to the hippocampal connection that is eliminated by the lesion, but rather to malfunctioning of the remaining intact normal brain. If such were the case, then complete isolation of the hippocampus through combined bilateral fimbria–fornix lesions and entorhinal lesions should have a different behavioural effect from either bilateral lesion alone. Again, such is not the case and this hypothesis can be rejected.

The data are clearly consistent with the disconnection notion, and are compatible with several different notions of cortical–subcortical interaction. On the one hand, the hippocampus may be acting as a throughput system, taking information from either the septum or entorhinal area, transforming it in some way, and passing it on to the next structure. Arguments can be made for both a cortical to subcortical system and a subcortical to cortical system (see Adey 1961) and the lesion data are not able to separate these alternatives. On the other hand, the hippocampus may be acting as an integrating system, comparing cortical and subcortical information, and determining its output on the basis of this comparison (Adey 1961; Douglas 1967; Vinogradova & Brazhnik, this volume). Since such a process would require simultaneous access to both cortical and subcortical information, interruption of either source of input would produce an impairment of behaviour. In either case, the picture is of an integrated septo-hippocampal system which requires simultaneous access to both cortical and subcortical areas in order to function normally.

HIPPOCAMPAL SUBFIELDS

In a series of experiments, Jarrard has examined the effects of lesions selectively disrupting the cell field CA1 and its projections through the alveus or the cell field CA3 and its projections through the fimbria (Jarrard 1976; Jarrard & Becker 1977). In general, CA3 and fimbria lesions produce behavioural changes that are similar to those found after total hippocampal damage while CA1 or alveus lesions have little effect. An important exception to this last statement is the acquisition of normal behaviour on the radial arm maze. In Jarrard's experiment, one group of rats was first trained to criterion performance on the radial arm maze, then given selective hippocampal lesions and tested for postoperative retention. In another group of rats, lesions were made before testing and the ability of the rats to learn the task was examined. CA3 lesions impaired both the acquisition and retention of the task. In contrast, CA1 lesions impaired only acquisition. These data imply that there is an important subdivision within the hippocampal system, with CA1 being necessary for the laying down of new memories but not for the retrieval of old ones, while CA3 is necessary for both (or at least for retrieval). These data suggest a means of relating the well-known anatomy and electrophysiology of the hippocampal system (see Andersen, this volume) to the data suggesting an important role of the hippocampal system in memory (see Weiskrantz, this volume).

BEHAVIOURAL FLEXIBILITY

In our usual testing procedure, the rat was rewarded only for the first choice of each arm. Consequently, he must be flexible in his responses to each arm. In each test, the first time the rat sees an arm he ought to run down it to get the food pellet, while every other time he sees an arm he ought not to run down it because there is no longer any food there.

Several experiments suggest that septo-hippocampal damage produces a selective deficit in tasks which require behavioural flexibility. For example, septo-hippocampal lesions produce a long-term deficit in spontaneous alternation (Douglas 1967; Johnson et al. 1977), a task which requires the rat to alternate his responses between the two arms of a T-maze. In contrast, similar lesions seem to have no effect on the acquisition of a simultaneous visual discrimination (Kimble 1963), a task in which the response to each stimulus is the same on every test. Likewise, rats with septo-hippocampal damage have extreme difficulty with a Differential Reinforcement of Low Response Rates (DRL) schedule (Schmaltz & Isaacson 1966; Johnson et al. 1977), an operant task which requires them first to press a bar to obtain food reward and then not to press the bar for a period of time following that reward. Rats with similar lesions, however, have no difficulty learning to obtain reward on a Continuous Reinforcement (CRF) schedule (Schmaltz & Isaacson 1966), a task in which a bar press is always correct. Other data (Kimble 1975; Olds 1972; Steward et al. 1977) also support the notion that the hippocampus is involved in behavioural flexibility.

To investigate the possible role of the septo-hippocampal system in behavioural flexibility, we altered the radial arm maze procedure so that the rat was presented with two different sets of arms. In one set, one pellet of food was placed at the end of each arm. On these arms, which I will call 'baited' arms, the rat was rewarded only for the first choice of each arm, as in our usual testing procedure. In the other set of arms, food was never found. On these arms, which I will call 'unbaited' arms, the rat was never rewarded for making a choice. The baited arms clearly require a flexible response; when first presented with a baited arm the rat should choose it, and when subsequently presented with that same arm he should not choose it. In contrast, an unbaited arm does not require a flexible response; since the arm never has food on it, the correct response for the rat is always to avoid it.

For this test, we used a 17-arm maze (Olton et al. 1977). Eight of the arms were baited arms, the other nine were unbaited ones. The arms were haphazardly mixed and organized in several different patterns. Rats were first trained preoperatively and then tested postoperatively. We obtained two different measures of performance. In both, the scores were transformed so that a score of 0 indicates chance performance, and a score of 100 indicates perfect performance. The first measure describes the ability of the rat to discriminate between baited and unbaited arms and is the probability of choosing a baited arm irrespective of whether the choice was the first (rewarded) one or a repeated (unrewarded) one. The second measure describes the ability of the rats to remember which of the baited arms they have chosen and is the probability of a correct response, given a response to a baited arm. Our previous experiments suggest that rats with fimbria–fornix damage ought to be impaired on this second measure because it is the same one which is used to describe choice behaviour in the eight-arm maze. The question is how lesions affect performance on the first measure, the probability of choosing a baited arm.

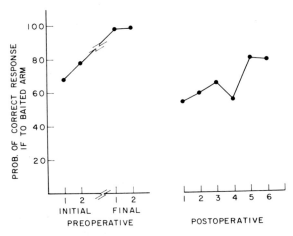

FIG. 3. The mean probability of a correct response given a response to a baited arm. This measure reflects the ability of rats to remember which of the baited arms had been chosen previously in that test. This measure is a percentage maximum performance score so that 0 reflects chance performance, 100 perfect performance.

If rats with fimbria-fornix damage have a deficit that is selective for flexible behaviours (rather than spatial discrimination *per se*), there should be no deficit on this measure. If, on the other hand, rats with fimbria-fornix damage have a more general deficit in spatial behaviours (independent of the behavioural flexibility dimension), there should be a deficit on this measure.

We currently have data from only three rats with successful fimbria-fornix lesions. Fig. 3 presents the probability of a correct response given a response to a baited arm, our usual measure of choice behaviour. Preoperatively, rats were relatively good at not repeating choices even during initial tests and with continued training came to perform almost perfectly. Postoperatively, there was a substantial impairment. During the first few days of postoperative testing, performance was actually worse than that during initial preoperative training and by the end of postoperative testing their performance was still worse than terminal preoperative levels. These data confirm the results from the eight-arm maze procedure demonstrating a substantial and enduring deficit in choice behaviour.

The interesting question, of course, is how rats performed with respect to the arms that never had food on them. Fig. 4 presents the probability of choosing a baited arm. Preoperatively, at the beginning of training, rats performed at approximately chance levels but by the end of training they had attained a stable high level of choice accuracy. Postoperatively, there was a substantial deficit for the first few test days. But initial postoperative

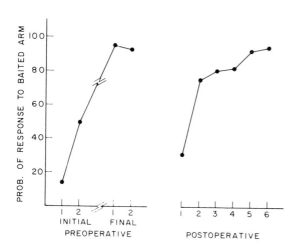

FIG. 4. The mean probability of choosing a baited arm. This measure reflects the ability of rats to respond appropriately to the unbaited arms and never choose them. This measure is a percentage maximum performance score; 0 reflects chance performance, 100 perfect performance.

performance was still above initial preoperative levels and the rats quickly recovered, so that within about 20 tests postoperative choice accuracy was approximately the same as terminal preoperative choice accuracy.

Although the predictions for this experiment are clear, the results unfortunately are somewhat mixed. On the one hand, there certainly was a deficit in discriminating among the baited and unbaited arms and rats made many errors postoperatively by responding to unbaited arms. On the other hand, that deficit was short-lived and the rats quickly recovered to preoperative levels. In light of the few rats in this study, any conclusions must be guarded. Nonetheless, the data here, and those from previous experiments, suggest that there is at least some role of the septo-hippocampal system in flexible, short-term memory behaviours that is independent of its role in spatial discrimination, but that this distinction must be quantitative rather than qualitative.

SPATIAL MEMORY

The experiments with normal animals demonstrate that rats develop a list of spatial locations which they have chosen during each test. Both the lesion data reported here and unit recording studies reported elsewhere (Olton *et al.* 1978*b*) suggest an important role of the hippocampus in this memory, but are consistent with interpretations based on sensory factors as well. To more explicitly test the notion that the hippocampus plays a role in spatial memory, we employed a stimulation procedure during performance on the radial arm maze to determine if that stimulation would produce retrograde amnesia (Olds & Olds 1961).

For this experiment, rats had bilateral electrodes lowered into the hippocampus near CA1. They were then trained to perform on the eight-arm radial maze. Each rat was allowed to make four choices and was then confined to the centre of the apparatus for approximately five minutes. He was released and rewarded for choosing the four arms not chosen before confinement. On control days, no stimulation was given. On test days, the hippocampus was stimulated for two seconds with sufficient current to generate a short electrophysiological (but not behavioural) seizure. About ten after-discharges occurred, followed for several minutes by a depressed EEG which subsequently returned to normal with theta rhythm of the usual amplitude during movement. After the EEG was normal, the rat was released from confinement and allowed to choose among the arms.

We have data from four rats at the present time, each given about 40 control tests and 20 stimulation tests. The data are clear. There was complete retroactive interference; after release from confinement the rats repeated the arms chosen before confinement as if they had never been there. On the other hand, there was no proactive interference; each arm chosen after confinement was not repeated after the first choice. This effect was consistent throughout the tests and showed no evidence of changing from the first test to the last. Also, stimulation of the amygdala, or the cingulate gyrus, even in current intensities sufficient to provoke extended behavioural seizures, had no influence on choice behaviour.

These data are important because they indicate that a normal hippocampal EEG is required for maintenance of the memory in the radial arm maze task (see also Livesey 1975). Stimulation was delivered while the rat was in the centre of the apparatus, after the choices has been made, and after the information about those choices had been stored in memory. Nonetheless, the stimulation completely disrupted choice behaviour on the arms chosen before confinement. Here, then, are data that are compatible with a memory interpretation, but not with a sensory discrimination interpretation of septo-hippocampal function.

CONCLUSIONS

In summary, I would like to suggest the following conclusions. (1) Lesions in the septo-hippocampal system have a profound and enduring effect on performance in the radial arm maze. (2) Lesions at the septal pole and the temporal pole of the hippocampus have an equivalent behavioural effect, especially when tested by the crossed-lesion procedure. (3) Lesions alter a number of behaviours besides choice accuracy, including response patterns, speed of running, and behavioural flexibility. (4) Normal hippocampal electrical activity seems to be required for retention of spatial memory in this task.

ACKNOWLEDGEMENTS

The research reported here was supported in part by Research Grant MH 24213 from the National Institute of Mental Health. I thank A. Black, J. Becker, G. Handelmann, L. Jarrard, J. O'Keefe, and J. Walker for assistance in developing the ideas presented here, and B. Lindeman and S. Williams for preparation of the manuscript.

References

Adey, W. F. (1961) Studies of hippocampal electrical activity during approach learning, in *Brain Mechanisms and Learning* (Delafresnaye, J. F., ed.), pp. 557–588, Thomas, Springfield, Ill.

Douglas, R. J. (1967) The hippocampus and behavior. *Psychol. Bull. 67*, 441–442

Geschwind, N. (1965) Disconnexion syndrome in animals and man, Part. 1. *Brain 88*, 237–294

Horel, J. A. & Keating, E. G. (1969) Partial Klüver-Bucy syndrome produced by cortical disconnection. *Brain Res. 16*, 281–284

Horel, J. A. & Keating, E. G. (1972) Recovery from a partial Klüver-Bucy syndrome in the monkey produced by a disconnection. *J. Comp. Physiol. Psychol. 79*, 105–114

Jarrard, L. E. (1976) Anatomical and behavioral analysis of hippocampal cell fields in rats. *J. Comp. Physiol. Psychol. 90*, 1035–1050

Jarrard, L. E. & Becker, J. T. (1977) The effects of selective hippocampal lesions on DRL behavior in rats. *Behav. Biol. 21*, 393–404

Johnson, C. T., Olton, D. S., Gage, F. H. & Jenko, P. J. (1977) Damage to hippocampal connections and behavior: DRL and spontaneous alternations. *J. Comp. Physiol. Psychol. 91*, 508–522

Kimble, D. P. (1963) The effects of bilateral hippocampal lesions in rats. *J. Comp. Physiol. Psychol. 56*, 273–283

Kimble, D. P. (1975) Choice behavior in rats with hippocampal lesions, in *The Hippocampus*, vol. 2. *Neurophysiology and Behavior* (Isaacson, R. L. & Pribram, K. H., eds.), pp. 309–326, Plenum Press, New York

Livesey, P. J. (1975) Fractionation of hippocampal function in learning, in *The Hippocampus*, vol. 2. *Neurophysiology and Behavior* (Isaacson, R. L. & Pribram, K. H., eds.), pp. 247–278, Plenum Press, New York

Mishkin, M. (1958) Visual discrimination impairment after cutting cortical connections between the inferotemporal and striate areas in monkeys. *Am. Psychol. 13*, 414–423

Mishkin, M. (1966) Visual mechanisms beyond the striate cortex, in *Frontiers of Physiological Psychology*, pp. 93–119, Academic Press, New York

O'Keefe, J. & Nadel, L. (1978) *The Hippocampus as a Spatial Map*, Oxford University Press, Oxford, in press

Olds, J. (1972) Learning and the hippocampus. *Rev. Can. Biol. 31*, 215–238

Olds, J. & Olds, M. E. (1961) Interference and learning in paleocortical systems, in *Brain Mechanisms and Learning* (Delafresnaye, J. F., ed.), pp. 153–183, Thomas, Springfield, Ill.

Olton, D. S. (1973) Shock motivated avoidance and the analysis of behavior. *Psychol. Bull. 78*, 450–456

Olton, D. S. (1978) Characteristics of spatial memory, in *Cognitive Aspects of Animal Behavior* (Hulse, S. H., Fowler, H. F. & Honig, W. K., eds.), Erlbaum Associates, in press

Olton, D. S. & Samuelson, R. J. (1976) Remembrance of places passed: spatial memory in rats. *J. Exp. Psychol.: Anim. Behav. Proc. 2*, 97–116

Olton, D. S. & Schlosberg, P. J. (1978) Food searching strategies in young rats: win-shift predominates over win-stay. *J. Comp. Physiol. Psychol.*, in press

Olton, D. S., Collison, C. & Werz, M. A. (1977) Spatial memory and radial arm maze performance of rats. *Learn. Motiv. 8*, 289–314

Olton, D. S., Walker, J. A. & Gage, F. H. (1978a) Hippocampal connections and spatial discrimination. *Brain Res. 139*, 295–308

Olton, D. S., Branch, M. & Best, P. (1978b) Spatial correlates of hippocampal unit activity. *Exp. Neurol. 58*, 387–409

Schmaltz, L. W. & Isaacson, R. L. (1966) The effects of preliminary training conditions upon DRL performance in the hippocampectomized rat. *Physiol. Behav. 1*, 1975–182

Steward, O., Loesche, J. & Horton, W. C. (1977) Behavioral correlates of denervation and reinnervation of the hippocampal formation of the rat: open field activity and cue utilization following bilateral entorhinal cortex lesions. *Brain Res. Bull. 2*, 24–48

Discussion

Weiskrantz: I wonder if one can draw the conclusions you have drawn without having a task in which one can vary the difficulty? For example, if it had been a four-arm maze instead of an eight-arm maze, perhaps you would have found a relative rather than an absolute deficit.

A further question concerns *lists* of choices. The eight-arm situation, or the *n*-arm situation, is one in which the animal has to remember a number of items. Was there a comparable test of non-spatial lists that the animals have been subjected to? That is critical for the interpretation.

Olton: Yes, I agree that this issue is critical for the interpretation. We are developing non-spatial analogues now. To the extent that the septo-hippocampal system is selectively involved in spatial memory, lesions here should have no effect in these non-spatial tasks. To the extent that the behavioural changes we see in the spatial procedures are due to non-spatial factors (such as loss of memory in general, failure to inhibit incorrect choices, etc.), lesions of the septo-hippocampal system ought to produce similar behavioural changes in both the spatial and the non-spatial procedures.

Weiskrantz: A further question is whether we are dealing with one deficit or more than one. Are we dealing with some kind of 'unthinking' rat *plus* spatial disabilities, or is this a unitary deficit?

A further point is the explanation of the below-chance performance—perseveration of responses to previously selected positions. That suggests an excessive anchoring to spatial cues which the damaged animal has some difficulty in overcoming.

Then, the stimulation experiment is in effect a retrograde amnesia paradigm similar to that of electroconvulsive shock. Such a paradigm provokes the old question of whether one is dealing with an interference with memory or with retrieval and performance. There is a large and contentious literature in that area, and logically the same questions of interpretation arise here.

Finally, you were cautious in restricting yourself to discussing extra-maze spatial cues as against maps, and I would simply say here that when we come to discuss models this distinction is surely a question of different *levels* of discussion rather than saying that one situation is a map situation and another is a spatial situation. It is a question of level of logical analysis rather than one experimental situation versus another.

Olton: First, the deficit induced by the lesion is not due just to the complexity of the task. If we use a four-arm maze the rats perform just as badly after fornix lesions, so the number of arms is not critical.

We found perseveration in the maze when we allowed the rats to choose freely but not when they were confined in between choices (Olton *et al.* 1978; D. S. Olton & M. A. Werz, unpublished). These data support the contention of Dr O'Keefe and Dr Black that choice perseveration, at least in this task, is due to response variables rather than stimulus ones. In other words, when a rat repeats a sequence of choices, he seems to do so because of the ballistic characteristics of his behaviour which I described in my paper, rather than because of any tendency to repeat spatial locations *per se*. I did not expect to find these results (Olton 1972), but the data are certainly consistent with this interpretation.

Grossman: In your very interesting experiment, you are playing on the well-known fact that animals with septal or hippocampal lesions can't resist doing something they have previously been rewarded for. In your test they are constantly being asked not to go to some place where they have found a reward earlier. I frankly find it difficult to see why one should attribute their poor performance to an interference with memory functions in general or spatial memory in particular, where a far simpler and more general 'disinhibition' hypothesis accounts for the results quite well.

Secondly, there are a number of old experiments, including some by John Flynn (Flynn & Wasman 1960) that showed that animals are perfectly capable of learning and performing conditioned responses during seizure-inducing electrical stimulation of the hippocampus. Clearly, animals can acquire (and recall) at least simple memories quite well where the hippocampus is non-functional.

Olton: You are correct on your first point. We have obtained data from several other procedures which demonstrate that rats have a strong 'win-shift' tendency (Olton *et al.* 1977; Olton & Schlosberg 1978). In other words, when rats have found food in a particular place, they are more likely on subsequent choices to go to some other place than to return to the original place. I think it is this tendency that makes the 'never-go' procedure so easy for them.

With respect to your second point, I need to emphasize that we are not looking at the ability of rats to learn during hippocampal stimulation, but rather their ability to retain information already learned (see also, for example, Livesey 1975; Soumireu-Mourat *et al.* 1975).

Winocur: We have some observations that relate to your paper. We have been using your eight-arm radial maze in a series of ongoing experiments involving rats with hippocampal and control lesions. We replicated the profound impairment that you observed and made similar observations on the running patterns. We also cued each arm differently with visual cues (for example, the floor of one arm was cued with black and white stripes, another with circles, etc.), and found a tremendous improvement in the hippocampally lesioned animals. They did not quite reach the level of the controls but their performance improved markedly. The cortical-lesion controls and sham-operated controls performed no differently from each other on either the spatial or the visual task. In the visual task the controls appeared to do slightly better than the hippocampally lesioned rats; in the spatial task they were very much better.

These experiments were all on postoperatively acquired behaviour. In another experiment, we preoperatively trained rats on the spatial version of the task and postoperatively tested them on the visual version, to see if there was any transfer of previously acquired strategies. There was no apparent transfer in the control groups. We were surprised to find, however, a substantial impairment on the visual task in the hippocampally lesioned group, after they had been preoperatively trained on the spatial task. This suggests that they were trying to use spatial cues postoperatively but were not doing it very well. This finding relates to the question of whether animals even attempt to deal with spatial cues after lesions to the hippocampus.

Black: Just a note of caution about the interpretation of Dr Winocur's last result. It is dangerous to consider a given task to be a spatial task or a non-spatial task. Animals can solve most tasks by employing different strategies; they might use spatial information or they might learn a sequence of stimulus–response associations, and so on. This is true of the eight-arm maze. That is, animals can use spatial strategy in this maze, or they can use a response sequence strategy (such as turning to the right whenever they exit from an arm and entering the next arm).

We have been doing experiments in the eight-arm maze in the room that we described earlier (p. 187) in which distant stimuli can be controlled and removed. (This experiment is being done in collaboration with G. Augerinos.) We compared the behaviour of two groups of rats. For one group, a number of stimuli were placed around the periphery of the room; for the second group, all stimuli were removed and the room was maintained as free of stimuli as possible. We found that rats in the impoverished-stimulus situation tended to use more stereotyped response sequences in obtaining food from each of the arms than rats in the stimulus-rich room. If Dr Winocur's experimental situation is like our impoverished situation, his rats may have been solving the problem using response sequence strategies rather than spatial information. The negative transfer from the preoperative situation to the postoperative cue learning situation that Dr Winocur described may, therefore, have been transfer from response learning to cue learning rather than from spatial to cue learning.

Weiskrantz: An important point you raise takes us back to an older approach suggested by Kimble (1969) and Kimble & Kimble (1970) of hypothesis-perseveration in a discrimination task. The argument was that hippocampally lesioned animals could adopt a hypothesis but were less willing to change from one hypothesis to another when the correct solution required it. Perhaps there is a persistence of a particular kind of hypothesis here, rather than an inability to use any particular one.

Livesey: We have experimental findings that are relevant to Dr Olton's work. In our published experiments (Livesey & Wearne 1973; Livesey & Bayliss 1975; Livesey & Meyer 1975) we used electrical (blocking) stimulation to the CA1 region or to the underlying dentate gyrus of the dorsal hippocampus of the rat as it learnt a simultaneous brightness discrimination. With CA1 stimulation we concluded that we were interfering with two separate processes; a consolidation process where, once the animal has learnt the task, the stimulation had no further effect on its performance; and a comparison or retrieval process where the animal was unable to learn the task and then, having learnt it without stimulation, again showed disruption with stimulation.

We found a striking difference between CA1 and dentate stimulation: CA1 stimulation interfered dramatically with acquisition; dentate stimulation did not interfere with acquisition of the response but interfered with relearning after the meaning of the cues had been reversed.

We speculated that the stimulated rats were performing as if they were working on a non-contingent 50% reinforcement schedule and were ignoring the cues and that the disruption related to attentive processes. Stimulated animals responded well but appeared not to be attending to the cues. Our findings were very similar to Dr Olton's in this regard.

For these studies we chose a simultaneous discrimination task in which the animal responded to one of two alternatives; that is, the task did not require active inhibition of response. The results led us to the view that the blocking stimulation interfered in some way with cue selection and that this involved detection of error (Livesey 1975). In recent studies (P. J. Livesey, J. P. Smith & P. Meyer, in preparation) we therefore decided to use a discrimination task which emphasized detection of error and the active withholding of a response to the negative cue—that is, a go/no-go discrimination task. In this task a lever press to the positive or go cue (a constant light) resulted in a reinforcement (a dipper full of milk). A lever press to the negative or no-go cue (a flashing light) during the 15 s no-go period resulted in resetting of the timer for a further 15 s. The end of each trial was signalled by a click of the apparatus, and the trials were presented in random order (Fellows 1967). There were 20 S^D and 20 S^Δ trials each day. We were concerned with the rats' responses during $S^\triangle$ trials, so the criterion of mastery of the task was 90% zero responses to $S^\triangle$ over two consecutive days.

In the first study rats were lesioned in either the CA1 or dentate gyrus in regions comparable to those where stimulation had been applied in the earlier studies. These animals were then required to learn the discrimination and then relearn it with cue reversal. There were eight rats in each of the CA1 and dentate lesion and control groups. In acquisition, controls and animals with dentate lesions did not differ significantly but animals with CA1 lesions showed a significant decrement in rate of learning ($U = 13$, $P < 0.05$). In reversal both dentate lesion animals ($U = 12$, $P < 0.05$) and CA1 lesion animals ($U = 1$, $P < 0.001$) were significantly retarded compared with the control animals.

In the second study, control animals were compared with animals learning the discrimination while receiving continuous stimulation (i.e. 0.5 s of stimulation every 3.0 s throughout the 40-trial period) to the CA1 region. There were 10 animals in each group. Eight of the control animals learned the task in 500 trials while only two of the stimulated animals reached criterion in this time ($U = 2$, $P < 0.01$).

It was observed that all animals quickly learned to respond in the presence of S^D. In the $S^\triangle$ period all animals, during the first few days of training, would press the lever rapidly for a while and would then extinguish the response. With the onset of the next $S^\triangle$ trial they would repeat this process. After some days this pattern changed and, after one or two bar presses, the animal would withhold its response until the next trial. Absence of reinforcement appeared to provide the cue for inhibition of response. After more days of training the normal animals then appeared to become aware of the cues. With change of cue from S^D to $S^\triangle$ the animal would start towards the lever, would look at (i.e. attend to) the light, and then swerve away. This was followed by the achievement of criterion performance.

For the stimulated animals that failed to reach criterion this transition did not take place, the animals continuing to respond as if they were using non-appearance of reinforcement as the signal to inhibit responding.

In summary, the use of the go/no-go task with lesioned animals produced a similar pattern of deficit to that observed in the stimulated animals learning simultaneous discrimination. With dorsal CA1 lesions or stimulation a deficit was observed in acquisition. This deficit was not evident in animals with dentate lesions or stimulation. Dentate lesion or stimulation did, however, result in a significant deficit in learning the reversal.

With stimulation of the CA1 region during acquisition of the go/no-go task it was evident that there was no defect in the rats' ability to withhold or inhibit a response. The deficit appeared to become evident at the stage when the control animals shifted from using absence of reinforcement as the cue for inhibition of response to use of the visual cues. Stimulated animals failed to utilize these cues—that is, failed to attend to them.

Vinogradova: To return to the alternatives between space analysis and memory, a very interesting analysis of this problem has been made by Susan Iversen (1976). She previously maintained that in primates after hippocampal lesions spatial orientation is deranged. She now thinks that memory is primarily deranged by hippocampal lesions. The essence of the problem seems to be the following. The hippocampus is really not necessary for the learning and retaining of traces in 'primitive' situations where the relations between stimuli and reinforcement are very simple. This kind of learning easily occurs without the hippocampus. But this structure is necessary in more complex learning situations, where cues are multiple and have complex patterns and where the straightforward relations between a stimulus and reaction are absent. This is exactly the case for learning in spatial situations, and it seems that memory for complex situations, such as all natural conditions are, is especially deranged by hippocampal lesions.

Some unique experiments were done in natural conditions in the USSR by Dr Krushinskaya (1966) which illustrate this. She worked in cedar birds *(Nucifraga caryocatactes)* which have a characteristic and complex behaviour in relation to their food supply. When the cedar nuts are ripe the birds make several thousand stores, each of small amounts of nuts, hidden over a wide area. During the winter they feed off these stores of nuts. The birds do not search for them; they just fly straight to their own stores, never touching stores made by other birds. Of course the landscape is changed in the winter in Siberia; there is deep snow and no foliage. Nevertheless the birds locate their food stores.

Dr Krushinskaya worked in the Altai Mountains for two years with these birds. They were kept in a home cage, and there was a large experimental cage with a natural landscape in it. Some birds received archicortical (hippocampal) lesions; other had lesions in the hyperstratum accessorium. These birds were put in the experimental cage and they made their small stores in various places, as normal birds do. Then they were taken to the home cage, and after some time (from 15 min to 24 hours) were returned to the experimental cage. With long time delays, whereas the birds with hyperstratum lesions went directly to their food stores, the hippocampally lesioned birds began to dig from one or other side of the cage, all the way across. But with short delays (15 min–3 hours) they were able to find their stores quite selectively. This could be interpreted as a loss of spatial map, but I think that it is the *memory* for the spatial map which is lost here. It is important to note that simple operant conditioned reflexes (lever pressing to light for a food reinforcement) were unimpaired in birds without the hippocampus.

References

Fellows, B. J. (1967) Chance stimulus sequences for discrimination tasks. *Psychol. Bull. 67*, 87–92
Flynn, J. P. & Wasman, M. (1960) Learning and cortically evoked movement during propagated hippocampal after discharges. *Science (Wash. D.C.) 131* 1607–1608
Iversen, S. D. (1976) Do hippocampal lesions produce amnesia in animals? *Int. Rev. Neurobiol. 19*, 1–49
Kimble, D. P. (1969) Possible inhibitory functions of the hippocampus. *Neuropsychologia 7*, 235–244
Kimble, D. P. & Kimble, R. J. (1970) The effect of hippocampal lesions on extinction and 'hypothesis' behavior in rats. *Physiol. Behav. 5*, 735–738
Krushinskaya, N. L. (1966) Some complex types of alimentary behaviour in cedar-birds after lesion of archicortex. *Zhurnal Evoluzionnoi Biokhimii y Fisiologii 11*, 563–568
Livesey, P. J. (1975) Fractionation of hippocampal function in learning, in *The Hippocampus*, vol. 2: *Neurophysiology and Behavior* (Isaacson, R. L. & Pribram, K. H., eds.), pp. 247–278, Plenum Press, New York
Livesey, P. J. & Bayliss, J. (1975) The effects of electrical (blocking) stimulation to the dentate of the rat on learning of a simultaneous brightness discrimination and reversal. *Neuropsychologia 13*, 397–407
Livesey, P. J. & Meyer, P. (1975) Functional differentiation in the dorsal hippocampus with local electrical stimulation during learning by rats. *Neuropsychologia 13*, 431–438
Livesey, P. J. & Wearne, G. (1973) The effects of electrical (blocking) stimulation to the dorsal hippocampus of the rat on learning of a simultaneous brightness discrimination. *Neuropsychologia 11*, 75–84
Olton, D. S. (1972) Discrimination reversal performance after hippocampal lesions: an enduring failure of reinforcement and non-reinforcement to direct behavior. *Physiol. Behav. 9*, 353–356
Olton, D. S. & Schlosberg, P. (1978) Food searching strategies in young rats: win-shift predominates over win-stay. *J. Comp. Physiol. Psychol.*, in press
Olton, D. S., Walker, J. A., Gage, F. H. & Johnson, C. T. (1977) Choice behavior of rats searching for food. *Learn. Motiv. 8*, 315–331
Olton, D. S., Walker, J. A. & Gage, F. H. (1978) Hippocampal connections and spatial discrimination. *Brain Res. 139*, 295–308
Soumireu-Mourat, B., Destratde, C. & Cardo, B. (1975) Effects of seizure and subseizure posttrial hippocampal stimulation on appetitive operant behavior in mice. *Behav. Biol. 15*, 303–316

Multivariate analysis of the septal syndrome

HOLGER URSIN, TORDIS DALLAND, BJØRN ELLERTSEN, THOMAS HERRMANN, TOM BACKER JOHNSEN, PETER LIVESEY, ZEENAT ZAIDI and HEGE WAHL

Institute of Psychology, University of Bergen, Norway

Abstract Since recent reviews of the behavioural effects of septal lesions agree that more than one explanatory concept is required, a multivariate analysis of the septal syndrome has been made. A total of 127 rats have been tested, 73 with septal lesions and 54 controls. The rats were tested in a standard test battery consisting of a residential maze, spontaneous alternation, spatial learning, approach/avoidance conflict and one-way active avoidance. Factor analyses reveal a complex change in the factor structure after septal lesions. None of the previous explanations of septal functions receives unequivocal support. The findings do not exclude the possibility that septal lesions interfere with a few general behaviour mechanisms or perhaps only one. However, if so, this factor does not explain as much of the variance as expected. Situational or test-dependent factors play a greater role in the variance. This is interpreted as an indication of insufficiency in the theoretical structure, or in conventional test designs.

In his review of 1966, McCleary concluded that the septal area was a nuclear complex. He said that he referred to it as an entity because there was no information available permitting any conclusions as to which anatomical structures were critical for the response inhibition. The research that followed demonstrated that there is indeed an anatomical organization within the complex, and this organization makes it very difficult to apply a single-concept model. All recent reviewers of the septal literature agree that more than one explanatory concept is needed (Fried 1972; Caplan 1973; Lubar & Numan 1973; Ursin 1976). They disagree about the numbers and names of these concepts.

Data from this type of research have more variance than is found for most biological phenomena. This variance could perhaps be used to our advantage. An adequate way of analysing how many explanatory factors really are needed to account for this variance requires multivariate methods.

In this report we have utilized this approach to explore the dimensions needed to explain the variance in a large study of rats with total or subtotal septal lesions.

These rats have been tested in a number of standardized tests. The resultant data matrix is both large and complex. One approach to the analysis is through correlational studies. These do not establish proofs of causality, but do indicate likely relationships that may then be further explored experimentally. A powerful tool for resolving the structure of relationships within a matrix of correlations is the factor-analytic method. Factor analysis provides a mathematical solution to the problem of locating a parsimonious combination of variables. If the elements—that is, the tests that make up the raw data for the matrix—have meaning and meaningful relation towards each other, then the experimenter may use this information to arrive at an interpretation of the nature of the factors. If a set of factors can be determined for normal animals on the basis of their performance on this battery of tests, then observation of the ways in which a similar factor solution derived for septally lesioned animals differs from that for the normal group may give us insights into important, perhaps fundamental differences that have developed as a result of the lesions. Furthermore, the relationship between such factors and particular types of lesions may help us gain insight in the anatomical organization of this complex from a functional point of view.

METHODS

A total of 210 male Møll Wistar rats (approximately 350 g body weight) has been tested in 'waves' consisting of 20–25 rats; controls and operated animals were mixed in each wave. After histology, 127 rats were accepted for multivariate analyses: 31 with large septal lesions, 42 with small, and 54 controls. For a rat to be included, histological criteria had to be met, and the test matrix had to be complete. The total duration of the experiment was three years. Rats were tested at all times of year.

Lesions were placed according to conventional electrolytic procedures. Histology was done without knowledge of the behavioural results. The extent of each lesion was drawn from projected images of the slides; the extent of the lesion was then transferred to pre-drawn diagrams. The size of each lesion was then evaluated as a percentage of the volume of each subnucleus.

Rats were allowed six days recovery after surgery. All animals were handled, and the test order was always the same (see Table 1). First the rats were placed for 22 hours in a plus-form *residential maze* (Barnett et al. 1966).

TABLE 1

Order of behavioural testing of rats with septal lesions and control rats

Day 1	Surgery
Days 1–6	Recovery
Days 6–12	Handling
Days 12–19	Residential maze
Days 19–24	Spontaneous alternation
Days 24–32	Spatial learning
Days 32–44	Approach/avoidance training
Days 44–48	One-way active avoidance

Recordings were made automatically of the number and duration of entries into each of four arms, extending from a central nest area. One arm contained food, one water, one an object (toy car), and one arm was empty. Three identical mazes were used; these were placed in a sound-insulated room and kept on a 12–12 hour light/dark cycle.

For *spontaneous alternation* a standard wooden T-maze was used (Dalland 1970). The rats were then tested for spatial learning in an open runway. The rats were trained to run down this runway for water. On the test day they were started from a new runway, added at right angles to the middle of the original runway, forming an open T-maze. A correct trial consisted of a straight run to the water without entering the wrong part of the original runway. A second trial was also given on the same day, the start arm then being placed on the opposite side of the original runway.

Approach/avoidance conflict was tested in another runway (360 cm; see Srebro et al. 1976). In this alley, the rats were rewarded with food. After stable approach had been established, the rats were shocked in the goal area. The location and behaviour of the rat was followed for 10 min (see Fig. 1).

One-way avoidance learning was tested in an unsignalled situation (Vanderwolf 1964; Werka et al. 1978). The animals were scored for acquisition, retention after 24 hours, and forced extinction (by punishing a previously correct response).

For the multivariate analyses a number of variables were selected. These are defined in Table 2. The computer programmes used for the analyses were all developed at the University of Bergen, Department of Psychometrics of the Institute of Psychology (Johnsen 1977). Some of the variables showed extremely skewed distributions and were therefore transformed before analysis using a log transformation ($x_t = \log(x+1)$).

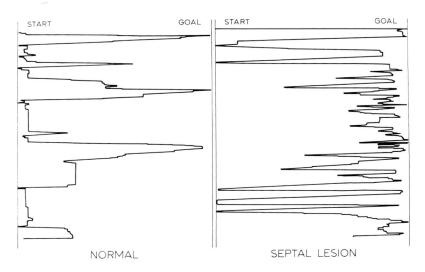

FIG. 1. Approach/avoidance in one normal rat *(left)* and one with septal lesions *(right)*. The location of the rat in the runway is recorded manually over the 10 min observation period. The rat is placed in the start area (top left), runs to the goal area (right) and receives shocks, leaves this area and runs towards the start area or approaches the goal again. The ordinate is a time axis, with start at top, and ending at the bottom of the graph after 10 min.

RESULTS

Discriminant analysis

As a first step in the analysis of the data from the two groups (normal rats and rats with septal lesions), a discriminant analysis was completed for the *residential maze*, the *approach/avoidance conflict*, and *one-way active avoidance*. The other two tests, spontaneous alternation and spatial learning, did not give rise to variables other than latencies which could be used for this purpose, and will only be dealt with sporadically in the following analyses.

The variables from each of the three situations were treated as a set of dependent variables which may be combined to differentiate between the two groups. A discriminant analysis may therefore be regarded as an extension of a normal one-way analysis of variance which again is an extension of the normal *t*-test for differences between two groups. The objective of the discriminant analysis was to provide a new variable which is a weighted sum of the original observations, and which gives a maximum difference between the two groups. The weights will also tell us to what extent each variable contributes to the total discrimination between the two groups. Finally, the method includes a test of the difference between the two groups

TABLE 2

Variables included in the factor analyses

Residential maze
WI	Water intake measured in grams
E1	Number of entries into food arm
E2	Number of entries into object arm
E3	Number of entries into water arm
E4	Number of entries into empty arm
T1–T4	Amount of time spent in each arm

Approach/avoidance
SHO	Total number of shocks taken in approach avoidance task
AVG	Avoidance gradient, measured in cm
CA	Total number of entries into conflict area
FLT	Total number of 'flights', defined as a direct run from the conflict area to the start box in less than 10 s
FAP	Number of fast approaches, defined as direct runs from the start box to the conflict area in less than 10 s
APT	Approach time, seconds from start of testing to the first shock
SHU	Number of shuttles, defined as runs from the start box to the conflict area and back to the start area in less than 30 s
FSF	Number of fast stops during flights. The delay is brief enough for the run to reach the flight criterion (less than 3 s)
SF	Number of all other stops during movements directed away from goal (shock) area
FSA	Number of brief stops during approach. The delay is brief enough for the run to reach the fast approach criterion (less than 3 s)
SA	Number of all other stops during runs towards the goal (shock) area
MIL	'Mileage': total distance run in the alley (measured in cm)

One-way active avoidance
TTC	Number of trials to criterion in one-way active avoidance
SHK	Number of shocks taken in one-way avoidance
NFA	Number of trials to first active avoidance
LT 1	Latency on first avoidance trial
LT 2	Latency on second trial
LTD	Difference between latency on first and second trial
RTN	Retention trials to criterion
FE	Forced extinction: number of punished trials before extinction of active avoidance

in respect to each variable regarded as a single dependent variable—that is, a simple t-test.

Residential maze. There was an overall significant difference between the two groups (Wilk's lambda $= 0.71$, $F(9/109) = 4.8$, $P < 0.001$). This difference was mostly due to four variables, which all showed a significant difference.

In a simple t-test, rats with septal lesions showed significantly more water intake ($P<0.05$), entries and time spent in the food arm ($P<0.01$ and 0.05, respectively) and entries in the water arm ($P<0.01$). The increase in entries is in agreement with previous data reported by Dalland (1975).

Approach/avoidance conflict. There was again a significant difference between the groups (Wilk's lambda = 0.71, $F(12/111) = 3.8$, $P<0.001$). The univariate t-tests showed highly significant differences on several variables. The septally lesioned group accepted more shocks than normals, had more visits to the conflict area, and more flights, fast approaches, and shuttles. Accordingly, the septally lesioned rats ran a longer distance (mileage) than normal rats in the alley. Their initial approach time was faster. (All P's <0.001.) These findings are in accordance with previous findings (Srebro et al. 1976).

One-way active avoidance. There was again a significant group difference (Wilk's lambda = 0.59, $F(6/121) = 13.9$, $P<0.001$). There were significant differences on all variables except retention. Rats with septal lesions took more trials, shocks and false avoidances to criterion ($P<0.001$ for all comparisons). The normal rats had a longer latency on Trial 1 ($P<0.001$) and septally lesioned rats had more resistance to extinction of the avoidance habit ($P<0.001$). Except for the difference in latency on Trial 1, the same findings have been reported by several authors (see Srebro et al. 1976).

Conclusion. Rats with septal lesions showed the expected changes in behaviour, which all constitute part of what has been referred to as the 'septal syndrome'. Therefore, our sample of rats exhibited the particular syndrome we were to analyse.

Factor analyses

The factor-analytic method employed was a component analysis, and should be regarded as a simple descriptive device for reducing the relations between a large group of variables to a smaller mutually independent set of factors. It must be stressed that the results depend on the inputs. They also depend on the type of data transformation that is performed, and the number of factors one accepts. The results reported here are from the matrix obtained following the data transformations effected to normalize the distributions.

For most analyses for the three test situations, three factors could be accepted following Kaiser's criterion (smallest eigenvalue larger than 1.0). We therefore decided to report three factors for all analyses. These three factors accounted for approximately 70% of the variance for most tests, with the maximum variance explained by the three factors being 88% for one test. The factors were rotated by a standard varimax procedure to simplify interpretation of the structure.

We first report factor analyses for each behavioural task separately, for normals and for rats with septal lesions. We then analyse selected variables from the three different tasks in an attempt to identify factors involving two or three of the situations.

Factor analyses for each test separately

Residential maze. For normals *Factor 1* (see Fig. 2) may be referred to as a 'Time' factor, loading on the time spent in each arm. *Factor 2* is an Activity factor. For normal rats, entries in the arms seemed fairly independent of the time spent in the arm; this is the reason for the differentiation between the first two factors. *Factor 3* is related to water intake, and, to our puzzlement, to the number of entries into the empty arm.

Rats with septal lesions showed changes in the factor structure with respect to the normals, most clearly for the group with large septal lesions ($n=31$, see Fig. 2, right column). We shall discuss this latter group. The combination of small and large lesions ($n=73$) showed only moderate changes in the same direction. *Factor 1* in rats with large septal lesions corresponds mainly to the Activity factor in normals, with an addition of the time spent in the food arm and in the empty arm. *Factor 2* is a new factor, related to object exploration and time spent in the empty arm. This Object factor is negatively related to entries into the food arm. Finally, we have a Water factor, *Factor 3*, which also loads on the time spent in the empty arm.

There are clear changes in the factor picture which relate to the septal syndrome. Septally lesioned rats have more entries than normal rats, but the organization of their Activity factor is only slightly changed, also involving the time spent in the food arm and the empty arm. There are two other changes in the factor structure suggesting changes in the organization of behaviour in rats with septal lesions. Their object exploration is changed, and also their behaviour related to water intake. Changes in water intake, object exploration, and increased activity are in accordance with several hypotheses on septal function (Ursin 1976).

Approach/avoidance conflict. In normal rats, *Factor 1* is an overall Activity

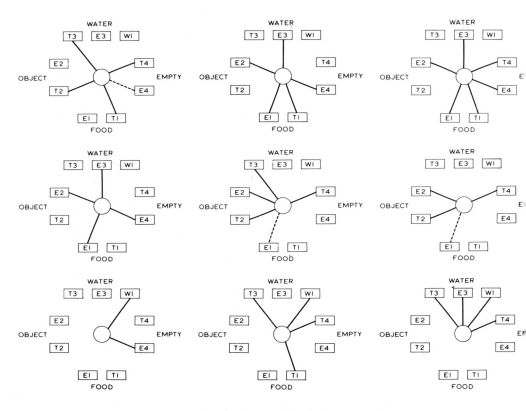

FIG. 2. Residential maze. Significant factor loadings for each factor: Factor 1 top row, Factor 3 bottom row. Normal animals left column, small septal lesions middle column, large septal lesions right column. Dotted line indicates negative loading.

factor (see Fig. 3). The correlations underlying the factor are generally significant and high. *Factor 2* is related to avoidance, fast stops during flight and fast stops during approach. It is negatively related to approach time, but this loading is weak. The leading variables are fast stops during flight (0.82) and the avoidance gradient (0.85). This factor, therefore, may be related to fear, but to the stops during flight rather than flight itself. Flight is one of the fixed action patterns available for 'fear'-motivated behaviour; the mutually exclusive alternative is referred to as 'freezing'. The stops during flight may reflect freezing also, since it is positively related to a high

NORMALS   SEPTALS

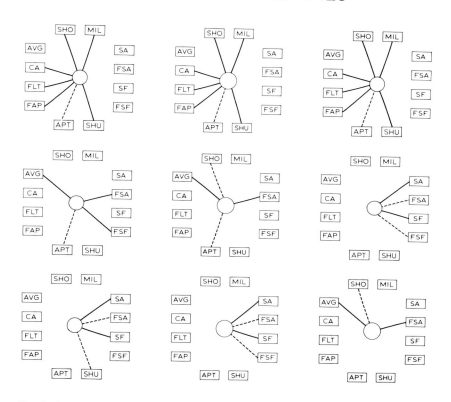

FIG. 3. Approach/avoidance conflict. Factor loadings for each factor (see Fig. 2 for details).

avoidance gradient. This hypothetical interpretation will be used hereafter. *Factor 3* is related both to shuttles and to the other 'stop' indicators; the long stops during flight and approaches relate negatively to shuttles and fast stops during approach. We will refer to this factor as the 'Stop' factor.

Rats with septal lesions have the same Activity factor as normals in spite of the fact that 'activity' is significantly increased. The septally lesioned rats seem to do what normals do, but more so. However, in the Stop factor and the Freezing factor there are clear changes. In rats with septal lesions, there is one factor which contains all the stop variables, with a negative relation between fast stops and long stops. The Freezing factor loses 'freezing', or fast stops during flights. Fast stops during approach, the number of shocks accepted (negatively) and the avoidance gradient remain in this 'Fear' factor.

This may suggest a change in 'freezing' or at least some change in the inhibitory aspects of fear-motivated behaviour.

There is no general loss of inhibition, but there are clear changes in the way stops relate to other variables. This analysis suggests the loss of a general modulating influence on the variables loading on activity. There are also important changes in the way avoidance and approach is organized. There is no loss of a single function like inhibition or freezing, but changes in the organization of behaviour related to these hypothetical functions.

One-way active avoidance. The factor structure for normal rats is very simple (see Fig. 4). There is an Acquisition factor (Factor 1), a Retention factor (Factor 2) and a factor related to the latency on Trial 1 (Factor 3).

The factor structure in the rats with large septal lesions is more complex. *Factor 1* is again an Acquisition factor, but it also loads on retention. *Factor 2* is the latency factor. *Factor 3* is a new factor. There is a clear change in the relationship between forced extinction and retention. They are negatively related in septally lesioned rats. Previous research has pointed out that the retention deficit observed in cats with septal lesions is not related to the passive avoidance deficit (McCleary *et al.* 1965).

Rats with septal lesions are defective in several of the variables in this test. These deficits seem related to two or three factors. One deficit is related to acquisition and retention, and one to acquisition and forced extinction, which may be related to 'passive avoidance'.

General factors

For a general explanation of the septal syndrome we should look for factors that contain variables from more than one test. An overall analysis of all variables reveals that a large number of factors may be included. When this is done, the factors are test-specific. For exploratory analyses, we have selected representative variables from different factors, and have made specific analyses to elucidate two particular problems. We have attempted to analyse the 'activity' phenomena, and also freezing and other types of inhibition. For these analyses, we have used the total group of septally lesioned rats ($n = 73$) (small and large lesions), even though the factor structure for this total lesion group tended to take a middle position between that of the large lesions and that of the normals.

Activity. We used the label 'Activity' both for Factor 1 in the residential maze, and for Factor 1 in the approach/avoidance conflict. There have been

MULTIVARIATE ANALYSIS OF THE SEPTAL SYNDROME

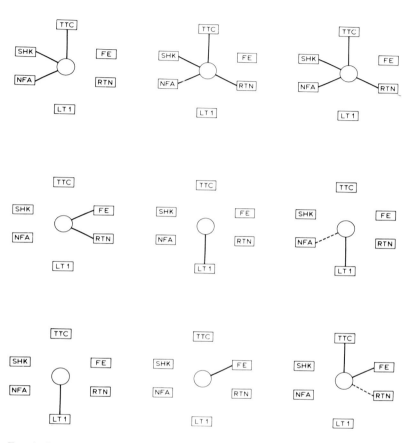

FIG. 4. One-way active avoidance. Factor loadings for normal rats and septally lesioned rats (see Fig. 2 for details).

numerous postulates of a general hyper-activity or hyper-reactivity in rats with septal lesions. However, our results indicate that these two factors may be completely independent. When only variables from these two factors were included in a separate analysis, they appeared as two independent and dominating factors, explaining together almost 60% of the variance. This is even more remarkable since in the first set of results septal lesions produced profound changes in all the leading variables on both factors, even if the factor structure remained unchanged.

362   H. URSIN *et al.*

Also, when other variables were added to the rotations, these two factors remained independent, and remained dominating variables with few relations with other variables.

'Freezing' and 'inhibition'. Loss of freezing and general loss of inhibition are prevalent interpretations of the septal syndrome, or parts of that syndrome. Freezing may be assumed to interact not only with shock-motivated behaviour, but might also interfere with start latencies. An analysis involving 14 variables selected from the approach/avoidance conflict and one-way active avoidance as well as latency variables from spontaneous alternation and spatial learning was therefore made (see Fig. 5). Five factors, all involving freezing or

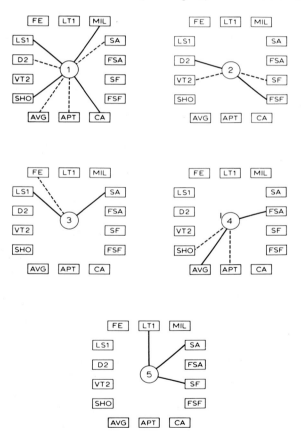

FIG. 5. Factor analyses of variables believed to represent different types of inhibition; septally lesioned rats only. Number in circle indicates factor number. Significant loadings: continuous line, positive; dotted line, negative loading.

inhibition, in one form or another, were evident. *Factor 1* involved the start latency in the spontaneous alternation situation, and variables from 'activity' and approach factors in the approach/avoidance conflict. *Factor 2* loaded on fast 'stops during flight' (FSF), for which our favourite interpretation is 'freezing'. It also loaded on latency on the second trial in the spontaneous alternation situation, and negatively with the number of vicarious trials and errors in that task. All four variables in the factor relate to different types of inhibition, but there was a negative relationship between the long-lasting stops during flight and vicarious trials and errors, on the one hand, and start latency and fast stops during flights on the other. There was no easy explanation apparent to us; the underlying inhibition mechanisms seem complex indeed. *Factor 3* loaded on stops during approach, the start latency in the spontaneous alternation situation, and negatively on forced extinction. Response inhibition or freezing or passive avoidance deficits may again be used as explanatory terms. This is also possible for *Factor 4* and *Factor 5*.

In other words, all the different factors in this analysis invited 'freezing', inhibition or 'passive avoidance' types of interpretation. Septally lesioned animals may indeed have some kind of profound change in freezing behaviour or inhibition, but it seems difficult to postulate a general or simple loss of freezing or inhibition as a total explanatory concept for the changes we observed. Srebro *et al.* (1976) found that the amount of freezing behaviour after large septal lesions depended on the task, and they did not find support for the hypothesis of any primary change in freezing or crouching behaviour.

Localization

Some variables correlated with lesions of particular nuclei. Time in the object arm and shocks accepted in the approach/avoidance situation correlated with lesions in the medial septal nucleus only. Entries into the water arm correlated with lesions in the dorsal septal nucleus only. Fast stops during avoidance correlated with lesions in the lateral septal nucleus only. Forced extinction of one-way active avoidance correlated with lesions both in the medial septal nucleus and the diagonal band of Broca. The number of shuttles in the approach/avoidance task correlated with lesions of the dorsal septal nucleus, medial nucleus and fornix. Retention of the one-way active avoidance correlated with lateral septal lesions, and lesions of the diagonal band of Broca and fornix. Variables like entries into the food arm, time spent in the empty arm, stops during avoidance, trials to criterion, shocks and number of false avoidances in the one-way active avoidance correlated with lesions in a number of structures.

One-way active avoidance deficit was not limited to any particular localization; this is in agreement with previous results (Grossman 1976; Srebro *et al.* 1976). 'Shuttles' depended on fornix lesions as well as septal lesions. Increased shuttling has been suggested to underly the improved Sidman avoidance and improved two-way active avoidance seen after both types of lesions (Grossman 1976).

However, even if this is the traditional way of analysing this type of data, such correlations should be regarded with some suspicion since we have so many correlations in this large matrix.

Factor loadings. An interesting alternative to this 'classical' neuropsychological approach, in which we analyse one variable at a time, is to investigate the relationships between factor loadings and brain structures. From the varimax rotated factor structure for each test discussed above,

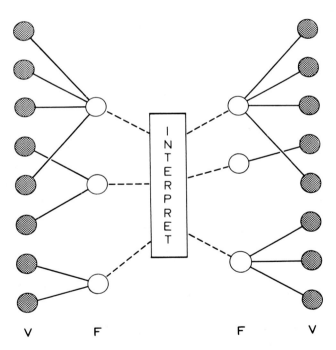

FIG. 6. Principles of the comparison between the factor analyses for normal animals and for lesioned animals (V, variable). The analyses produce factors (F, open circles) which are then compared and interpreted by us. This interpretation may be excluded and replaced by a mathematical comparison between the factors, producing sets of factors with optimum correlation (canonical analysis).

factor scores were computed for each rat. These scores were then correlated with the extent of lesion of the various septal subnuclei.

Fig. 7 illustrates the logic for this operation. Instead of trying to understand the factor structure by theoretical considerations (see Fig. 6), we now consider only the factor structure for rats with septal lesions, and correlate the factor loadings with septal lesions.

When this analysis is made for all rats with septal lesions ($n = 73$), the Activity (entry) factor from the residential maze correlated with medial septal lesions only ($P < 0.05$). The object exploration factor correlated with

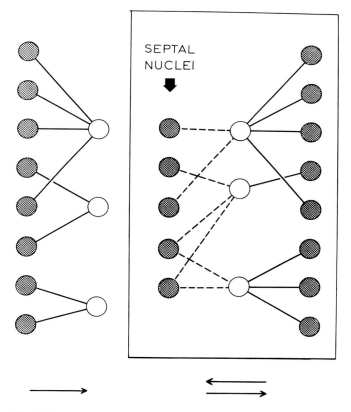

FIG. 7. Instead of an interpreter intermediate between the factors in normal animals and septal animals we have now included the septal nuclei. When the analyses include the septal nuclei, and the relationship between groups of nuclei and variables, we hope to be able to understand the difference from the normal animals better. In the analyses presented in this paper we always go from the variables to the factors; it is, however, also possible to predict variables from the types of lesion produced (indicated by arrows at bottom of figure).

all types of lesions, and the water factor did not show any significant correlation with any particular type of lesion (see Table 3). For the approach/avoidance conflict, the Activity factor correlated significantly with lesions in the dorsal nucleus and the medial nucleus ($P<0.05$ for both) and also fornix ($P<0.01$). The second factor, the avoidance gradient factor, did not correlate with any particular lesion type. The 'Stop' factor correlated significantly with lesions in both the lateral septal nucleus ($P<0.01$) and medial septal nucleus ($P<0.05$). For one-way active avoidance the Acquisition factor correlated significantly with all types of lesions ($P<0.01$), the latency factor did not correlate with any particular lesions, while the forced extinction factor correlated significantly with lesions in the dorsal septal nucleus.

When only rats with large septal lesions were analysed ($n=31$) the picture is different; there is only one significant correlation. The Activity factor in the approach/avoidance conflict correlated significantly with fornix lesion ($P<0.05$). This may be due to the small degree of variance in this group with very large lesions.

TABLE 3

Localization of septal factors (Factor loading analysis)

	Residential maze	Location	Approach/ avoidance	Location	One-way active avoidance	Location
Factor 1	Activity	MS	Activity	DS-MS	Acquisition	All
Factor 2	Explore	All	Fear change	None	Forced extinction (PA)	DS
Factor 3	Water intake	None	Stop	LS-MS	Escape	None

MS, DS and LS: medial, dorsal and lateral septal nuclei. PA: passive avoidance.

Canonical analysis. So far, we have only used one lesion type at a time for our comparisons between lesions and behaviour. It is also possible to combine the lesions. This has been done by a canonical analysis, which identifies the relations between two sets of variables. This is achieved by defining pairs of new variables, one from each set (lesions and behaviour), which are (1) simple functions of the variables within each set *and* (2) maximally correlated with each other. This may be regarded as a type of simultaneous factor analysis where the solution for each set depends on the solution for the other set. In this way, we hope to identify common structures across sets of variables.

For this analysis, we have used the material consisting of all the septal lesions ($n = 73$). For the residential maze, two roots were accepted, explaining a total of 79% of the variance. Root 1 correlated with entries in the food arm, entries in the water arm and negatively with time spent in the empty arm. This root correlated significantly with all categories of lesions, indicating that no fixed localization exists for important aspects of the activity variable in the residential maze. The other root correlated negatively with entries in the water arm and positively with fornix lesions. No ready explanation is apparent to us.

For the approach/avoidance conflict again two roots were obtained, explaining a total of 77% of the variance. Root 1 correlated with variables from the Activity factor in this test, and with lesions in the dorsal septal and medial septal nucleus as well as with fornix lesions. The other root correlated with the Stop factor; this root correlated with lesions in the lateral, dorsal and medial septal nuclei. It is generally accepted from many models that 'stops' or inhibitions are related to septal nuclei. In our case, we do not find any particular localization within the septal nuclei, but a strong correlation between these nuclei and this type of phenomenon. This is the same finding as we reported for the single correlations; this analysis points to the possibility that the stop mechanisms are tied together and all relate to all septal nuclei.

Two roots were also accepted for the one-way active avoidance, explaining 57% of the variance. Root 1 correlated with the Acquisition factor; as expected, this correlated with all lesion types. This confirms our previous conclusion. Root 2 correlated with forced extinction and lesions in the fornix. The importance of this fibre bundle for extinction and 'passive avoidance' phenomena has been dealt with in other reports.

CONCLUSION

This paper is our first report on an application of multivariate analysis to neuropsychological data. The multivariate methods have been used purely as descriptive devices. The results should be regarded as preliminary, and the next stage in this type of analysis should utilize model testing based on tests of specific hypotheses about relations between different groups of variables. So far, we have only studied the relations between the variables in respect to simple linear models, and simple group differences. More complex relationships may easily be conceived, and should also be pursued in future work with multivariate techniques.

We believe that some conclusions may be reached already at this stage. Our findings do not exclude the possibility that septal lesions interfere with a few or perhaps only one general behaviour mechanism like inhibition or sensitivity or something else. However, if so, this mechanism does not explain as much of the variance in our conventional test situation as we expected. Situation- or test-dependent factors play a much greater role in the variance. We interpret this as one indication of insufficiency either in our theoretical structure, or in our test design. Either we are not testing directly enough for 'the septal function', or the organization of the behaviour influenced by septal nuclei is really as complex as our multivariate tests suggest.

Multivariate analyses are complex in respect to interpretation. We do not claim this approach to be the exclusive method in neuropsychological research. However, we do believe that the complexity of the septal syndrome can only be unravelled by methods of analysis which consider this complexity.

ACKNOWLEDGEMENTS

The research reported here was sponsored by the Norwegian Research Council for Science and the Humanities. The data-processing of the raw data from the residential maze was done by the EDB-Section of the Medical Faculty, University of Bergen.

References

BARNETT, S. A., COCKROFT, A. L. & SMART, J. L. (1966) An artificial habitat for recording movement. *J. Physiol. (Lond.) 187*, 15–16

CAPLAN, M. (1973) An analysis of the effects of septal lesions on negatively reinforced behavior. *Behav. Biol. 9*, 129–167

DALLAND, T. (1970) Response and stimulus perseveration in rats with septal and dorsal hippocampal lesions. *J. Comp. Physiol. Psychol. 71*, 114–118

DALLAND, T. (1975) *The Nature of the Perseverative Behavior of Rats with Septal and Rats with Dorsal Hippocampal Lesions.* Doctoral dissertation at University of Bergen

FRIED, P. A. (1972) Septum and behavior: a review. *Psychol. Bull. 78*, 292–310

GROSSMAN, S. P. (1976) Behavioral functions of the septum: a re-analysis, in *The Septal Nuclei* (DeFrance, J. F., ed.), pp. 361–422, Plenum Press, New York

JOHNSEN, T. B. (1977) Kontroll av standardprogrammer. Mimeograph, Dept of Psychometrics, Institute of Psychology, Bergen

LUBAR, J. F. & NUMAN, R. (1973) Behavioral and physiological studies of septal function and related medial cortical structures. *Behav. Biol. 8*, 1–25

MCCLEARY, R. A. (1966) Response-modulating functions of the limbic system: initiation and suppression, in *Progress in Physiological Psychology*, vol. 1 (Stellar, E. & Sprague, J. M., eds.), pp. 209–272, Academic Press, New York

MCCLEARY, R. A., JONES, C. & URSIN, H. (1965) Avoidance and retention deficit in septal cats. *Psychon. Sci. 2*, 85–86

SREBRO, B., ELLERTSEN, B. & URSIN, H. (1976) Deficits in avoidance learning following septal lesions in the albino rat. *Physiol. Behav. 16*, 589–602

URSIN, H. (1976) Inhibition and the septal nuclei: breakdown of the single concept model. *Acta Neurobiol. Exp. 36*, 91–115 (Memorial Paper in Honor of Jerzy Konorski)

VANDERWOLF, C. H. (1964) Effects of combined medial thalamic and septal lesions on active-avoidance behavior. *J. Comp. Physiol. Psychol. 56*, 31–37

WERKA, T., SKÅR, J. & URSIN, H. (1978) Exploration and avoidance in rats with lesions in amygdala and the piriform cortex. *J. Comp. Physiol. Psychol.*, in press

Discussion

Gray: I want to congratulate Dr Ursin, because I know the enormous difficulties of doing this kind of analysis with human personality data. He is absolutely right in saying that in principle this is a tool which not only can decide, but was devised for the purpose of deciding, whether one is dealing with a unitary difference between two groups of individuals or a number of differences. This is the way to do it; but, as he pointed out, there are formidable difficulties in practice. You need a large number of subjects. In this case, you are applying your analysis to a large number of subjects which have to have lesions, and they have to go through a large number of complicated tests. Whereas, in collecting human personality data, typically you take a group of people off the street and bring them in for rapid paper and pencil tests, so you can at least get the data relatively easily. I am impressed by the power with which nonetheless you are able to tell the difference between the loss of a factor, and a change in a factor pattern; but I wonder whether the problem of putting subjects through many sequential tests is nonetheless too confounding for practical purposes, in this particular case.

Dr Rawlins has found that the behaviour of septally lesioned rats can change with repeated exposure to different test conditions, and indeed septally lesioned rats can learn, given enough time, to overcome a particular deficit (Rawlins 1977). Suppose your normals and lesioned rats in test 1 of your series behave differently, as they do. Suppose that their behaviour on test 2 is a function not just of the lesion, but of their behavioural experience in test 1. That means that in test 2 your two groups may differ not because the lesion has altered their behaviour in the way you observe, but as a consequence of the different ways they have already behaved on test 1. And this will go on, through your sequence of tests, to become an ever-more complicated problem.

Ursin: I agree that this is a complication, but I do not see how you can get round this problem, and I do not think it confounds the results obtained with this particular test sequence. We have used our standard battery of tests in the order we always run such experiments. We proceed from non-traumatic tests gradually to tests requiring shocks, the test with most shocks (one-way active avoidance) being the last. I wouldn't balance the test order, because that would really worry me. If I had the animals in the one-way active avoidance test first, and then put them in the residential maze, I would not know how to interpret the data. We know that normal and lesioned rats will be different in the active avoidance task; septal lesion rats receive more shocks. That could affect later approach data, but there are no shocks until the approach avoidance test in our design.

Rawlins: The experiments (Rawlins 1977) Dr Gray referred to did not in fact use shock. Having found that medial septally lesioned subjects perseverated in extinction in the alley, I next ran them in a Skinner box using a progressive ratio schedule; initially they did perseverate, but this did not last. When their performance was once more comparable with that of the control subjects, I retrained them in the alley and then re-extinguished them. They no longer perseverated in the alley at all. It seems possible that experience with a sparse schedule of reinforcement in the Skinner box changed the animals' response to non-reward in the alley, so it may not be only tasks employing shock which affect performance on subsequent tasks.

Weiskrantz: The methodological problem that arises is the effect of order of testing, and the fact, going back to the Kimble kind of hypothesis (Kimble 1969; Kimble & Kimble 1970), that lesioned animals do not easily shift their strategies. This is an important consideration whenever animals are put in a series of situations. As almost all experiments have a preliminary 'shaping' phase, the problem may be a widespread one.

Livesey: One important point is that before the analysis was undertaken, Dr Ursin looked at the results of individual tests and these conformed well with the sort of pattern found by other people.

Weiskrantz: Having done this analysis, what is the message that you would like to leave us with, Dr Ursin? Is it that the effect of septal lesions is incredibly complicated and McCleary has to be split into a number of different persons? Or, that we have some additional insight into the way the septum is organized with respect to a limited number of groups of behavioural tasks? The first message comes through clearly, that you can embarrass various people if you try to generalize from their positions, but on the positive side, what should people working on septal lesions take away from this for further guidance?

Ursin: I don't know the answer. If these results embarrass anyone, they definitely embarrass me as much as anyone else. I have accustomed myself to the thought that single-concept models of the septal syndrome have to be revised. However, I hoped we should end up with a limited number of factors that I could have handled. The results are nowhere near the factors I have predicted myself (Ursin 1976). For instance, when we did the last analysis of inhibition, I had hoped to come up with, say, three or four types of inhibition only. But the message so far with this analysis is that there are too many factors for me to handle in any available theoretical framework.

Gray: This is where the question of whether the factors are correlated becomes very important. If these different inhibition factors are themselves correlated you may simply be seeing too many trees for the wood, because each of the individual factors reflects test-specific influences, but nonetheless they all also share a common variance. If they are correlated factors, you can explore the correlations between them. If they are uncorrelated factors because you made them uncorrelated, you obviously cannot do that. But if they come out empirically as orthogonal factors (that is, uncorrelated), then the problem is indeed the way you have posed it, and that would be worrying.

Ursin: The analysis we have run is one where we ask for independent (orthogonal) factors. One way to go may be to be less rigid in asking for independent factors. The easiest argument against our results is to say that we selected bad tests. However, they are representative of fairly standard tests, I think. But the tests have to be refined in one of two ways. One is to be more and more specific on the type of psychological dimensions involved. I see promising developments there. The other way is to look at the elements of behaviour. Let us go further back to ethology, as Dr Black was suggesting, looking at smaller units of behaviour (see Lubar *et al.* 1973).

Olton: I have a question which has to do with your data matrix. As I understand it, on your F1 factor in the plus-shaped maze, rats with septal lesions were more active than controls. In your approach/avoidance tests, rats with septal lesions were also more active than controls. Then you said that those two behavioural changes were not related to one another at all. That seems an important conclusion, namely that behavioural changes in two different tasks which look to be the same are in reality due to two entirely different mechanisms.

Ursin: This may be an important characteristic of septal lesions. The behaviour of the rats with lesions is stimulation-specific in many situations. For instance, Köhler (1976*a, b*) finds this for orienting responses, habituation and exploration. Exploration in one situation does not mean the same as exploration in another situation, and these behaviour types are differentially sensitive to septal and hippocampal lesions.

Azmitia: I gather, Dr Ursin, that you started out wanting to find out what sort of behavioural test would help one to understand the function of a specific anatomical structure, and you have been very precise and sophisticated in your analysis of, and approach towards, the behavioural tests and their interpretation; but your attitude towards the anatomy underlying the behaviour seems casual. It doesn't appear strange to me that when an answer is found it is an unclear answer, since the questions asked were not as rigorous on the anatomical side as they were on the behavioural side of the study. Methods are available now by which you can destroy the cells in a particular brain area without disrupting the fibres of passage. There are also methods by which the efferents and afferents of an area can be determined and studied. Therefore it is possible to be as sophisticated in the anatomy as you are in the behaviour.

Ursin: We have applied an 'equipotentiality' principle for each subnucleus; this is necessary since we have to end up with a quantification of the lesion —a number which may be entered into the matrix.

DeFrance: In defence of Dr Ursin's strategy, I think it does make good sense. In the first place, physiologists and anatomists tend to get caught up in the detail, which may not have a functional significance. If you look at the septal complex in Golgi stains, one is amazed at the tremendous number of possible interactions between fairly distinct anatomical regions. For example, the dendrites of the medial septal cells extend right over into the lateral septal region. Therefore, if you make a lesion in the lateral septal region, how specific is it?

Weiskrantz: The point is not so much the precision of measurement as the degree of independence of those particular structures. Your measures of cells in different places may indeed be independent, Dr Azmitia, but anatomically they are not independent. I think Dr DeFrance's point is that one is dealing on the anatomical side with structures which are themselves correlated with each other to a certain extent.

References

KIMBLE, D. P. (1969) Possible inhibitory functions of the hippocampus. *Neuropsychologia 7*, 235–244

KIMBLE, D. P. & KIMBLE, R.I. (1970) The effect of hippocampal lesions on extinction and 'hypothesis' behavior in rats. *Physiol. Behav. 5*, 735–738

KÖHLER, C. (1976a) Habituation of the orienting response after medial and lateral septal lesions in the albino rat. *Behav. Biol. 16*, 63–72

KÖHLER, C. (1976b) Habituation after dorsal hippocampal lesions: a test dependent phenomenon. *Behav. Biol. 18*, 89–110

LUBAR, J. F., HERRMANN, T. F., MOORE, D. R. & SHOUSE, M. N. (1973) Effect of septal and frontal ablations on species-typical behavior in the rat. *J. Comp. Physiol. Psychol. 83*, 260–270

RAWLINS, J. N. P. (1977) *Behavioural and Physiological Correlates of Limbic System Activity.* Unpublished D. Phil. Thesis, Oxford University

URSIN, H. (1976) Inhibition and the septal nuclei: breakdown of the single concept model. *Acta Neurobiol. Exp. 36*, 91–115

A comparison of hippocampal pathology in man and other animals

L. WEISKRANTZ

Department of Experimental Psychology, University of Oxford

Abstract Bilateral damage to the medial temporal lobe, including the hippocampus, in man is associated with a severe amnesic syndrome. It is still not clear whether the hippocampus (or its output pathways and related target projection sites) is the critical structure in producing this syndrome, especially as more severe learning deficits in animals are found with lesions in the anterior inferotemporal cortex than with hippocampal lesions *per se*. But the problem of trying to relate memory deficits in man and animals depends on the characterization of the amnesic syndrome itself. It was originally thought to be a failure of input into long-term memory store or a failure of consolidation. Medial temporal and hippocampal lesions in animals do not produce results that fit such a characterization. On re-examination of the human syndrome, however, for which some of the evidence is reviewed, it appears that the amnesic patients can learn and remember over long intervals if certain testing paradigms are used. The results are more readily matched to some of the results of hippocampal lesion studies in animals. Two main classes of current theories of the amnesic syndrome are discussed. A somewhat different approach is suggested here, based on the dissociation between human amnesic subjects' commentaries and their objective performance, which suggests a dissociation between levels of processing rather than a failure on any particular level.

Undoubtedly one of the spurs to the study of hippocampal function in animals was the dramatic clinical finding in the 1950s that bilateral medial temporal lobe damage, including the hippocampus, produced a crippling and enduring amnesic syndrome in human patients (Scoville & Milner 1957; Penfield & Milner 1958). Such reports arrived on the scene, it may be recalled, against a background of considerable scepticism engendered by Lashley's monumental work suggesting that there are no local regions of the brain that are absolutely critical for learning (given adequate sensory reception) or memory. The clinical reports of patients such as H.M. (in whom bilateral medial temporal resection was carried out for relief of severe epilepsy) appeared to indicate, as we shall see, that a particular region was essential for the consolidation of long-term memory. Actually patients with very similar memory impairments had been known to neurologists for many years. These were occasional patients suffering from encephalitic disease or more commonly from Korsakoff psychosis and Wernicke's disease, among other aetiologies. But the work of Milner and colleagues concentrated attention on the importance of the hippocampus in the human amnesic syndrome. This was especially so not only because the hippocampus was obviously implicated in the bilateral medial temporal lobe resections, but also because later reports of unilateral temporal lobe damage indicated that selective amnesic states could be produced, such that left temporal lobectomy impaired verbal memory and right temporal lobectomy impaired memory for complex visual and auditory patterns. In these mild but selective amnesic cases, there is a correlation between the extent of hippocampal involvement and the severity of the impairment on the relevant memory tasks (evidence by Corsi reviewed by Milner 1970, 1971).

Given this impetus some 20 years ago, a note of caution may still be desirable. It is still not established with certainty that the hippocampus is the critical structure in the medial temporal lobe complex that causes memory impairments when damaged, or that amnesia follows inevitably from bilateral pathology in it. Increasing hippocampal damage probably also correlates with increasing damage to the overlying white matter and surrounding neocortex. Or it may be that a combination of hippocampal and other damage is crucial. It seems important to stress these possibilities in the context of the present paper, dealing as it does with comparisons of animals and man. Animal work in the 1950s was also proceeding at a rapid pace, and a few years before the reports by Milner and co-workers it was established that bilateral inferotemporal cortical lesions in the monkey produce a severe and enduring learning impairment (Mishkin 1954; Mishkin & Pribram 1954). Interestingly, one of the control lesions that in that early work produced little or no impairment was a bilateral hippocampectomy! The impairment in the monkey was known to be specifically visual (material learned through the auditory or tactile modes was unaffected) whereas the human amnesic impairment is multi-modal. Nevertheless, Iversen and I (1970) set out to test the effects of inferotemporal lesions (some in combination with hippocampal damage) on long-term memory for tasks that could be learned very rapidly. This work did produce interesting results; there was indeed a striking impairment in long-term retention, but it emerged that this was because of enhanced interference phenomena in memory rather than a failure of storage as such (Iversen 1970). We will return to that theme soon. But it is worth noting that the analysis of the inferotemporal deficit is now much further advanced, and we know that it can be fractionated into at least two components (cf. review, Weiskrantz 1974). More interestingly, it recently has emerged that damage to the most anterior region of the inferotemporal cortex, including the temporal pole, causes an extremely severe learning deficit in monkeys, more severe than any we have seen before. We still do not know whether this far anterior focus is modality-specific. Such a region might be disconnected by damage to white matter overlying the hippocampus, and Horel & Misantone (1974, 1976) have shown that a disconnection of the anterior temporal lobe in the monkey can have severe consequences for visual discrimination learning. And so we should still be cautious in assuming that the medial temporal amnesic deficit in man is due to the hippocampal component alone and in trying therefore to relate the human clinical syndrome to studies of restricted hippocampal damage in animals.

Having sounded this note of caution, we still have a serious difficulty in relating findings from human and animal investigations, because even those animal studies that included bilateral medial temporal lobe lesions, modelled on the surgery carried out on patients such as H.M., have yielded findings apparently in serious disagreement with those from the human clinic. Indeed, Milner was herself an author of just such an animal study (Orbach *et al.* 1960). But such an apparent disagreement forces us to turn to the problem of the amnesic syndrome itself, leaving aside the question of the locus of the critical pathology, because the precise form of its characterization determines the type of animal paradigm that we try to design, and thereby determines whether the animal results are judged to be discrepant or not.

The general clinical description of amnesic patients such as H.M. is well known (Talland 1965; Zangwill 1966; Milner 1966). Their most striking inability is in remembering incidents in daily life, even after a gap of just a few minutes, and even when the incident (such as seeing the experimenter) is experienced on several occasions. This difficulty need not be accompanied by any impairment in intelligence provided it is based on measures that do not demand the bridging of a gap of more than a few minutes (in which rehearsal is prevented). For example, they can carry out mental arithmetic and other intellectually demanding tasks, and they need have no difficulty in reciting back strings of digits or in other short-term memory tasks (Baddeley & Warrington 1970). At the same time, they have no difficulty in conversing about at least some events from early life, and their verbal skills are unaffected. This general description applies not only to patients like H.M. but also to patients suffering from Korsakoff's syndrome and to some postencephalitic patients. This is not to say that all such patients are identical in all respects;

given the complexities of Korsakoff pathology, for example, involving often both mamillary bodies as well as medial dorsal thalamus and other structures (Victor *et al.* 1971), patients with different aetiologies are entitled to different overall spectra of symptoms, such as the presence or absence of confabulation, but in the long-term human memory tests that we have been using we can find no essential differences between different aetiologies of the amnesic syndrome (for example, see Winocur & Weiskrantz 1976).

In confronting these patients informally in the clinical situation, the conclusion is practically irresistible that they cannot form a durable record of their new experiences. In terms of the familiar form of serial two-process models, one is tempted to say that they cannot transfer information from a short-term store (lasting well under a minute) to a permanent long-term store or—remaining detached from the question of one- or two-process models and whether, in the latter case, the processes are serial or parallel—to say that there is a failure of 'consolidation' of memory. Milner (1968) was the earliest proponent of this general type of interpretation.

Given such an orientation, it should prove impossible for an animal with the same disability to learn and remember events over long intervals. The dilemma is that animals with medial temporal damage or with restricted hippocampal damage can learn and remember, at least under conventional conditions of laboratory testing; indeed sometimes they learn more rapidly than control animals. This is not to say that they are unimpaired in other ways, of which more later. But when animal tests are deliberately designed to separate the short-term and the long-term memory components, monkeys with inferotemporal or inferotemporal plus hippocampal lesions can retain information over the long term provided the tasks are administered in such a way as to minimize interference from other tasks (Iversen 1970). It also appears that the forgetting curves of such animals are not different from those of controls, which they certainly should be (Weiskrantz 1971; Dean 1976). It was this disappointing set of results that led us to conclude that our own search for the amnesic syndrome in animals had failed. Either the same structures are differently organized in monkey and man, or we were not directing equivalent questions to the two species. Dr Elizabeth Warrington, of the National Hospital, and I decided to re-examine the human amnesic syndrome more closely. Meanwhile, newer lines of animal enquiry by other workers give rise to fresh optimism about the essential identity of organization, but such optimism stems in part from a re-examination of the human deficit itself by a variety of investigators.

The first conclusion that emerged in our own work on human subjects was that a consolidation hypothesis was inadequate. Under certain conditions there was evidence of excellent learning and retention over long intervals by amnesic subjects. There was an early strong hint of this in a verbal free-recall task when we analysed the subjects' errors; many of them were words that intruded from lists of words the subjects had seen on previous days (Warrington & Weiskrantz 1968b). But the clearest evidence came from experiments using the technique of 'cued recall'. In this type of experiment the subject is not asked explicitly to say what he remembers or whether he can recognize an item, but instead he is given partial information and asked to identify the whole item. For example, he may be given fragments of a picture or of a word (Warrington & Weiskrantz 1968a) or the first two or three letters of a word (Weiskrantz & Warrington 1970a), or a semantic hint (Warrington & Weiskrantz 1971, 1974), or the first of a pair of rhyming words (Winocur & Weiskrantz 1976). One can conduct perfectly straightforward learning and memory experiments using these methods. With repeated exposure of the more complete items, it is found that the subject requires less partial information to identify the whole item, and this can be plotted in the form of a conventional learning curve. The adequacy of the cue can be tested at varying intervals after learning, and this can be plotted as a conventional forgetting curve. Under appropriate conditions, amnesic subjects show learning that is not significantly slower than controls, and their memory is often remarkably durable over hours or days (Weiskrantz & Warrington 1970b). We have frequently repeated this kind of study and found it to be robust. That it is not restricted to our patients alone is suggested by Milner's observation with H.M., in whom retention over an interval of one hour and also over four months has been reported with the use of the cued recall method for pictorial material (Milner et al. 1968; Milner 1970). Indeed, much earlier Williams (1953) had demonstrated a beneficial effect of offering 'prompts' to amnesic subjects. A related type of result is provided by A. D. Baddeley & N. Brooks (personal communication) who find that amnesic subjects retain their improved performance in solving a particular jig-saw puzzle with repeated practice and in arranging scrambled words into a meaningful sentence.

A second point that emerged was that the beneficial effects of cueing could not readily be ascribed to its merely being an easier method for all subjects, amnesic and normal alike. If that were so, all subjects should be aided by the cues to a comparable degree. But amnesic subjects are differentially aided by cueing so that they can remember more or less normally with the appropriate set of cues, whereas their performance with a more conventional measure (e.g. yes/no recognition) is much poorer; that is, there is a significant interaction between groups of subjects and the method of retention testing (Warrington & Weiskrantz 1970, 1974).

The next two points to have emerged carry positive rather than negative implications. We can ask whether cueing is especially beneficial at any particular stage of learning or retrieval. The answer is that cueing is critical only when it is used at the stage of testing for retention. It works even when the material is presented during acquisition in the conventional form of standard lists of whole items, provided retention is later tested by cueing. If retention is tested without cueing, whether or not cueing was used at the acquisition stage, the amnesic subjects perform very poorly (Warrington & Weiskrantz 1970). This carries the strong implication that the amnesic deficit lies with mechanisms beyond the initial input of the items into storage, and obviously suggests a difficulty with retrieval.

Finally, the cue is not an automatic and assured device for successful retrieval; it depends on how many items can be matched by the cue and the order in which these items were exposed to the subject. Under certain conditions amnesic subjects are actually severely impeded by the use of cues that are deliberately designed to be ambiguous. Suppose, for example, the amnesic subject is first presented with a set of items and his retention is tested by appropriate cues. Under these conditions, as we know, his retention will be good. Now soon afterwards we present him with a second list of items which can be matched by the same cue as the first set. When his retention is now tested he will show a dogged persistence in producing the items from the first set instead of the second set, far more dogged than control subjects. We have called this the 'prior learning effect' (Warrington & Weiskrantz 1974). The purest example arises when the two successive lists of items are the *only* sets that can be matched by the cues, as for example with words like COTTON, MOAN, etc., followed by COTTAGE, MOAT, etc.; COTTON and COTTAGE are the only common words matched by the cue COT, and MOAN and MOAT are the only ones matched by MOA. Here the amnesic subject will persist in saying COTTON, MOAN, etc. in response to their respective cues, whereas the control subject will rapidly switch to the more recently presented items. A replication of this effect can be seen in Fig. 1. The effect may well be an analogue of discrimination reversal in animals, which is often impaired by hippocampal lesions. A rather elegant similar example was provided by Winocur, who succeeded in demonstrating good paired-associate learning by amnesic subjects when the pairs of words were constrained by a rule common to the whole list, such that all pairs are, for example, similar in meaning or similar in sound. Such good learning is, in itself, remarkable because all investigators agree that conventional paired-associate learning (i.e. not rule-constrained) is disastrously poor with amnesic subjects. But it was also found that if a second set of pairs was presented sharing both the

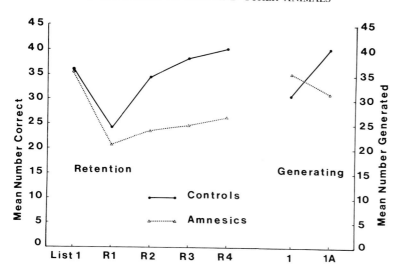

FIG. 1. 'Reversal learning.' For each recall cue (the initial three letters) there were only two English root words available as possible responses. Subjects were first taught one set of words (List 1) and then were given four trials with the alternative set (R_{1234}). The same cues were used on all five trials. Panel at right (Generating) indicates number of times each word was offered by subjects after end of reversal condition when asked to provide two words to each cue. (From Warrington & Weiskrantz 1978.)

same constraining rule and also the same initial words for each pair, the amnesic subjects persisted strongly in producing the first associate rather than switching to the second (Winocur & Weiskrantz 1976). The implication of these findings is that the amnesic subject is bedevilled by items that happen to gain prior entry into his store, and these persist and thereby interfere with retrieval of later items. A further implication of all the findings with cueing is that the cues may indeed exercise their beneficial effect precisely when they serve to constrain items that would otherwise intrude inappropriately as errors.

In trying to relate these findings to those with animals subjected to medial temporal lobe or selective hippocampal damage, it is easy enough to find examples of enhanced interference phenomena among successive tasks (Winocur & Mills 1970; and see review by Iversen 1976), and of dogged persistence in learning and extinction (Kimble 1969; Kimble & Kimble 1970), in discrimination reversal (Douglas & Pribram 1966), and in tasks such as DRL (differential reinforcement of low rate of responding) (Schmaltz & Isaacson 1966). But there are also some difficulties. First, many of these deficits are not specific to hippocampal damage as such but may also be seen after, for example, orbital frontal lobe damage. In fact, this problem may be lessened somewhat (but not completely) by our introductory comments urging caution about the assumed critical role of the hippocampus in the amnesic syndrome; there are anatomical links between the anterior inferotemporal cortex and the dorsal medial nucleus of the thalamus (Klinger & Gloor 1960), which in turn projects to the whole of the prefrontal cortex. Secondly, there are studies of hippocampal and/or fornix lesions that appear to suggest a specifically spatial function for the hippocampus. For example, Mahut has found that spatial discrimination reversal is impaired whereas non-spatial reversal tasks are normal (Mahut 1971, 1972; Mahut & Zola 1973). Another striking example comes from the work of Olton (this volume) who has devised a radial arm maze for studying 'spatial memory' in rats; hippocampal lesions severely impair performance in this apparatus. Superficially, at least, spatial deficits appear to be orthogonal to the cluster of deficits that are more readily matched to the human amnesic syndrome. But at least we can say that the animal studies no longer need be bedevilled by their failure to find a consolidation type of deficit: it appears that the human deficit also is not one of consolidation. Whatever specific explanation prevails, there seems a reasonable hope that it will apply equally well both to animals and to people.

What are the explanatory possibilities? These probably have been considered in more detail at the human level, given the richness of theoretical developments in the area of human memory and the speed with which new experiments can be done with human subjects. Indeed, one wag has suggested that there are now more theories of the human amnesic syndrome than there are amnesic subjects. We have reviewed various of the theoretical possibilities elsewhere (Weiskrantz & Warrington 1975), but for present purposes they can be put into two broad categories: those that assume some specific impairment of information processing, and those that assume that one out of several types of memory systems is defective.

Milner's consolidation hypothesis is an example of the first type: it assumed a failure of transfer into long-term store. A variant might be that long-term traces are established but decay abnormally rapidly. Other examples of processing hypotheses are those that assume that information enters storage without difficulty but is encoded inappropriately, leading to excessive interference amongst stored items or inappropriate or faulty retrieval. The best known example of such a hypothesis comes from the work of Butters & Cermak (1974) who have argued that Korsakoff patients, although they may well be capable of encoding semantically . . . may prefer to rely on an acoustic or associative level of encoding'. The suggestion derives from experiments in which they either used different types of cues for recall or compared verbal and nonverbal forms of stimulus items to be recognized after an interval. Perhaps the neatest example comes from their use of Wicken's technique of release from proactive interference which suggests that Korsakoff patients do not react to subtle changes in semantic categories in successive lists. Much of Butters and Cermak's work depends upon testing after very short intervals and they rest heavily on the claim that Korsakoff patients are impaired in short-term memory tasks, a finding with which not everyone is in agreement (Baddeley & Warrington 1970). A rather different form of an encoding hypothesis comes from Baddeley, based on the finding that amnesic patients are quite able to take advantage of word clustering based on taxonomic category, but are impaired in their use of clustering based on visual imagery (Baddeley & Warrington 1973).

At the other end, perhaps, of the information processing sequence can be found various interference hypotheses. These are based rather directly on the evidence for enhanced intrusions and interference phenomena as noted above. Butters & Cermak (1974), among others, have also recorded an increased sensitivity of their amnesic patients to proactive interference, although their interpretation of this sensitivity, as we have noted, is not formulated in terms of a primary enhancement in interference *per se*. Warrington and Weiskrantz have been most closely associated with the suggestion that inappropriate traces cannot be suppressed, or they are more slowly unlearned or forgotten. Recently we have put the suppression form of such a hypothesis to a severe test and found it wanting (Warrington & Weiskrantz 1978). In particular, false positive responses do not intrude excessively on the first occasion when they might be expected to do, but only emerge gradually with repeated opportunities; nor are interference effects between successive lists found in excessive amounts when the lists do not have to be differentiated by the amnesic subject. Thus, although the beneficial effects of cued recall are highly robust and amnesic subjects are differentially aided by such methods, and while increased levels of interference can be readily observed, there is not strong support for the view that cues work by constraining inappropriate response alternatives *per se*, although other interference hypotheses still remain to be tested.

Explanations of the second type, that assume an impairment in one out of several memory systems, are of two main forms which curiously balance each other. The first is Gaffan's (1972, 1974) suggestion that amnesic patients are impaired in discriminating degrees of familiarity of an item, while associative links can still be formed and stored. Recognition is impaired because all items tend to appear more nearly equally familiar (or unfamiliar) than for the normal subject. Gaffan has supported his hypothesis with some ingenious animal experiments which deserve to be replicated. The hypothesis has interesting implications and it also fits clinical impressions, as well as some, but perhaps not all, of the experimental findings with amnesic subjects.

One may also ask what the theoretical basis of adequate familiarity discrimination itself might be; it has been argued by some that recognition itself, at least in part, is mediated by associative retrieval mechanisms (Anderson & Bower 1972; Mandler *et al.* 1969; Warrington & Ackroyd 1975). And here we find examples of hypotheses that are, in this way, complementary to the familiarity hypothesis. The argument is that particular forms of associations cannot be established by amnesic subjects and that this failing in turn leads to faulty recognition. Thus it is argued, for example, that retrieval often depends on background contextual cues associated with the items to be remembered, a position advocated by Winocur who interprets impairments in hippocampectomized rats in such terms (Winocur & Mills 1970; Winocur & Breckenridge 1973). Related is the 'spatial anchoring' or 'cognitive map' hypothesis of Nadel & O'Keefe (1974) which, taking its lead from single neuron recordings, suggests that the hippocampus provides a spatial framework or map with which events can be associated and within which they can be ordered. This, too, is an intriguing hypothesis, although neurological patients with documented spatial disorders are by no means typically amnesic, and a 'topographical amnesia' can occur in isolation from other aspects of global amnesia (De Renzi *et al.* 1977). Nor, of course, are all hippocampus units spatial. But the hypothesis and findings from which it is derived are provocative and influential and, as we have already seen, one of the major effects of hippocampal lesions in animals appears to occur in spatial learning and/or reversal tasks, whether or not these have anything to do with amnesia.

One curious but recurring feature of the amnesic syndrome tempts one to strike out along a rather different theoretical line from those theories we have just been considering, which are all cast in terms of a blockade or weakness in a single information processing sequence or of a blockade of a particular form of memory. By now it is known that amnesic subjects can learn and remember a variety of tasks when they are tested in particular ways. The only person who remains unconvinced about this is the amnesic subject himself, who persists in failing to acknowledge that his performance is based on specific past experience, or may occasionally confabulate about such a basis. This is nicely illustrated by an anecdotal account made some 66 years ago by the Swiss psychologist Claparède (1911) of an alcoholic

Korsakoff patient. He pricked the patient's finger with a pin, and afterwards she withdrew her hand when he moved his hand near hers. When he questioned her about why she did so, she answered 'Isn't it allowed to withdraw one's hand?' He persisted with the questioning, and she said 'Perhaps there is a pin hidden in your hand'. And when he asked why that should be, she said 'Sometimes pins are hidden in hands'. But, commented Claparède, she never recognized the 'idea of pricking as a memory' (cf. translation by MacCurdy 1928).

Now this is apparently an example of avoidance conditioning. We wondered whether we could condition amnesic patients using a classical procedure, and if so whether they would display a similar lack of acknowledgement of their memory as did Claparède's patient. We used eyelid conditioning, which is not seriously aversive, rather than pricking with a pin or delivery of electric shocks. The results on two subjects, one a post-encephalitic case and the other an alcoholic Korsakoff patient, are shown in Figs. 2 and 3 (Weiskrantz & Warrington 1978). With both subjects there was unmistakable evidence of conditioning, with retention over both the 10-minute rest pauses and over an interval of 24 hours. We also found the same striking lack of verbalized acknowledgement of their demonstrable memories as did Clapa-

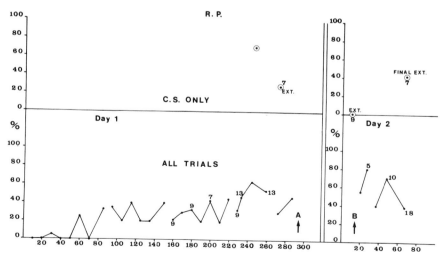

FIG. 2. Percentage of conditioned eyelid responses as a function of trials for subject R.P. Top panel shows 'probe' trials and extinction trials, in which no air puff was delivered. Bottom panel shows conditioning and all other trials combined. Interruptions in graph indicate occurrence of 10-minute rest breaks. Numbers above some points refer to occasions in which percentage is based on other than units of 10 trials. Arrows refer to stages at which interviews were recorded (see text). (From Warrington & Weiskrantz 1978.)

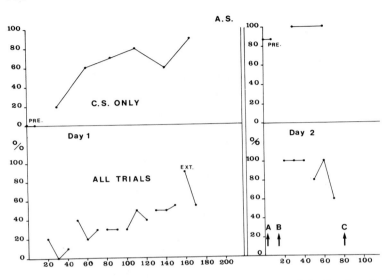

FIG. 3. Percentage of conditioned eyelid responses for subject A.S. Details as for Fig. 2. (From Weiskrantz & Warrington 1978.)

rède, even when the subjects were invited to comment within minutes on their experience while remaining seated in front of the apparatus and even when they subsequently displayed good evidence of having been conditioned. Despite the distinctive appearance of the apparatus it was never mentioned as being associated with the delivery of an air puff to the eye, itself presumably an uncommon experience. One of the earliest examples of long-term retention in amnesic patients was for motor skill learning as found by Milner (1962), who similarly reported that while the patient H.M. was able to learn mirror-drawing over a three-day period, yet each day found H.M. completely unaware that he had done the task before despite good savings from day to day. Sidman et al. (1968) also reported a striking dissociation between H.M.'s learning of a visual discrimination and his failure to describe the learned contingencies.

The sheer variety of tasks with which amnesic subjects by now have been shown to display long-term retention makes it increasingly difficult to conclude that there is a failure of consolidation of particular types of material or an insensitivity to particular encoding demands. In addition to the evidence for eyelid conditioning, amnesic patients show good retention of motor skill learning (Milner 1962; Corkin 1968), visual discrimination learning (Sidman et al. 1968), recall of pictures and words with a variety of cues (Warrington &

Weiskrantz 1968a, 1974; Weiskrantz & Warrington 1970a), retention of learning of anomalous pictures in the McGill Anomalies Test (Warrington & Weiskrantz 1973; and A. D. Baddeley & N. Brooks, personal communication), rule-governed verbal paired-associate learning and retention (Winocur & Weiskrantz 1976), retention of facilitation of stereoscopic perception of random dot stereograms (Ramachandran 1976, and personal communication), retention of the McCulloch colour-grating illusion over 24 hours (Warrington & Weiskrantz, unpublished observations), and retention of facilitation of solving jig-saw puzzles and for arranging words into sentences (A. D. Baddeley & N. Brooks, personal communication).

Whatever all of these tasks have in common in which good learning and retention have been demonstrated, it appears that the patients themselves on one level do not have access to their memories; nor is it necessary as a practical requirement that they have such access in order for their memories to be tested and demonstrated objectively. None of the current theories of the amnesic syndrome appear to have focused on the striking dissociation between the subjects' commentaries and their objective performance, which suggests a dissociation between levels of processing rather than a failure on any particular level. It may be that the dissociation is akin, in this respect, to that between performance and awareness noted in 'blind-sight' (Weiskrantz et al. 1974). But this general orientation towards the problem poses an even bigger problem in terms of our main theme: how would we study such dissociations in an animal model for the amnesic syndrome, and how would we ourselves recognize such a dissociation in an animal when it occurred?

References

ANDERSON, J. R. & BOWER, G. H. (1972) Recognition and retrieval processes in free recall. *Psychol. Rev. 79*, 97–123

BADDELEY, A. D. & WARRINGTON, E. K. (1970) Amnesia and the distinction between long- and short-term memory. *J. Verb. Learn. Verb. Behav. 9*, 176–189

BADDELEY, A. D. & WARRINGTON, E. K. (1973) Memory coding and amnesia. *Neuropsychologia 11*, 159–165

BUTTERS, N. & CERMAK, L. S. (1974) The role of cognitive factors in the memory disorders of alcoholic patients with the Korsakoff syndrome. *Ann. N.Y. Acad. Sci. 233*, 61–75

CLAPARÈDE, E. (1911) Récognition et moïite. *Archives de Psychologie (Genève) 11*, 79–90

CORKIN, S. (1968) Acquisition of motor skill after bilateral medial temporal-lobe excision. *Neuropsychologia 6*, 225–265

DEAN, P. (1976) Effects of inferotemporal lesions on the behavior of monkeys. *Psychol. Bull. 83*, 41–71

DE RENZI, E., FAGLIONI, P. & VILLA, P. (1977) Topographical amnesia. *J. Neurol. Neurosurg. Psychiatr. 40*, 498–505

DOUGLAS, R. J. & PRIBRAM, K. H. (1966) Learning and the limbic system. *Neuropsychologia 4*, 197–220

GAFFAN, D. (1972) Loss of recognition memory in rats with lesions of the fornix. *Neuropsychologia 10*, 327–341

GAFFAN, D. (1974) Recognition impaired and association intact in the memory of monkeys after transection of the fornix. *J. Comp. Physiol. Psychol. 86*, 1100–1109

HOREL, J. A. & MISANTONE, L. J. (1974) The Klüver-Bucy syndrome produced by partial isolation of the temporal lobe. *Exp. Neurol. 42*, 101–112

HOREL, J. A. & MISANTONE, L. J. (1976) Visual discrimination impaired by cutting temporal lobe connections. *Science (Wash. D.C.) 193*, 336–338

IVERSEN, S. D. (1970) Interference and inferotemporal memory deficits. *Brain Res. 19*, 227–289

IVERSEN, S. D. (1976) Do hippocampal lesions produce amnesia in animals? *Int. Rev. Neurobiol. 19*, 1–49

IVERSEN, S. D. & WEISKRANTZ, L. (1970) An investigation of a possible memory defect produced by inferotemporal lesions in the baboon. *Neuropsychologia 8*, 21–36

KIMBLE, D. P. (1969) Possible inhibitory functions of the hippocampus. *Neuropsychologia 7*, 235–244

KIMBLE, D. P. & KIMBLE, R. J. (1970) The effect of hippocampal lesions on extinction and 'hypothesis' behavior in rats. *Physiol. Behav. 5*, 735–738

KLINGER, P. & GLOOR, P. (1960) The connections of the amygdala and of the anterior temporal cortex in monkeys. *J. Comp. Neurol. 115*, 333–369

MACCURDY, J. T. (1928) *Common Principles in Psychology and Physiology*, Cambridge University Press, London & New York

MAHUT, H. (1971) Spatial and object reversal learning in monkeys with partial temporal lobe ablations. *Neuropsychologia 9*, 409–424

MAHUT, H. (1972) A selective spatial deficit in monkeys after transection of the fornix. *Neuropsychologia 10*, 65–74

MAHUT, H. & ZOLA, S. M. (1973) A non-modality specific impairment in spatial learning after fornix lesions in monkeys. *Neuropsychologia 11*, 255–269

MANDLER, G., PEARLSTONE, Z. & KOOPMANS, H. S. (1969) The effects of organization and semantic similarity on recall and recognition. *J. Verb. Learn. Verb. Behav. 8*, 410–423

MILNER, B. (1962) Les troubles de la mémoire accompagnant des lésions hippocampiques bilatérales, in *Physiologie de l'Hippocampe*, Centre National de la Recherche Scientifique, Paris

MILNER, B. (1966) Amnesia following operation on the temporal lobes, in *Amnesia* (Whitty, C. W. M. & Zangwill, O. L., eds.), pp. 109–133, Butterworths, London & Washington, D.C.

MILNER, B. (1968) Preface: material specific and generalized memory loss. *Neuropsychologia 6*, 175–179

MILNER, B. (1970) Memory and the medial temporal regions of the brain, in *Biology of Memory* (Pribram, K. H. & Broadbent, D. E., eds.), Academic Press, New York

MILNER, B. (1971) Interhemispheric differences and psychological processes. *Br. Med. Bull. 27*, 272–277

MILNER, B., CORKIN, S. & TEUBER, H.-L. (1968) Further analysis of hippocampal amnesic syndrome: 14 year follow-up study of H.M. *Neuropsychologia 6*, 215–234

MISHKIN, M. (1954) Visual discrimination performance following partial ablations of the temporal lobe. II. Ventral surface vs. hippocampus. *J. Comp. Physiol. Psychol. 47*, 187–193

MISHKIN, M. & PRIBRAM, K. H. (1954) Visual discrimination performance following partial ablations of the temporal lobe. I. Ventral vs. lateral. *J. Comp. Physiol. Psychol. 47*, 14–20

NADEL, L. & O'KEEFE, J. (1974) The hippocampus in pieces and patches; an essay on modes of explanation in physiological psychology, in *Essays on the Nervous System* (Bellairs, R. & Gray, E. G., eds.), pp. 367–390, Clarendon Press, Oxford

ORBACH, J., MILNER, B. & RASMUSSEN, T. (1960) Learning and retention in monkeys after amygdala-hippocampus resection. *Arch. Neurol. 3*, 230–251

PENFIELD, W. & MILNER, B. (1958) Memory deficit produced by bilateral lesions in the hippocampal zone. *Arch. Neurol. Psychiatr. 79*, 475–497

RAMACHANDRAN, V. S. (1976) Learning-like phenomena in stereopsis. *Nature (Lond.) 262*, 382-384

SCHMALTZ, L. W. & ISAACSON, R. L. (1966) The effects of preliminary training conditions upon DRL 20 performance in the hippocampectomized rat. *Physiol. Behav. 1*, 175-182

SCOVILLE, W. B. & MILNER, B. (1957) Loss of recent memory after bilateral hippocampal lesions. *J. Neurol. Neurosurg. Psychiatr. 20*, 11-21

SIDMAN, M., STODDARD, L. T. & MOHR, J. P. (1968) Some additional quantitative observations of immediate memory in a patient with bilateral hippocampal lesions. *Neuropsychologia 6*, 245-254

TALLAND, G. (1965) *Deranged Memory*, Academic Press, New York

VICTOR, M., ADAMS, R. D. & COLLINS, G. H. (1971) *The Wernicke-Korsakoff Syndrome*, F. A. Davis, Philadelphia

WARRINGTON, E. K. & ACKROYD, C. (1975) The effects of orienting tasks on recognition memory. *Memory and Cognition 3*, 140-142

WARRINGTON, E. K. & WEISKRANTZ, L. (1968a) A new method of testing long-term retention with special reference to amnesic patients. *Nature (Lond.) 217*, 972-974

WARRINGTON, E. K. & WEISKRANTZ, L. (1968b) A study of learning and retention in amnesic patients. *Neuropsychologia 6*, 283-291

WARRINGTON, E. K. & WEISKRANTZ, L. (1970) Amnesic syndrome: consolidation or retrieval? *Nature (Lond.) 228*, 628-630

WARRINGTON, E. K. & WEISKRANTZ, L. (1971) Organisational aspects of memory in amnesic patients. *Neuropsychologia 9*, 67-73

WARRINGTON, E. K. & WEISKRANTZ, L. (1973) An analysis of short-term and long-term memory defects in man, in *The Physiological Basis of Memory* (Deutsch, J. A., ed.), pp. 365-395, Academic Press, New York & London

WARRINGTON, E. K. & WEISKRANTZ, L. (1974) The effect of prior learning on subsequent retention in amnesic patients. *Neuropsychologia 12*, 419-428

WARRINGTON, E. K. & WEISKRANTZ, L. (1978) Further analysis of the prior learning effect in amnesic patients. *Neuropsychologia*, in press

WEISKRANTZ, L. (1971) Comparison of amnesic states in monkey and man, in *Cognitive Processes of Nonhuman Primates* (Jarrard, L. E., ed.), pp. 25-46, Academic Press, New York & London

WEISKRANTZ, L. (1974) The interaction between occipital and temporal cortex in vision: an overview, in *The Neurosciences, Third Study Program* (Schmitt, F. O. & Worden, F. G., eds.), pp. 189-204, The MIT Press, Cambridge, Mass. & London

WEISKRANTZ, L. & WARRINGTON, E. K. (1970a) Verbal learning and retention by amnesic patients using partial information. *Psychon. Sci. 20*, 210-211

WEISKRANTZ, L. & WARRINGTON, E. K. (1970b) A study of forgetting in amnesic patients. *Neuropsychologia 8*, 281-288

WEISKRANTZ, L. & WARRINGTON, E. K. (1975) The problem of the amnesic syndrome in man and animals, in *The Hippocampus*, vol. 2: *Neurophysiology and Behavior* (Isaacson, R. L. & Pribram, K. H., eds.), pp. 411-428, Plenum Press, New York

WEISKRANTZ, L. & WARRINGTON, E. K. (1978) Conditioning in amnesic patients. *Neuropsychologia*, in press

WEISKRANTZ, L., WARRINGTON, E. K., SANDERS, M. D. & MARSHALL, J. (1974) Visual capacity in the hemianopic field following a restricted occipital ablation. *Brain 97*, 709-728

WILLIAMS, M. (1953) Investigation of amnesic defects by progressive prompting. *J. Neurol. Neurosurg. Psychiatr. 16*, 14-18

WINOCUR, G. & BRECKENRIDGE, C. B. (1973) Cue-dependent behavior of hippocampally damaged rats in a complex maze. *J. Comp. Physiol. Psychol. 82*, 512-522

WINOCUR, G. & MILLS, J. (1970) Transfer between related and unrelated problems following hippocampal lesions in rats. *J. Comp. Physiol. Psychol. 73*, 162-169

WINOCUR, G. & WEISKRANTZ, L. (1976) An investigation of paired-associate learning in amnesic patients. *Neuropsychologia 14*, 97-110

ZANGWILL, O. L. (1966) The amnesic syndrome, in *Amnesia* (Whitty, C. W. M. & Zangwill, O. L., eds.), pp. 71-91, Butterworths, London & Washington, D.C.

Discussion

Winocur: Professor Weiskrantz has shown that the memory performance of brain-damaged amnesics can be substantially improved by cueing patients at recall with partial information about the material to be remembered. Our own work with amnesics has also been concerned with the use of cueing techniques to compensate for memory deficits but we have focused on the manipulation of cues extrinsic to the tasks themselves. For example, in one experiment (with Dr Marcel Kinsbourne, Hospital for Sick Children, Toronto, Ontario), we found that, although Korsakoff amnesics were characteristically poor in recalling a previously presented list of paired-associate words under standard testing conditions, their memory improved significantly when contextual conditions associated with the original learning and recall were made highly distinctive. The results are presented in Fig. 1. In original learning, the paired-associate lists, which consisted of 12 semantically related stimulus and response words (e.g. battle—army), were presented four times followed 60 seconds later by a recall test. Forty-eight hours later,

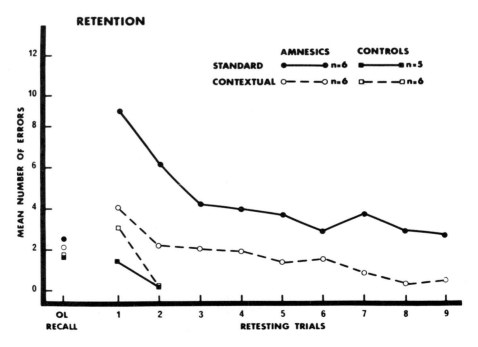

FIG. 1. (Winocur). Error scores of amnesic and control groups during recall and re-learning of a semantically related paired-associate list under standard and distinctive contextual cue conditions.

subjects were returned to the same room for nine re-learning trials. In the standard condition, all testing took place in a small room with sound and illumination levels normal. In the contextual condition, all testing was conducted in a room illuminated only by a bright red lamp and with taped music playing in the background. The main points to be drawn from Fig. 1 are that there were no differences between amnesics and controls in the initial recall test, that amnesics generally were inferior to controls in re-learning both lists, that amnesics were better at recalling and re-learning the list under the contextual conditions, and that cue manipulations did not affect the performance of the controls.

Our data generally indicate that amnesics' performance is consistently better in tasks which can be directly associated with distinctive external cues. In another set of experiments, Dr Kinsbourne and I used a transfer of learning paradigm developed by Professor Weiskrantz and myself (Winocur & Weiskrantz 1976) in which subjects are initially taught a list of semantically

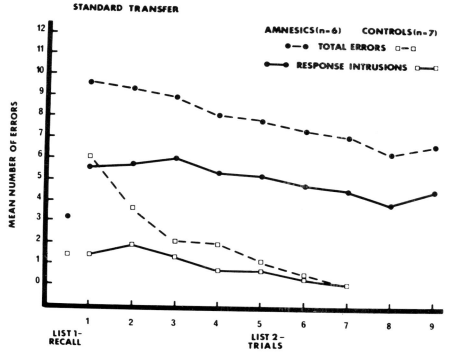

FIG. 2. (Winocur). Error patterns of amnesic and control groups in standard transfer of learning test.

related paired-associate words (e.g., peace–tranquil) and subsequently required to learn a second list consisting of the original stimulus words to which new similarly related response words are to be associated (e.g., peace–calm). Amnesics typically have no difficulty in learning the original list but are severely impaired in learning the second list. As can be seen from Fig. 2, this impairment is related to a tendency to provide the previously correct List 1 words in response to the familiar stimuli. Such response intrusions are, of course, consistent with the well-known vulnerability of amnesics to interference phenomena. We have found, however, that if the two lists are presented under contrasting contextual conditions, of the sort previously described, response intrusion errors are dramatically reduced and List 2 learning is significantly improved. Fig. 3 compares List 2 learning by amnesics and controls in a standard condition where both lists are presented in a normal, familiar setting and in a contextual-shift condition where List 1 is presented in the presence of highly distinctive cues and List 2 is presented in a normal but contrasting setting. An acquisition-only condition is also represented where amnesic and control groups were taught a list under standard testing conditions without any prior training. Fig. 3 shows that amnesics benefited significantly from the contextual shift and in fact there was no statistical difference between amnesic's performance in the contextual shift and acquisition-only conditions. As for the controls, their best performance, as expected, was in the acquisition-only condition where there was virtually no interference from past experience and they were unaffected by cue variations.

These results are part of a general pattern emerging in our research which, we believe, relates the amnesic syndrome at least in part to a fundamental misuse of information in high-interference situations where cues ambiguously signal different responses. Furthermore, they are in direct accord with Kinsbourne & Wood's (1974) proposal that amnesics suffer a deficiency in the use of contextual cues at retrieval.

For the most part, amnesic subjects in the experiments just described have come from populations of Korsakoff patients, generally believed to have extensive damage to the limbic system and particularly in the dorso-medial thalamus and mamillary bodies. These structures are known to have important hippocampal connections and indeed some investigators, notably Brion (1969), have related the amnesic component of Korsakoff psychosis to hippocampal damage. Certainly in its important features, the memory loss pattern associated with Korsakoff patients does not differ significantly from that reported by Milner (1967) and others for cases of bilateral hippocampal damage. Accordingly, we began to examine the effects of contextual manipulation on the behaviour of animals with hippocampal damage in an attempt to uncover possible points of convergence between the animal hippocampal and human amnesic syndrome.

Although the hippocampus has been conceded a central role in mediating memory in humans, attempts to demonstrate similar memory loss in hippocampally damaged animals generally have been unsuccessful. This, however, may be due to significant procedural differences in testing memory across

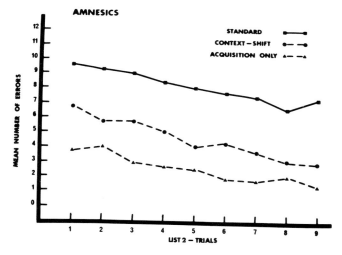

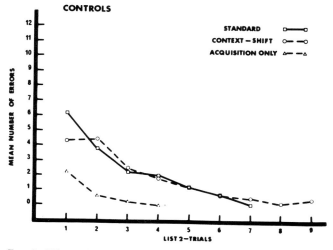

FIG. 3. (Winocur). Comparisons of errors made by amnesic and control groups in learning a semantically related paired-associate list under standard transfer, contextual-shift transfer, and acquisition-only test conditions.

TABLE 1 (Winocur)

Mean scores for hippocampally lesioned and control groups of rats before and after a shift in contextual cues (AB, BA) or no shift (AA, BB)

	Hippocampal lesion			Cortical lesion (control)			Operated controls		
	Original learning	Re-learning	% savings	Original learning	Re-learning	% savings	Original learning	Re-learning	% savings
Shift:									
AB	40	16.7	53.2	34	4.0	86.7	47.5	4.0	90
BA	40	12.5	56.5	52	6.0	88.6	37.5	2.5	93.8
$\bar{X}$	40	14.3	55.1	43	5.0	87.6	42.5	3.8	91.9
No shift:									
AA	46	2.0	95	37.5	10.0	70.8	44	0	100
BB	33.3	1.7	91.7	38.0	4.0	90.0	41.7	0	100
$\bar{X}$	39.1	1.8	93.2	37.8	6.7	81.5	42.7	0	100

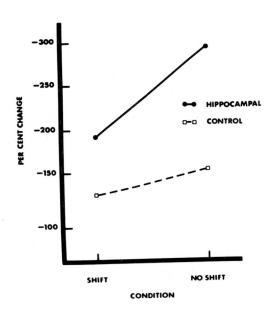

FIG. 4. (Winocur). Performance of hippocampally lesioned and control groups of rats on a reversal learning task expressed as a percentage of the original discrimination learning.

species. In animal experiments, special care is usually taken to ensure that the original learning and retention conditions are as similar as possible. Moreover, the distinctive character of the novel learning situation and the absence of distraction between test sessions facilitate the retrieval process by minimizing interfering influences. This contrasts with human testing where the potential for interference is far greater because of the use of well-known material in highly familiar settings. Long inter-test intervals filled with routine activities in the same general setting and normal psychological variation undoubtedly contribute additional sources of interference at the time of recall.

We thus reasoned that if the apparently normal memory of hippocampal animals is a function of the standard practice of ensuring minimal variation between original learning and retesting conditions, their performance may be adversely affected by procedures which vary the cues across tests. This hypothesis was examined in an experiment by Janet Olds and myself in which we tested groups of hippocampally lesioned rats and controls on a simultaneous discrimination habit and several days later tested recall for that discrimination either under the same conditions or under highly contrasting conditions where as many as possible visual, auditory, and tactile contextual cues were varied. The results were quite clear (see Table 1). All groups learned the original discrimination equally well and there were no differences in recall when background stimuli were not varied. However, the hippocampally damaged animals showed a clear memory impairment in the contextual shift condition.

These results emphasize the increased dependence of hippocampally lesioned animals on the presence of distinctive conditioned cues for effective functioning. This characteristic may also help to explain the disproportionate amounts of interference associated with hippocampal damage in tasks (e.g., passive avoidance learning, extinction, reversal learning) where experimental contingencies change despite the presence of familiar stimuli. Impairments on these tasks may reflect the failure of hippocampally damaged animals to redefine the significance of functional stimulus elements and identify a new contextual situation. If animals with hippocampal lesions are basically insensitive to subtle changes in experimental conditions, it follows that manipulations which contrast the stimulus conditions associated with the different task components should compensate for this deficit and contribute to a reduction of negative transfer. This hypothesis was tested in a reversal learning paradigm, involving a simultaneous discrimination task in which contextual stimuli were varied between the original learning and the reversal test. Hippocampally lesioned rats were found to be much worse than controls in reversing the previously learned discrimination when contextual conditions were constant throughout testing. However, the difference was eliminated when the original and reversed stimulus relationships were presented in a highly contrasting surround (see Fig. 4). These results may be compared with a transfer of learning performance of human amnesics and suggest that hippocampally damaged animals similarly benefited from the presence of salient contextual stimuli which dissociate incompatible tasks and facilitate appropriate response selection.

Although research aimed at establishing parallels between the animal and human syndromes is still in its early stages, at this point the contextual hypothesis has been shown to be useful in describing similar aspects of the respective deficit patterns. It may, in addition, provide the basis of a unifying theoretical framework regarding the behavioural function of the hippocampus across species.

Gray: We know that there will be no more patients like H.M., and there aren't many now. To what extent can we regard as relevant to the septo-hippocampal system, or the hippocampus in general, data from other kinds of amnesic patients?

Weiskrantz: There are still some surgical cases around, unfortunately. One of our cases, for example, had been given a unilateral temporal lobectomy for relief of epilepsy which resulted in the amnesic state; one assumes that the other temporal lobe was pathological. Some of Penfield's early cases were of this sort. But there are two issues. First, to what extent are all patients who superficially show the amnesic syndrome similar in detail psychologically? There is some dispute about this; Dr Winocur and I have looked at patients with different aetiologies on one particular type of long-term memory test and we did not find differences. Dr Warrington and I also have not noted differences on our cued recall memory tests between amnesic patients with different aetiologies. These subjects are entitled to differ in many other ways because of varying pathologies, but on long-term memory, we see no qualitative differences.

Secondly, the anatomical basis is also controversial. It was formerly argued that in the alcoholic Korsakoff syndrome, the common pathological structure was the mamillary bodies. That has been challenged by Victor (1964), who has evidence of four cases of degeneration of the mamillary bodies without the amnesic syndrome but with the other neurological defects associated with Wernicke's disease. He has argued that the medial dorsal nucleus of the thalamus has the closest association with amnesia. That takes us out of the hippocampus, but it is one reason why I am interested in species differences, because of the old evidence for a fornix–medial dorsal nucleus projection.

It may be that such a projection is more prominent in monkeys than in rodents. So the anatomical picture with regard to the hippocampus and amnesia is not clear. The pathology suggests that in Korsakoff syndrome the medial dorsal nucleus as well as the mamillary bodies degenerate. There are a few cases of fornix lesions in man in which the amnesic syndrome is reported. There is a little post-mortem evidence. We are now getting post-mortem material on two of our Korsakoff cases whose memory impairments were studied very intensively, but we don't know the histology yet.

Vanderwolf: It's important to consider that different behaviours may have different neurological bases whether they are learned or not. With respect to eyelid conditioning, there is a paper by Marquis & Hilgard (1936) showing that removing the occipital cortex has no effect on conditioning of this kind. Norman *et al.* (1974) have shown that decerebrate cats show good eye-blink conditioning.

Weiskrantz: I am not altogether surprised that we get eyelid conditioning in a brain-damaged patient; what struck me was the dissociation between the clear long-term learning and retention of conditioning, on the one hand, and the subject's commentary on it, on the other. In my paper I also mentioned earlier examples of such a dissociation.

Vanderwolf: Yes, and Dr Grossman mentioned earlier that in hippocampal seizures you can get good conditioning of leg flexion, but for another behaviour that involves running around, you get amnesia.

Gaffan: There are many controversies in this area; fortunately we don't need to be concerned with them all. I personally don't believe that there *is* an amnesic syndrome. In other words, I think it is doubtful whether there is an identifiable clinical entity which people agree on and which recurs in different patients.

Weiskrantz: I not only disagree with that; I disagree with it strongly!

Gaffan: I base this on disagreements between laboratories.

Weiskrantz: Patients with complex pathology, which is the usual situation, are entitled to a range of deficits including short-term memory deficits, migraine, ataxia, or what have you. There are two important points. On the specific kind of long-term memory tests that we are looking at, the amnesic patients behave similarly. The second question goes right through the analysis and interpretation of clinical material. This is *not* how complicated and varied a pathological picture you can see, but what can be dissociated from what—what features are essential for the designation of the amnesic syndrome and which features are optional, as it were. The Boston group and the National Hospital group agree, I understand, that while the Boston patients often have short-term memory deficits this is not a necessary feature seen in all amnesic patients. Dr Warrington and her colleagues have argued that short-term and long-term memory impairments are clearly dissociable and that amnesic patients can be perfectly normal on short-term memory measures (Baddeley & Warrington 1970; Warrington & Shallice 1969; Shallice & Warrington 1970; Warrington & Weiskrantz 1973). The fact that the two sets of scores are correlated may help the factor analyst, but it doesn't help the neuropsychologist who wants to discover what the absolutely critical element is in a syndrome.

Gaffan: It is important to know whether some things can be dissociated from other things, but I don't think we have to accept amnesia as a recognizable, recurring syndrome. If you show, as you have done, that some people can be observed with virtually no evidence of long-term memory retention in tests of recognition memory, whereas in tests of short-term memory with distraction they perform at normal levels, that is an important fact about memory, and we have no need to worry where the lesion is or whether it is a recurring syndrome, and whether there is always the same anatomical cause.

Secondly, this symposium is on septo-hippocampal interactions and there are cases where the post-mortem evidence indicates that the fornix which connects those structures has been damaged, and these patients have severe difficulties with memory. I am not saying that any patient coming with what some psychologists would say is the amnesic syndrome and others, general dementia, or any person who has difficulties with memory, necessarily has damage to the hippocampal system. We do need to say that there are three reported cases (two with post-mortem evidence) where the body of the fornix has been damaged bilaterally, and no other structure is damaged bilaterally. All three patients had extreme memory difficulties (Hassler & Reichert 1957; Brion *et al.* 1969; Heilman & Sypert 1977). There are no cases where people have claimed that fornix transection has resulted in a different behavioural pattern of impairments. The controversy arises from the fact that some people claim to have cut the pre-commissural descending columns of the fornix deeply and observe no psychological effects at all.

Grossman: Isn't there a case on record of congenital absence of the fornix, and the woman's ability to learn and remember was normal?

Weiskrantz: The agenesis cases introduce interesting developmental issues. The same arises in agenesis of the corpus callosum, and it may be difficult to generalize from the development of a brain lacking a structure to the damage of that structure in the adult brain.

Grossman: You said that your hypothesis would be hard to test in animals. A possible test would be as follows. You can train an animal to do something to gain access to another task, which is then rewarded. If the animal learns the first task, it must have a memory of the fact that the second task will be rewarded. It has to have access to that memory.

Molnár: I did this experiment (Molnár 1965; cited in Molnár 1973). We had a big cage with three identical feeders attached to the centre of three walls, but we used only one of them. The animals (cats) were trained to sit quietly in the centre of the cage. On hearing CS1 the cat had to go outside to a smaller cage. When the animal stepped on the floor of the small cage (i.e., pushed the pedal there), CS1 went off and CS2 occurred; the animal could then go to the feeder placed opposite the enclosure to receive food. So, as Dr Grossman has just said, if the animal learns to approach the 'detour' part of the task (i.e. to approach the pedal), it must have a memory of the fact that the second task will be rewarded. Further, we had two groups of animals (Fig. 1, p. 398). In Group I, SC1 and CS2 both came from the top of the large cage; in Group II, CS2 came from above the correct feeder. After hippocampal lesions, the cats are hyperactive, roaming around instead of sitting quietly and waiting for the CS1, as they were before the operation. If they heard CS1 they continued their incessant walking and came to the enclosure (and stepped on the pedal) only by chance. Their subsequent behaviour showed interesting differences between the two groups. If CS2 came from above the positive feeder (Group II), they went immediately to collect their reward. If, however, CS2 came from above the top of the cage they continued moving and visited all the feeders repeatedly. This latter observation supports the notion of the key role of the hippocampus in spatial orientation, also (cf. O'Keefe & Black, this volume).

Weiskrantz: In fact, the experiment suggested by Dr Grossman may not be a good model for the human experiment. I have in mind two different meanings of 'access'. There is access in the conventional sense of retrieval, where some partial cue or associated piece of information provides a route to stored information and aids in its being extracted. I would not be surprised if access in this sense still operated in an animal with, say, section of the fornix, and that he might show adequate learning of a two-compartment serial task.

But I want to introduce a second meaning of 'access'. It refers to *monitoring* of our own behaviour and our own past experience. There are various routines that can be carried out skilfully at one level which we do not monitor; indeed, the more skilled the action the less likely is monitoring to occur, as in driving a car under undemanding conditions. We can be certain that the skill depends on past experience and on quite specific access to quite a large store of information, and therefore there is excellent access to this information in the first sense of access, but not in the same sense that allows us to stand back, as it were, and offer a commentary as the routine action

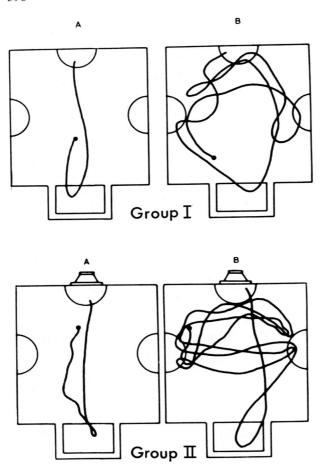

Fig. 1. (Molnár). Effect of bilateral, subtotal hippocampal ablation on performance in a chained alimentary task. A: before and B: after hippocampal lesion.

Group I: both CS1 and CS2 presented from the top of the cage.

Group II: CS2 transferred to just above the feeder to be approached. Movement of a tiny lamp attached to the head of the cat was photographed in red background light.

unfolds. We tend to offer commentaries only for a restricted range of actions, whether learned or not; presumably, those that might benefit from them. Indeed, nothing interferes with driving so much as being asked to make a detailed commentary about how one actually drives! Access, in my second sense, is a commentary at one level about what is happening at another level. I am suggesting that the amnesic patient, or the animal with corresponding pathology, has lost a particular monitoring capacity, or his monitoring system has become disconnected. It is relatively easy to understand this concept of loss of monitoring in cases of 'blind-sight'; there are patients with damage to the visual cortex who can be shown to be capable of making visual discriminations but who report no awareness of 'seeing' as such. I am suggesting that perhaps the amnesic patient can be shown to remember (as indeed he can) but reports no awareness of remembering (or, in some cases, a false awareness of remembering far more than he is entitled to remember). This is why I am attracted to the notion of the hippocampus as an interface between cerebral neocortex and older, lower brain structures. This is not to say that an interface may not impose certain transformations of its own in order to make the transfer of information back and forth more efficacious. An interface is not necessarily merely a relay.

One can ask what the advantage of such a monitoring system might be. One obvious advantage is that you can have memory not only in the sense that your past experience can affect your current behaviour, but you can think about your memories. The notion of the hippocampal animal as 'unthinking' may be very apt. When we think about our memories, we can order them in space and time; we can compare one with the other; we can manipulate them cognitively. The amnesic person is unable to do that and he is severely disabled and has to be hospitalized, perhaps not because he lacks the memory but because he lacks access to it in a manipulatable form.

Similarly, in 'blind-sight', even though there is at least some measure of form discrimination, spatial localization and control of eye movements by external stimuli, the subject thinks he is guessing and, in the conventional use of the word 'see', he doesn't see. He is unable to use that visual information to operate cognitively on his visual world, to manipulate his visual imagery. So there are two meanings of access: in animals I am not sure how to get at the difference operationally, but David Gaffan's dissociation between recognition and associative memory comes closest to it.

Gaffan: May I just say that the most dramatic memory impairments in fornix-transected monkeys are in picture recognition memory, where the performance of normal monkeys is very impressive (Gaffan 1977); the delayed matching task with objects is not as well performed by normal animals and is almost certainly very contaminated with associative processes, since the objects are rewarded at acquisition. But the results from this task (Gaffan 1974) do indicate a very specific memory impairment, rather than any general visual or motor disorder, since the same animals who were impaired in matching were perfectly able to perform a purely associative task with the same objects.

O'Keefe: Dr Weiskrantz, I want to question your assumption that in the normal person in your conditioning situation there is only one memory trace—that which underlies the conditioned response. This leads you to conclude that when you question a person about his knowledge concerning the conditioning experiment the answers he gives are dependent on some part of his brain gaining access to the part which contains the sole memory trace. It seems to me equally possible that there are two independent memory traces stored in different parts of the brain, one underlying the conditioning and another separate, say episodic, memory system which contains the information that in certain circumstances the person had certain experiences.

Weiskrantz: One reason why I stressed the bizarre character of the conditioning apparatus was because episodic memory was not necessary for correct identification. The subject had never seen the apparatus and did not have to remember when or even whether the experience had occurred, because he was sitting in front of the apparatus all the time; in the same way that a cue in the verbal learning situation can provide non-episodic access to the item, the appearance of the apparatus could serve as a cue in a non-episodic sense for that situation; yet it doesn't seem to do so. Also, the kinds of learning and retention we now know that amnesic subjects can display are many and include situations where both semantic and phonetic rules are applied and can be shown to be operating, and include a variety of forms of verbal learning. It is not just crude reflex conditioning that is retained but a variety of kinds of learning with no single characterization, as I tried to argue in my paper. The important factor is the nature of the retrieval cue, not the nature of the original learning, or so I believe.

Vanderwolf: Can the patients learn to find their way round the hospital?

Weiskrantz: Typically, no, although this can be a matter of degree. However, it is worth repeating that there are cases with topographic memory difficulty, typically associated with temporal-parietal lesions of the right hemisphere. They have severe difficulty in finding their way round and in learning spatial mazes. But they have no amnesia for non-spatial events, and would not generally be said to be 'amnesic' in the clinical sense.

Swanson: In patients whose corpus callosum has been cut for medical reasons, when a visual stimulus is presented to the right hemisphere and produces a reaction, if the patient is asked to say why he did it, he can't explain why. Is there any similarity between this and what is seen in your patients?

Weiskrantz: That is an example of a horizontal disconnection, stopping interaction between the two hemispheres. Dr N. Geschwind has argued that several neurological defects can be explained in terms of interruption of cortical–cortical connections. I have considered this possibility in visual 'blind-sight' cases—that it could be a disconnection from verbal commentary. Professor R. W. Sperry kindly allowed me to test his commissurotomy patients in the way we tested our first visual case. The 'split-brain' patients are qualitatively different from ours. That is why I suggest that we consider vertical rather than horizontal disconnection for these cases of amnesia; and, after all, we have evidence that one thing the hippocampal system does is to transfer information from one level to another.

Rawlins: I have only met amateur human amnesics, so I am not sure what sort of conditions best demonstrate amnesia in man. Professor Weiskrantz said that it's not easy to see how to demonstrate a disconnection of the monitoring system in animals. However, it is clear that there are tasks in which animals reliably show a deficit after hippocampal damage. John O'Keefe is able to characterize these tasks by appealing to a (sometimes intuitive) spatial component in them. Jeffrey Gray tries to characterize them by referring to the consequences of the animal's behaviour, particularly when the behaviour leads to non-reward or punishment. Can you, Professor Weiskrantz, suggest a pleasing classification of tasks in which your monitoring system needs to be effective? Can you discriminate between your classification and Dr O'Keefe's, or between yours and Dr Gray's?

Weiskrantz: If you take the kinds of characterizations that come from Jeffrey Gray's group, from the spatial hypothesis group, and perhaps from David Gaffan's evidence on recognition failure, the appropriate situation appears to be one in which an animal cannot depend merely on one piece of information to provide the answer out of a number of stored alternatives; he must order the events in relation to each other, not only spatially but temporally, or in terms of one or another cognitive system. He must search in memory, not simply activate a well-established link. The spatial maze is one situation in which a search may be necessary, but I am unhappy about the notion that a spatial factor is essential because the human amnesic subject has a massive non-spatial memory difficulty and, as I have said, spatial memory loss can be dissociated from non-spatial memory loss. Of course, it may be that the rat hippocampus is functionally different from or more restricted than man's. Dr Gray's characterization of the prototypical sensitive task is one in which the animal *temporarily stores* information in order to give him a much needed chance to pause and reconsider. There are two parts to that suggestion. According to my view: pausing and reconsidering, perhaps; but temporary storage, no. Amnesic patients have no difficulty in considering transient events intelligently and their short-term memory can be quite normal. The hippocampus, despite evidence of the possibility of persistent neurophysiological changes—frequency potentiation, for example—is not, in my view, the place to look for storage as such. It is the place in which to look for a sorting out of possible stored alternatives available at a different level (and if such a sorting out isn't required, then the hippocampus can be dispensed with).

Rawlins: Do you think, then, that the effects of hippocampal damage seen in rats are solely a consequence of the way in which we test rats? There are few good, complex situations in which we can test them which are free of spatial elements.

Weiskrantz: There seem to be two ways of proceeding, and no doubt both ways will always be followed. One is to ask, for example in relation to David Olton's residential maze, is it 'lists' of items to be remembered or is it 'space'? One continuously dissects the behavioural tasks that are sensitive to septo-hippocampal lesions and tries to sort out what are the absolutely essential features and what are not. What I would *like* to be able to suggest, as a second approach, is some way to study vertical disconnection phenomena in animals, but we have no precedents and will have to find new methods.

Lynch: Is it possible that cues enter memory systems at different levels of the nervous system, where they become mixed with what is going on *at* that level? When you are mixing a cue with a response in the eye-blink situation, that may be happening way down the neuroaxis, but the cue is also entering the nervous system at other levels—cortical levels—and being integrated with the types of processing going on at those levels. The hippocampus may be necessary for that mixing process further up. Would that idea work as well as your verbal integrator?

Weiskrantz: I'm not sure, but so long as one is just dealing with, say, eye-blink conditioning, we can say that decorticate animals can do that. Probably the isolated eyelid could do it! But you can get complex verbal learning in amnesic subjects with the right cueing conditions, and this *must* depend on cortical mechanisms.

Lynch: But that does require cueing, whereas the eye-blink does not.

Weiskrantz: The conditioned stimulus is a cue.

Black: You asked about situations that would enable us to demonstrate a difference between the use of lists and the use of spatial information. It seems to me that we could do this if we found a situation that requires the animal to employ information about spatial relationships among stimuli.

Even if one demonstrates that there is a deficit in spatial information processing as a result of hippocampal lesions, one can't conclude that the deficit is specific to spatial information processing, unless you also demonstrate that there is no deficit in the processing of non-spatially organized information.

I don't think that that has been done yet. We have been focusing on the spatial situation first, so that we know the situation well for which we want to construct non-spatial analogues.

Weiskrantz: It depends on one's interest. David Gaffan has been working on recognition memory, which is often a case, almost by definition, of temporal order, where the subject must say whether an event did or did not occur before or, more commonly in man, whether it occurred more recently than the alternative.

Black: I thought that under certain circumstances animals with lesions of the hippocampus seem to do quite well on that task (M. Mishkin, unpublished data).

Gaffan: As general comments, firstly I very much appreciate the strength of Dr Olton's results; he is using a spatial recognition memory task and asking the rat to remember a list of spatial locations; and at a choice point he has to discriminate between a relatively familiar location where he has already been, and mustn't now go, and a relatively novel situation where he *hasn't* recently been, and he must now go there. In the monkey we do the same experiment with a list of objects (Gaffan 1974) or pictures (Gaffan 1977) in a completely non-spatial task where the animal has only to remember whether he has seen one or other object more recently. I am very happy that you also find recognition memory impairment in the rat.

To go to patients, there are gross impairments of the same kinds of recognition memory tasks, and I am not clear how Professor Weiskrantz's hypothesis explains these deficits. When the subject responds in, say, a forced-choice recognition with pictures of faces, in the most literal sense he is not asked to comment but just to point to the picture he has seen. I find it hard to believe that if you took such a patient and gave him food like Dr Olton's rats for doing a spatial or any other memory recognition task, his deficit would disappear. So I think we should make a comparison between the deficits in rats and monkeys, and the deficits in patients.

Finally, if you make a slightly different form of explanation in terms of the commentary notion and say that the commentary is not the patient's 'gossip' about what he is doing, but, rather, an elaboration of the memory trace, that is a more subtle internal hypothesis. That would be fine for explaining deficits in picture recognition or face recognition. But if the monkey needs to have access to a memory trace, reorganize it, think about it, in order to do a memory task well, and that is why he has impaired recognition memory for objects, I find it hard to believe that he is not doing equally complicated and difficult memory operations in comparable tasks in which he is not impaired, namely associative tasks where he must remember whether an object was rewarded or not. Equally complex control processes are surely required in the two kinds of task.

Weiskrantz: The word 'gossip' is a bad one because it suggests an epiphenomenal kind of chit-chat between the subject and experimenter. In fact the most important 'gossip' is the dialogue the subject has with himself. In a recognition task, the subject is trying to work out something about the order and arrangement of events, where there are several possible permutations. In the typical associative discrimination task, that degree of ordering and permutation is not required; it is usually a binary situation, even though the stimuli may be complex.

Black: I have some questions about the recognition memory hypothesis. In the eight-arm maze there is a deficit among animals with hippocampal lesions when the relevant stimuli are distant from the apparatus. Recently, Gordon Winocur has shown that there is no, or at least very little, deficit when a distinctive stimulus is placed on the floor of each of the eight arms of the maze. This suggests that animals with lesions of the hippocampus can perform successfully on a recognition task if the appropriate stimuli are available. (See p. 345.)

Gaffan: This must be clarified. Are the lesioned animals impaired relative to the controls on the visual conditions?

Winocur: I can't answer that in terms of statistics; lesioned animals made more errors than controls on the visual task, after preoperative training on the spatial task. The data have not yet been statistically analysed and I cannot say if the difference is significant.

Weiskrantz: There *is* a confounding between associative and recognition memory; there is a clear association within any one day between a particular visual stimulus and reinforcement.

Black: Isn't that true of spatial cues?

Weiskrantz: Exactly: there are both recognition and associative elements in the spatial situation. It is possible that what you are calling a spatial situation favours one factor more than the other, and conversely, in the visual case, it's the other way around.

Andersen: I would like to return to your model, Dr Weiskrantz: we perhaps agree that the eye-blink reflex need not be the same process as the thing which produces the memory trace of it. Would you comment on the following experiment by Peter Matthews' group (Goodwin *et al.* 1972)? In a normal, blindfolded subject, one of his biceps is vibrated to produce a tonic vibration reflex in response to the Ia afferent impulses. He is instructed to move the other arm to copy the position of the vibrated arm. He manages to do so, although with a certain lag, but is not aware of it—and becomes surprised at the position of the arm when he opens his eyes.

Gray: Your point is that this kind of disconnection takes place in the normal subject?

Andersen: Yes, and not necessarily associated with hippocampal activity.

Weiskrantz: There is no doubt about this. It would be a very inefficient nervous system that had its monitoring system always connected to the whole stream of its activity. We are *not* aware which eye we use in a particular situation. We don't know anything directly about the state of our lenses or many other bodily adjustments. In skilled behaviour such as driving a car, the more skilled you become, the less able you are to give a commentary on what movements you have been making recently. It is efficient *not* to have that kind of information.

Gaffan: In understanding memory we should not overlook the fact that in patients, the best evidence, as I mentioned before, is that lesions of the body of the fornix, interrupting connections between septum and hippocampus, produce a memory deficit but they do not produce defects in response to non-reward or difficulties related to spatial tasks, which are produced by quite different brain lesions.

The evidence in man suggests that we should concentrate on memory, because the ability to distinguish novelty is not simply something used to control the orienting reflex or something that produces behavioural inhibition; it is something which is a memory output and can be used for many different purposes and for subtle memory tasks. You can train an animal to respond with either approach or withdrawal to a novel stimulus, or whatever you like, but he nevertheless has the ability to distinguish the novel from the relatively familiar input, and that is an aspect of memory.

There is one aspect that I would like to ask the physiologists about. We can't, of course, make any rational model other than a purely speculative one that will go directly from synaptic events to the behavioural response of an animal or a person in a memory task, in detail, at this stage. However, in a memory task, what we can do in psychology is to make calculations about the distributions of the two classes of stimuli which we are asking subjects to discriminate, namely things he has seen and things he hasn't seen, in a recognition task. We can calculate a d' index and other similar indices on the performance, which are defined in terms of the variance of the noise distribution and the variance of the signal distribution, and the difference between their means. These enable us to predict how the substrate of the memory trace should behave in terms of possible distributions of signal and noise which are consistent with the discrimination of the whole organism (Gaffan 1978). I would like to see more detail on the form of the decline of long-lasting changes, either the potentiation changes, or the decline of

Dr Vinogradova's kind of decreases and increases. I would like to see this in terms of the mean and the variance and the higher moments and the shape of the distribution; then I could say if those were consistent with normal memory performance.

References

BADDELEY, A. D. & WARRINGTON, E. K. (1970) Amnesia and the distinction between long- and short-term memory. *J. Verb. Learn. Verb. Behav.* 9, 176–189

BRION, S. (1969) Korsakoff's syndrome: clinico-anatomical and physiopathological conditions, in *The Pathology of Memory* (Talland, G. A. & Waugh, N. C., eds.), Academic Press, New York

BRION, S., PRAGIER, G., GUÉRIN, R. & MME TEITGEN (1969) Syndrome de Korsakoff par ramollissement bilatéral du fornix. *Revue Neurologique 120*, 255–262

GAFFAN, D. (1974) Recognition impaired and association intact in the memory of monkeys after transection of the fornix. *J. Comp. Physiol. Psychol.* 86, 1100–1109

GAFFAN, D. (1977) Monkeys' recognition memory for complex pictures and the effect of fornix transection. *Q. J. Exp. Psychol.* 29, 505–514

GAFFAN, D. (1978) Measurement of trace strength in memory for pictures. *Q. J. Exp. Psychol. 30*, in press

GOODWIN, G. M., McCLOSKEY, D. I. & MATTHEWS, P. B. C. (1972) Proprioceptive illusions induced by muscle vibration: contribution by muscle spindles to perception. *Science (Wash. D.C.) 175*, 1382–1384.

HASSLER, R. & REICHERT, T. (1957) Über einen Fall von doppelseitiger Fornicotomie bei sogenannter temporaler Epilepsie. *Acta Neurochirurgica (Wien)* 5, 330–340

HEILMAN, K. M. & SYPERT, G. W. (1977) Korsakoff's syndrome resulting from bilateral fornix lesions. *Neurology 27*, 490–493

KINSBOURNE, M. & WOOD, F. (1974) Short-term memory processes and the amnesic syndrome, in *Short-Term Memory* (Deutsch, J. A., ed.), Academic Press, New York

MARQUIS, D. G. & HILGARD, E. R. (1936) Conditioned lid responses to light in dogs after removal of the visual cortex. *J. Comp. Psychol.* 22, 157–178

MILNER, B. (1967) Memory disturbance after bilateral hippocampal lesions, in *Cognitive Processes of the Brain* (Milner, P. M. & Glickman, S., eds.), Van Nostrand, Princeton

MOLNÁR, P. (1973) The hippocampus and the neural organization of mesodiencephalic motivational functions, in *Recent Developments of Neurobiology in Hungary*, vol. IV (Lissák, K., ed.), pp. 93–173, Akadémiai Kiadó, Budapest

NORMAN, R. J., VILLABLANCA, J. R., BROWN, K. A., SCHWAFEL, J. A. & BUCHALD, J. S. (1974) Classical eyeblink conditioning in the bilaterally hemispherectomized cat. *Exp. Neurol.* 44, 363–380

SHALLICE, T. & WARRINGTON, E. K. (1970) Independent functioning of verbal memory stores: a neuropsychological study. *Q.J. Exp. Psychol.* 22, 261–273

VICTOR, M. (1964) Observations on the amnestic syndrome in man and its anatomical basis, in *Brain Function*, vol. II: *RNA and Brain Function: Memory and Learning* (Brazier, M. A. B., ed.), pp. 311–340, University of California Press, Berkeley & Los Angeles

WARRINGTON, E. K. & SHALLICE, T. (1969) The selective impairment of auditory verbal short-term memory. *Brain 92*, 885–896

WARRINGTON, E. K. & WEISKRANTZ, L. (1973) An analysis of short-term and long-term memory defects in man, in *The Physiological Basis of Memory* (Deutsch, J. A., ed.), pp. 365–395, Academic Press, New York & London

WINOCUR, G. & WEISKRANTZ, L. (1976) An investigation of paired-associate learning in amnesic patients. *Neuropsychologia 14*, 97–110

Final general discussion

FREQUENCY POTENTIATION AND MEMORY

Gray: May we take up the question of frequency potentiation again? Material has been presented on this (by Dr Andersen and by Dr DeFrance) and it has been much studied at the neurophysiological level, but I don't know any data that indicate that it has a role in normal functioning, or anything to do with memory, for which it is often used as a model. Dr Rawlins has done experiments which suggest that this kind of stimulation is not very powerful in controlling an animal's behaviour.

Rawlins: My experiments (Rawlins 1977) were not in fact designed to investigate frequency potentiation as such; the primary aim was to establish whether the lamellar organization of the perforant path would enable rats to use stimulation, whose excitatory effects would be restricted to a fairly narrow lamella within the dorsal hippocampus, as a discriminative stimulus. Initially I attempted to train the subjects to choose one arm in a Y-maze when stimulated through one perforant path electrode, and to choose the other when stimulated via another perforant path electrode. The electrodes had been positioned under electrophysiological control to ensure that, while both of them would elicit responses visible using an 'on beam' recording electrode, the areas maximally excited by each of the stimulation electrodes were differently located along the longitudinal axis of the hippocampus (Rawlins & Green 1977). Having failed to train rats to respond appropriately to stimulation of this kind in an appetitive task in the Y-maze, I tried to train them to use perforant path stimulation as a discriminative stimulus in an appetitive task in the Skinner box. Here they had to press one of two levers; the appropriate lever was signalled by the location of the stimulation electrode (a discrimination of lamellae) or by the frequency (4 or 12.5 Hz) of the stimulation delivered via one perforant path electrode (frequency discrimination). Again, there was no evidence of learning, though the stimulation did not interfere with the discrimination if the correct choice was at the same time indicated using a stimulus light located in the appropriate lever.

Lastly, I attempted to train a further group of animals to use perforant path stimulation of 4 or 10 Hz as a conditioned stimulus in classically conditioned suppression of bar pressing in the Skinner box. Again, suppression was observed if both intrahippocampal stimulation and an exteroceptive stimulus were presented at the same time; it was also seen if the exteroceptive stimulus alone was used; but it was not seen when perforant path stimulation alone was used. Furthermore, no sign of a behavioural response was visible when the stimulation was first delivered to the unanaesthetized subject, although a clear hippocampal response was visible through the chronically implanted recording electrode. Thus, it was impossible to demonstrate any behavioural effect of restricted perforant path stimulation at all; given this, I was unable to demonstrate any significance of frequency potentiation.

Weiskrantz: The question is at two levels: can the intact whole animal use this information, in the same way as Doty has asked about stimulation at various loci in brain; and, secondly, will it serve as an isolated model, say, for plasticity in some region or systems in the brain, whether or not they have any relation or access to other parts of the brain or the brain as a whole? Those are two different questions. I can imagine many things changing in the animal—say, synaptic conductivity, which would be crucial for learning—to which an animal might not be able to respond as a specific event in a conditioned suppression test.

Gray: Either way my question remains: what evidence is there that frequency potentiation either has a role in normal function, or is a good model for plasticity? I accept the evidence that long-lasting potentiation can occur in the intact animal, but what does that tell us about the role of the hippocampus in normal functioning?

Lynch: Douglas (1977) has put different trains of stimulation into the perforant path of hippocampus in the rat and finds that the best parameters clearly mimic certain of those that Jim Ranck has described, namely four pulses separated by 3–5 ms, spaced apart, like the complex-cell behaviour already described. These are excellent conditions for inducing potentiation in the intact animal. This makes us ask what frequencies mean—what these burst patterns mean. We assume when we see these patterns of firing that there is information exchange, but potentiation may give us cause to think that the synapses are being adjusted in synaptic strength, and this may be done by using specific frequency patterns.

Weiskrantz: May I introduce an important distinction here? There are a variety of ways in which any system can change its sensitivity. The most obvious examples are adaptation and fatigue, and we don't get excited talking about those changes when we discuss memory, but we do get more interested when we hear about manipulations of the kind that Dr Vinogradova has carried out, where the properties of the stimulus are changed and one shows that dishabituation is very specific to the kind of input that produced the original habituation. If, for example, frequency potentiation is done with a particular frequency or pattern of stimulation, is the change then specific to that pattern and not to other patterns? Can you train it to a specific input stimulus or are you simply changing the sensitivity of the system?

Andersen: To me, those two possibilities are exactly the same.

Weiskrantz: So you would call muscular fatigue 'memory'?

Andersen: Neither more nor less than I call long-lasting facilitation, memory. It is a plastic, long-lasting change. At the moment we don't know the molecular basis for these events. But it has now been shown that long-lasting facilitation is specific to the used line, and that it is sensitive to the frequency of the used line. But my main point is that at present we should not use frequency potentiation as a model of memory; it is a model of synaptic change—the very small building blocks from which the complex thing we call memory is constructed.

Ursin: When we talk about behavioural effects of stimulating electrically through the hippocampus we deal with three different phenomena. The first is the blocking or 'jamming' type of stimulation that Peter Livesey has been using; the second is the semipathological sensitization of the kindling effect, which may or may not be related to frequency potentiation; the third is the type that Dr Rawlins is describing, where hippocampus stimulation is used as a cue. We must keep these three things separate. The relevance of sensitization for learning and memory is indeed questionable. In any conditioning experiment one has to check whether it is the repetition of the stimulus which produces the change (sensitization), or whether the changes are due to the temporal contiguity of two or more stimuli (conditioning, learning). As I understand him, Dr Gaffan has been saying that the sensitization model could be used for memory models, but you have to add certain conditions. Can he amplify this?

Gaffan: I think we have to accept long-lasting potentiation as a basic trace phenomenon which, when further elaborated, could be used as a memory trace which will be useful in the kinds of memory task that are impaired by a lesion of the hippocampal system. I don't see what further evidence one could have that an electrophysiological cellular phenomenon is important in natural behaviour except of this kind; that when you make a lesion of the structure that it is taking place in, you get changes in the appropriate kind of behaviour, and that there is a correlation between the electrical recording and what is needed to explain the behaviour of the normal animal. This correlation must be exact and, as I said before, we need more precise data on the time course of the decline of potentiation and any other long-lasting physiological changes, including not simply the means of various measures but also their distribution. If you find a good correlation, what else can you do? How do you know that John O'Keefe's spatial units have anything to do with the perception of space? We only assume that if one shows that a lesion impairs that kind of task, and the unit responds appropriately, the animal is using it for something.

O'Keefe: It seems to me that the physiologist can show the psychologist what sort of synaptic changes he ought to look for during learning. In the learning of a recognition task, for example, you could try to show that there are changes in hippocampal synapses of the type that Per Andersen has shown in the slice after potentiation.

Gaffan: A point that has been raised in relation to long-lasting potentiation in the hippocampal system is that the same process might occur in other parts of the brain. But that is important, because there are many aspects of memory in which animals with fornix section are not impaired (including remembering that an object has been accompanied by non-reward!). So it is fortunate that it is *not* unique. Secondly, it has been said that this is not an associative memory; it is not like a Hebb synapse; all it can tell you is that this input had occurred before. But since the kind of memory task on which a fornix-sectioned animal is impaired is exactly the type where the animal must tell if something has occurred before, that is not an objection. The other problem raised is the time course. The evidence from the intact animal is less good than that from brain slices, but there is good reason to think that these changes last for at least 48 hours. Isn't it time we became less shy of trying to bring together the physiology and psychology? If a change lasts for 48 hours, that is certainly long enough for memory from my point of view, as a psychologist.

Secondly, there is a rapid decrease in the effect which has been thought to be inconsistent with a long-term memory system. We need not be worried by this; there is also a rapid decrease in memory in the first minute, as Professor Weiskrantz and I have shown in recent unpublished experiments on object recognition memory in monkeys.

Segal: It is relevant to this question that I have found cells in the entorhinal cortex of rats which fire rapidly after a conditioned stimulus, at the rate of 10–15/s for up to 10 seconds (Segal 1973). I also found that the responsiveness of cells in the hippocampus to the conditioned stimulus would depend on this. In other words, the closer the next conditioned stimulus to the previous one which evoked this increase in entorhinal cortex cellular activity, the larger the response of the hippocampal cells to the conditioned stimulus. This effect was blocked by cutting the perforant path between the entorhinal cortex and hippocampus, so it looks like a case of natural frequency potentiation that causes an increase in responsiveness of the hippocampus to an external stimulus.

DeFrance: It is too early to be disappointed that we cannot find tight correlations between the potentiation phenomena that we see and overt behavioural changes. This merely stimulates us to think about possible mechanisms for long-term changes in cellular excitability. Intuitively, however, it does seem reasonable that learning involves a change in cellular excitability.

Weiskrantz: Any mechanism underlying learning must involve an enduring change at some point in the system: the question is whether this is sufficient, not just necessary, for a model of learning. The difficulty that psychologists see when they distinguish between, say, muscular fatigue and learning is that even in the typical simple memory situation it is not just a change that is important but a *comparison*, between previous and present events. That comparison is commonly of a matching sort, 'this is the same event or a different event' or 'this is an event which has a certain significance because it is the same event which in the past had certain consequences', as you have in conditioning situations. That is what you cannot ask a muscle, or the adapted eye. The muscle doesn't know that its state has altered, nor does the retina know when it is dark-adapted. That kind of consideration is the kind of elaboration I think is needed if an observed change is to satisfy criteria for 'memory'.

Andersen: I don't think anyone has claimed these as memory, but we hope they may be building blocks for memory.

Vinogradova: Hesse & Teyler (1976) showed the disappearance of long-lasting potentiation after electroshock in the rat hippocampus. They regarded this as equivalent to the erasing of short-term memory traces by electroshock in animals.

Ranck: One way to begin to determine the psychological significance of some physiological mechanism is to study the electrophysiology of behaviourally identified neurons. In theory one could determine which cells had a change in firing *associated* with learning and then see if the electrophysiological properties of the cell had changed, and in what way. Technically this is possible now.

There is a central problem in all studies of single cells and learning. If learning has occurred the behaviour, or potential behaviour, of the animal has changed. Therefore, a neuron may change its firing during learning, but still be associated with the same behavioural events. In a sense this neuron has not changed during the learning. In my own work, I always assume the most conservative hypothesis, and look for the characteristics of the cell which are constant, even though learning may be occurring.

Weiskrantz: Perhaps it would be fair to conclude (in answer to Jeffrey Gray's original question) that frequency potentiation has not been shown in any behavioural sense to be of significance in memory, but that it provides a possible way of looking at one element in any such change that would be necessary.

FURTHER COMMENTS ON THE HIPPOCAMPAL COGNITIVE MAP THEORY

O'Keefe: I would like to add two points to the discussion of the cognitive map theory in our paper (O'Keefe & Black, this volume). The first concerns the way that hippocampal place units might be connected together to form a map of an environment, and the second concerns the way that these maps are initially built and how they are subsequently modified.

As we discussed in our paper, there are cells in the hippocampus which represent places in an environment. Conversely, a place in an environment would be represented by a group of these place cells, and the whole environment would be represented by all of these groups taken together. Such a collection of place representations would not constitute a map, however. For a map one also requires a mechanism which connects together these place representations into a spatially ordered framework. The framework must specify which places are next to each other and the directions and distances between places. One way that part of the framework might be provided is by neighbouring place cells representing neighbouring parts of an environment, a topographically isomorphic representation. So far, the physiology suggests that this is not the case, since two neighbouring place cells recorded from the same microelectrode position are as likely to represent two distant parts of an environment as two neighbouring ones (O'Keefe 1976).

An alternative possibility is that there is a mechanism which shifts the focus of excitation within the hippocampus as a function of the animal's displacement during movements through the environment. Thus as an animal moved in an environment this system would calculate how far and in what direction that movement would take the animal. This information about the spatial parameters of the movement would be transmitted to the hippocampus where, in combination with the activity of the place cells representing the animal's position at the start of the movement, it would activate the place cells representing the part of the environment where the movement landed the animal. A different movement starting from the initial position might result in a different spatial displacement and would result in the activation of a different set of place cells. We have proposed that hippocampal theta activity might be a reflection of the operation of the spatial displacement mechanism and have suggested one way in which it might work (O'Keefe & Nadel 1978).

A second point concerns the motive which leads an animal initially to build a map of an environment and subsequently to update that map when it no longer agrees with that environment. We believe that maps are built and maintained out of the purely cognitive motive of curiosity, out of a desire to know about the places in its world and the things they contain. Curiosity is generated by the mapping system itself. When there is a mismatch between the sensory input which the hippocampus receives when the rat is in a place and the sensory input which it expects to receive in that place, a subset of place cells which we have called misplace cells signals the mismatch. Reference to Fig. 2 of our paper (p. 186) might give us some insight into this misplace mechanism. If one compares the firing of the cell during those trials when one or more of the sensory inputs have been removed (F-I) with the firing when all cues are present (A-D), there is a clear increase. Thus, for example, in H and I the cell continues to identify the place in the environment but does so with an increased number of spikes. We think that this increased firing acts as a misplace signal and that when it reaches a certain level it disrupts the ongoing behaviour and triggers exploration. Exploration, then, is viewed as information-gathering behaviour which is directed towards those sensory inputs in an environment which are either not represented or are inaccurately represented in the animal's map of that environment. Its purpose is to correct the inaccurate representation and to bring it into line with the new sensory information. We suppose that when an animal first enters an unfamiliar environment there is no representation of that environment in the hippocampus and therefore a maximal firing in the misplace cells. As the rat explores and place representations are built up in the hippocampus, the firing of these cells would decrease and in some cases cease all together. These suggestions have not been directly tested yet, but they might account for the fact that the highest amplitude theta activity is often seen when an animal first enters a novel environment.

Black: We had difficulty in convincing others that our cognitive map model did relate to reality, largely because what is predicted on the basis of this model is often very similar to what is predicted on the basis of a simple stimulus–response learning model. The problem arises from the following confounding. If you are going to provide the animal with a situation in which it can form a map of the environment during exploration, you must provide it with stimuli in its environment. Once you have provided such stimuli, one can always argue that the animal is using them as cues which it approaches or avoids rather than as components of a spatial map. In short, whenever you create an environment that is rich in stimuli in order to allow the animal to form a cognitive map, you also have an environment in which the animal can learn to move about by approaching or avoiding particular cues. It seems to us that one way out of this bind is to employ situations that require the use of information about spatial relationships. That is what we were trying to do in the experiments described in our paper.

Weiskrantz: It seems to me that there is a cognitive problem about the cognitive map itself! How does one go from a spatial cognitive map to the kinds of non-spatial cognitive operations implicated both by hippocampal recording and studies of the effects of lesions? When you go from rat to man, that becomes a serious problem.

There is a further question, which arises from lesion studies, of whether there are not a number of deficits on non-spatial tasks that require explanation, and, looking at the spatial deficits in detail, whether they are best characterized in terms of a cognitive map system or in terms of a behavioural inhibition system or, indeed, in terms of some other model. So, are there alternatives to this spatial model on the basis of evidence from lesions and from electrophysiology? But the first question is the cognitive one: how can you extend this model to a broader range of cognitive operations?

Livesey: I want to rush in here where angels fear to tread, or at least where Dr O'Keefe didn't tread in his paper, to look at what we might mean by a cognitive map and then put an alternative proposition to the one Dr O'Keefe proposed. The reference to angels is not entirely innocent; it encapsulates what I would picture as a cognitive map. That is to say, the angel knows where he or she (or it) is, and it knows where in that particular instance it does *not* want to be. In other words, there is purposive behaviour involved. The notion of a cognitive map is therefore one where the organism knows the environment and knows what it is going to do in relation to that environment. We can picture an animal moving through a familiar environment as a spot of light, which is the animal's record of where it is. But that in itself isn't, in my view, the whole picture of a cognitive map. I think the animal is engaged in some purposive behaviour and knows its objective, and as it moves through that environment there is an expectation of the next set of stimuli it will encounter and a marking-off of those stimuli. If the animal is distracted, or fails to follow the pathway, there is a mismatch which is registered. If the animal gets into a new part of the environment another type of mismatch is registered. If an animal goes to a particular place where it expects to find a reward, or whatever its goal is, and doesn't find it, again you get this registration of mismatch.

The notion I wanted to develop is that while there *is* undoubtedly a cognitive map, it is not of the type that John O'Keefe has been talking about. I find it hard to conceive of single cells acting as memory cells, just as I have always had problems with the notion that a single cell constitutes a discrete memory store or 'unit of memory'. An image or representation of the environment is undoubtedly developed within the nervous system but I believe the hippocampus is not the locus.

Many of the behavioural situations that have been related to hippocampal function do not involve spatial mapping of the sort referred to by Dr O'Keefe. In our own experiments the animals were involved in simultaneous discrimination of visual cues independent of spatial location (Livesey & Bayliss 1975; Livesey & Meyer 1975; Livesey & Wearne 1973), or successive or go/no-go discrimination at one spatial location (P. J. Livesey, J. P. Smith & P. Meyer, in preparation). Jim Olds, in a series of brilliant studies, showed that many hippocampal cells respond with anticipatory changes in firing rates to signals that the animal has learned to identify with food or water, the nature of the response being determined by the motivational state of the animal (Olds & Olds 1961; Olds 1967; Olds & Hirano 1969; Olds *et al.* 1969). In these studies the animals were stationary and in 'an expectant waiting mode'. Thompson (1976) working with the nictitating membrane response in the rabbit also demonstrated very rapid conditioning of unit responses in the hippocampus in immobilized animals.

The anatomical location of the hippocampus also argues against such a function. It sits astride two major information-processing systems, the posterior neocortex which deals with environmental information and the brain core structures that deal with information about internal states, e.g. changes in need levels with consequent modulation of drive states, rewards and punishments. On this very basis one would expect the hippocampus to be involved in relating the two sets of information. My view is that it functions to facilitate the development of representations of the effectiveness of interactions between organism and environment and in the utilization of representations so laid down. In particular, I would argue that the hippocampus has evolved as a system especially sensitive to situations where an anticipated outcome fails to eventuate—that is, in the detection of error or changes in the meaning of relationships that have already been established.

The work of Jim Ranck (1973) and Olga Vinogradova (1975) also strongly supports the view that we are looking at a comparator system and not a spatial map and I believe O'Keefe's findings also conform with this interpretation.

If the hippocampus is concerned with expectancy and with direction of attention when expectations fail to materialize, then it must be particularly sensitive to change. For this to be so the anticipated event or match must first be signalled. Both Olds (Olds & Olds 1961) and Thompson (1976) have observed very rapid changes in responsiveness in hippocampal cells as an association is registered. Thompson commented that 'Because the response develops within a very few trials of training it may well be the earliest neural indication in the brain that learning is occurring'.

O'Keefe appears to be demonstrating the process of establishment of familiarity and expectancy within a spatial context. We know from other work in our laboratory (J. A. Bell & P. J. Livesey, in preparation) and from Mackintosh's work (Mackintosh 1973) that the context or environmental setting within which, for example, a cue–response–reward relationship is established, can be crucial for later recognition of the significance of the cue. If the cue is then presented out of context the significance is lost. It therefore seems likely that O'Keefe's 'location' cells are signalling an expectancy in a spatial context and this is a similar phenomenon to Olds' 'anticipatory firing' and Ranck's 'approach-consummate' cells. O'Keefe's finding of 'mismatch' cells, that responded when the animal was anticipating an event at a location and this anticipation was not realized, is consonant with Ranck's 'approach-consummate-mismatch' cells. While a spatial representation or map of the environment is essential for these cognitive processes to be effected, this does not mean that the hippocampus holds such a map, but rather that it utilizes it.

Finally, let me mention a topic which we have neglected in this symposium —the relation between the septo-hippocampal system and emotion. One of the major input/output systems of the hippocampus is with the brainstem core and structures that have been shown to monitor needs and to mediate motivational and emotional states, yet the significance of this relationship has scarcely been touched upon.

The rewarding effects of brain stimulation to the septum have been demonstrated in rats (Olds 1956) and man (Heath 1964) and for the CA3 region in the rat (Ursin *et al.* 1966). Reward and punishment effects have been observed from hippocampal stimulation in man (Heath 1964).

Olds (1969) demonstrated anticipatory firing of neurons in CA3 and concluded that the firing was in expectation of rewards associated with food or water. The approach-consummate cells described by Ranck (1973) would seem to be in this category. Jeffrey Gray, in his theoretical construct generated by his findings, refers to frustrative non-reward and an anxiety model which directly implicates emotional state. In his view the hippocampus is particularly sensitive to the absence of expected reinforcement.

In my expectancy model I have talked of the generation of expectancy and resultant match or mismatch. But expectancy of what? Expectancies are generated in relation to reinforcers or, more directly, I believe, to the affective components of reinforcers—that is, rewards and punishments. We have a strong hint, then, that the hippocampus is directly involved in the monitoring of reinforcement processes and that these may, in fact, constitute the affective components of our experience. This is an aspect of septo-hippocampal function that calls for our attention in the future.

FUNCTIONS OF THE SEPTO-HIPPOCAMPAL SYSTEM IN MAN AND OTHER ANIMALS

Grossman: It seems to me that there is an intuitively obvious truth relevant to the general question of the function of the septo-hippocampal system that we have so far disregarded, namely that this system almost certainly does not have one single function. It is patently absurd to take any large component of the brain and to try to attribute to it a single function, whether spatial mapping, some sort of retrieval function, behavioural inhibition, affect, or whatever. None of us would take this approach to, say, the thalamus or hypothalamus. Yet, we have been doing just that with respect to the septo-hippocampal system which is at least as complex. The behavioural effects of hippocampal lesions are as diverse as those observed after destruction of any other major component of the brain, if not more so. The worst we could do is to argue whether it has one function *or* another. What we should do instead is to try to reach agreement on the nature of the multiple functions the septo-hippocampal system appears to have, and to discuss whether they might be related in some sensible way; in order to discover whether there is some reason why those functions are located together in this part of the brain. The hypothalamus, for example, has many different functions, but they have some relationship to each other. We must not forget, in this context, that one tends to find what one looks for. Those of us who have been convinced that inhibitory processes that affect all reactions to non-reward and, perhaps, punishment are an important aspect of the functions of the septo-hippocampal system have tended to observe behaviour in situations designed to bring out such influences. Others have similarly stressed experimental situations that emphasize some aspects of memory, spatial organization, and so on. I would like to suggest that we look for common elements rather than differences in our observations as well as our interpretations.

Black: Let me comment on that point. When we use terms such as 'spatial', we don't imply a single function. The term encompasses a variety of functions, and the point is to provide a rationale for grouping them together.

In addition, I think that the attempt to limit the number of functions that one ascribes to a particular brain structure is important from a research strategy point of view. It forces one to at least attempt to reduce the number of different functions that one ascribes to the structure. I think that it is much easier to take what might be called a more slothful approach, and simply list a variety of functions for particular structures that correspond to the variety of behavioural tests that one carries out. Different structures will, of course, involve different numbers of functions. For example, the anatomical and behavioural data suggest that the septum may involve more different behavioural functions than the hippocampus.

One further point. In listening to the discussions, it has struck me that there is a common feature to most of the theoretical notions that have been expressed at this symposium. Many share the idea that hippocampal circuits are involved in what might be called match–mismatch information processing —that is, in detecting certain types of environmental changes. Dr Vinogradova's theory clearly makes this assumption; the spatial theory of O'Keefe & Nadel (1978) states that the hippocampus is involved in detecting changes in spatial cognitive maps; Dr Gray's hypothesis argues for the role of the hippocampus in detecting environmental changes, as does Dr Gaffan's. These hypotheses do differ, of course, with respect to the importance that they assign to this match–mismatch system, but at least it's a common feature.

Gray: I should like to try to reconcile what Abe Black has just said (with which I fully agree) with what Peter Grossman said previously. It seems that all of us (with the possible exception of Dr Weiskrantz!) have used the notion of a comparator. We haven't actually given historical credit where it is due here. This idea comes out in Sokolov's (1960) original notion of the orienting reflex, of the neuronal model of the stimulus and of the matching of the model against the incoming information. What Dr Black just said about the use of the comparator for different kinds of input is critical, and can offer the solution to the problem posed by Dr Grossman. The actual septo-hippocampal circuits that we have discussed may be used by different kinds of input in different circumstances. The ones that have taken my attention have been situations in which what is being compared is the presence or absence of a reward. In those circumstances certain kinds of behaviour are appropriate and the septo-hippocampal system has an output to appropriate mechanisms (about which we know nothing) which produce that kind of behaviour.

Spatial behaviour seems to be an interesting case in which a lot of mapping of expected and actual outcome is important for what the animal is doing. There I would suppose that input is from slightly different places and output is to slightly different structures.

Again, with list memorizing and list production, there are somewhat different inputs and different outputs. The common feature is there, and the differences are because of the different integrations with different, other, inputs and outputs, about which we know very little.

I would like to pass this back to the anatomists, because one of my own difficulties in thinking of the septo-hippocampal system as controlling behaviour is the problem of how it communicates with motor systems. We know almost nothing about that.

Winocur: If we are to agree on the match–mismatch or comparator system as a major mechanism, we should also note that it is not a system or mechanism that is obliterated by hippocampal damage. As Dr Weiskrantz has shown (see his paper, pp. 373–406), it is much more inefficient, but a certain amount of matching and comparison *can* be undertaken if the patient or the animal is helped with additional information—additional cueing. I have studied this both in humans and in other animals, and when we create analogous situations between the testing paradigms used in each, we find surprisingly similar results. I think we can reconcile the animal and human data.

Weiskrantz: An earlier study by Kenneth Spence (1966) is relevant. He looked at eyelid conditioning with and without 'awareness' by human subjects, and recorded extinction rate. He did this because of a discrepancy between human and animal results on eyelid conditioning. With 'awareness', human subjects extinguish more rapidly than animals; without 'awareness', human subjects are closer to animal subjects. Looking at the extinction pattern in the amnesic subject may indeed be a way of inferring whether he has 'awareness'; in other words, there is already a paradigm that was deliberately devised just for this purpose of comparing man and animals.

Olton: Dr Gray's questions should not be overlooked—namely, what the anatomical connections are from the septo-hippocampal system to behaviour, and how the rest of the system functions where these connections are not present.

Vanderwolf: On Dr Gray's point, we know that the reticular formation in the midbrain is essential for a large variety of behaviours; after lesions these animals are unable to walk or to right themselves (Lawrence & Kuypers 1968). Grantyn et al. (1973) have shown that if the hippocampus is stimulated in various fields, one obtains excitatory postsynaptic potentials in midbrain or pontine reticular neurons. This would be one means by which hippocampal output could have a direct influence on reticulospinal mechanisms and hence on behaviour.

Swanson: Although we are left with the impression that the hippocampus acts as a comparator in associative and cognitive functions and so on, little attention has been paid here to equally important mechanisms underlying effects referred to by Dr Grossman and Dr Ursin, showing that the septo-hippocampal system profoundly affects the autonomic nervous system, the regulation of the neuroendocrine system, and more complex components of behaviour which involve these latter systems, such as motivation and emotion. The hippocampus has a unique structure and a unique set of inputs and outputs, and an understanding of its functions must include a consideration of what the precise nature of its inputs are and what part of the neuraxis they influence through its output. Present data suggest that sensory information is the most important input into the septo-hippocampal complex; in fact we know almost nothing about what type of information this is. It is not primary or even secondary sensory information; it is most probably polymodal sensory information, with the exception of more direct olfactory inputs to the entorhinal area. A second major source of inputs arises in the brainstem, including the hypothalamus and thalamus.

What does this tell us about the function of the hippocampus? Other parts of the brain, for example the striatum and deep pontine gray, receive widespread cortical inputs as well as diverse inputs from the brainstem. So the unique properties of the hippocampus must lie in the organization of its cortical and ascending inputs. At present, it is primarily the fault of anatomists that it is not possible to bridge the gap between structure and function in the hippocampal system. This can only be done by characterizing the organization of its inputs more clearly, especially with regard to the associative and cognitive functions of the hippocampus.

For example, the place system hypothesis of hippocampal function deals with electrophysiologically defined 'place units'. To what extent do such units form an anatomically defined map of the environment in the hippocampus, in a manner analogous to the organization of visual, auditory, and somatosensory units in their respective primary sensory cortices? Furthermore, how and to what extent is the topographical information arriving in the primary sensory cortical areas transferred to the entorhinal area? We know from the work of Jones & Powell (1970) that there is a polysynaptic relay of information from primary to secondary somatosensory cortex and then into prefrontal and temporal association cortices. This appears to be the simplest way by which somatosensory information is likely to gain access to the hippocampus. At the level of the association cortices there is also secondary visual and auditory input, probably to the same cells in some cases. Thus, the output of such a neuron to the hippocampus may be fired by a variety of sensory stimuli. Anatomically, the topography of projections from the primary and secondary sensory cortices, and how much of it is preserved and enters the hippocampus, needs to be determined.

Vinogradova: The crucial point is what kind of information goes to the hippocampus. I won't discuss the septal inputs, as they are better known, but I want to describe the sensory responses of neurons in the entorhinal cortex which were investigated in our laboratory by Dr Stafekhina (Stafekhina & Vinogradova 1976). There are considerable differences between these responses in the medial and lateral entorhinal areas (28 A and B). In 28 A, in the rabbit, about 70% of the units respond to various kinds of sensory information (in various combinations), but especially to visual and auditory stimuli. In the majority of units responses are of the 'specific', patterned type, and their patterns vary with change of a stimulus. For example, tones of various pitches may evoke in one and the same unit specific responses of quite different patterns. The same unit may respond also to visual and somatic stimuli by different types of responses. Thus, in 28 A a high level of convergence is observed with preservation of information coding in the output signals of neurons. Such characteristics of responses can be seen only in the highest association areas of the neocortex.

The situation is quite different in area 28 B. Here only 25% of units respond to visual and auditory stimuli. The responses are very sluggish, with latencies up to 200–400 ms, and diffuse. However, these units respond perfectly to somatic stimuli. They may respond selectively to tactile stimulation of one paw, eyelid, ear, etc. by short-latency responses. We also tested in 28 A and B the effects of electrical stimulation of the frontal and posterior convexity of the neocortex. We found a high proportion (34%) of units in the medial entorhinal responding to posterior neocortical stimulation by time-locked driving with short latencies (13–36 ms). There were no responses to stimulation of the frontal neocortex, or just weak diffuse effects. Exactly the opposite happens in the lateral entorhinal cortex where 42% of cells respond by short-latency (11–18 ms) driving to stimulation of frontal, but not of posterior cortex. Thus, there is a real double dissociation. This distinct localization in the entorhinal cortex is interesting in view of Van Hoesen's data on the differentiation of its neocortical inputs and on differentiated terminations of pathways from sub-fields of the entorhinal cortex to the hippocampus (Van Hoesen & Pandya 1975).

Swanson: To turn to the internal circuitry of the hippocampal formation, work summarized by Dr Andersen (1975) has emphasized that the hippocampus has a very rigid spatial organization, which he characterizes as essentially lamellar, or transverse to its longitudinal axis. Given the inputs described, one would like to know what the computer inside is doing with it. The internal circuitry is extremely complicated and not fully understood. However, the simple trisynaptic circuit that Dr Andersen described can be used as a starting point. Thus, pathways can be traced out of the entorhinal area which innervate both the dentate gyrus and Ammon's horn. There is, therefore, a direct entorhinal input to field CA3, and one relayed to it through the mossy fibre system. The mossy fibres also activate field CA4, however, which makes things much more complex, since the latter projects quite widely (in the longitudinal sense) to the dentate gyrus, Ammon's horn, and the subiculum on both sides of the brain. So it would seem that there are rather direct projections (one cannot say 'activation' from this kind of study) to wide parts of the dentate gyrus, Ammon's horn, and the subicular complex on both sides of the brain after the activation of a small region in the entorhinal area. Furthermore, since field CA1 projects back to the entorhinal area (L. W. Swanson, J. M. Wyss & W. M. Cowan, unpublished work 1977), the whole circuit may be started off again as a result of the initial stimulus. It is conceivable that information entering the entorhinal area from the neocortex, for example, may end up circulating up and down both sides of the hippocampus to some extent.

Andersen: There is also inhibition, which cuts across the lamellae and modifies the activity. Because what matters is activation, not mere projections.

Gray: Those connections must be there for a purpose; they can't just be being turned on and off again for no good reason!

Swanson: The important feature of hippocampal topography is the longitudinal organization of each of its cortical fields, with each field in turn having its own characteristic pattern of longitudinally organized afferent and efferent connections.

Gray: We ought now to be trying to put together the anatomy (and physiology) with the behavioural ideas we have discussed. If we take the notion of the septo-hippocampal system as a comparator seriously, it should *not* get regular, fixed sensory input or perceptual information. The cells responsive to many different kinds of stimulation in the entorhinal cortex and then in the hippocampus must be *potentially* responsive but not all the time; that is to say, only some of this information must be allowed into the septo-hippocampal system as and when required for the match–mismatch mechanism to work.

Weiskrantz: Are there not two stages here? In order for there to be a match–mismatch, there has to be a generous supply of input information; a further gate then occurs as to what happens when there is a mismatch. Where are you going to put the filter?

Gray: One of the interesting things anatomically is that the output from the hippocampal formation goes back, via the subiculum, both down to the midbrain and out to the entorhinal area. Is it possible that the septo-hippocampal system controls its own inputs, both about novelty and about familiarity (to use Olga Vinogradova's distinction), in such a way that it can itself control the information coming in at both ends? It can't be receiving all the sensory world under all circumstances all the time.

Zimmer: One should also remember that there are projections from the thalamic nuclei to the hippocampus itself (Herkenham 1976) and to the retrohippocampal areas (Shipley & Sørensen 1975). Secondly, when considering the septo-hippocampal output, recent work by Herkenham & Nauta (1977) shows that some of the limbic output converges with pallidal output on the lateral habenula, which then in turn projects to the brainstem.

Finally, on phylogeny, Dr K. Kawamura (personal communication) has been studying the connections of the different association cortices, combining his own results with what is available from the literature. In lower mammals, such as the cat, there are projections from these cortices through the cingulum to the entorhinal area, whereas in primates (monkey) there appear to be additional connections to these areas from the phylogenetically younger frontal parts of the cerebral cortex. Accordingly, going from one species to another one will have to consider qualitative as well as quantitative differences in the hippocampal input.

Molnár: May I remind you here that Douglas (1967) has a nice terminology. He says that there are two basic inputs to the hippocampal formation: predigested (or entorhinal) and motivational (or septal) information may reach the structure. We should take into consideration the motivational input through the septal nuclei.

Endre Grastyán's work, showing a correlation between the orientation reflex and the hippocampal functions, should perhaps also be recalled here (Grastyán 1959; Grastyán *et al.* 1959, 1966; Grastyán & Vereczkey 1974). He said that orientation is accompanied by hippocampal theta rhythm and that the function of the hippocampus is to inhibit this reflex.

The point is that while concentrating on the theta mode of action, we are inclined to forget that the hippocampus might also be desynchronized. As to the functional state of the hippocampus during theta and desynchronization, respectively, an increasing amount of evidence suggests that during the so-called 'pacemaker action' of the septal input, what actually happens is that the septum, instead of pacemaking, periodically filters out parts of the cell activity, elicited by the excitatory entorhinal inputs, impinging on the dendrites of pyramidal cells (cf. Molnár 1973). This means that we should take into consideration both the theta rhythm and the desynchronization when trying to decipher the functional meaning of electrical events in the hippocampus. Probably the latter lies with the more active state, and during desynchronization you really find suppression of thalamic evoked potentials elicited by electrical pulses to the reticular formation. Further, if you elicit movements in animals by stimulating the motor cortex, simultaneous hippocampal stimulation will inhibit these movements (Grastyán 1959).

A final comment to Dr Vanderwolf; you are right, Grantyn & Grantyn showed a direct output from the hippocampus to the reticular formation, but they also showed that there is a direct input from the limbic midbrain area to the hippocampus, which is, most significantly, inhibitory in nature (Grantyn & Grantyn 1972).

Srebro: There is a general impression from this discussion that the major part of the septum is just a shadow, or appendix, of the hippocampus! We should remember that the septum is a very complex and heterogeneous structure. It has extensive projections to other cortical areas besides the hippocampus as well as a wealth of connections with the amygdala, hypothalamus, thalamus and the midbrain structures. It has sometimes been called 'an interface' between the phylogenetically old structures of the mesencephalon and diencephalon and the newer cortical structures of the telencephalon. This discussion has not brought out the fact that physiological and behavioural functions related to the hippocampus may be only a small part of the septal involvement in CNS processes. Still, it should be acknowledged that this symposium has provided abundant evidence that the integrity of the medial septum is necessary for the normal functioning of the hippocampus.

Andersen: We should also remember Olga Vinogradova's and Andersson & Santini's (1972) data that there is a large and effective somatic input to the pyriform cortex. This may be an important input to the entorhinal area. It comes predominantly from the skin and from the joints and has a relatively short latency (20 ms, in the cat).

We should also not forget that the bulk of activity is completely intrahippocampal and only a small fraction comes out. With regard to output, we should remember that the number of efferent axons increases dramatically in evolution, so that in man the hippocampus has a much larger output than in the monkey, where the hippocampal output is again much greater than in the rat. Finally, of the output that I have studied, the thalamic cells receiving fimbrial output are driven as effectively as the lateral septal nuclei.

So any scheme must include on the afferent side, short-latency skin and joint afferents, and on the efferent side, the thalamic output.

Vanderwolf: I just want to add here that the idea that theta activity has something to do with novelty has surely been disproved by a lot of data now.

Gray: This brings up the problem of the relationship between data acquired by electrophysiological recording methods and data obtained by intervention methods. Dr Vanderwolf and his group have seen extremely good correlations between movement and theta activity in the rat. He agrees that this correlation breaks down in other species such as the rabbit and cat. It seems a counsel of despair simply to attribute this to species differences in the meaning of theta, and not to seek further explanations. In any case, when one turns to intervention methods (including lesions and other methods of abolishing theta activity), there is no causal relationship in the rat between theta and movement. It is more reasonable to suppose that the correlation between theta and movement is greater in the rat than in the rabbit or cat because, during exploratory behaviour, the rat moves and sniffs around while the rabbit and cat remain still and use their eyes. Theta activity is present when all three species actually move; but when they are exploring their environment, whether you see movement or not turns on the way each species explores. The relationship between theta and exploratory behaviour—the collecting of information—is still very good, just as Grastyán (Grastyán *et al.* 1959) said a long time ago.

Black: I don't think that the correlations break down; they are still here, but they become more complicated. That is, the correlation between movement and hippocampal theta cells occurs in all of these species which display hippocampal theta activity; in addition a second type of theta activity, atropine-sensitive theta, which is not correlated with movement and whose behavioural correlates do not seem to be worked out clearly, also occurs; there seem to be large differences among species with respect to the prominence of this second type of hippocampal rhythmical slow activity (RSA). For example, it occurs with great frequency in rabbits and very infrequently in rats.

With respect to your last point, as I mentioned earlier, animals take in information and process it very well when they are not producing theta activity.

* * *

Weiskrantz: This may be a good point at which to conclude the discussion. I think we all agree that it has been a provocative and productive symposium. I am impressed by two things: first, that out of the undoubted complexity and welter of information there is nevertheless some common ground on which certain sign-posts can be placed; no doubt at the next meeting the sign-posts will be somewhat different and more detailed, but at least there has been some convergence. Second, unlike practically any other area of brain function, the septo-hippocampal system is an area in which not only are anatomists, physiologists, neurochemists and psychologists working towards a common goal, but they are interested in each other's data and consider that they have a bearing on their own discipline. This is both unusual and encouraging.

References

ANDERSEN, P. (1975) Organization of hippocampal neurons and their interconnections, in *The Hippocampus*, vol. 1: *Structure and Development* (Isaacson, R. L. & Pribram, K. H., eds.), pp. 155–175, Plenum Press, New York

ANDERSSON, S. A. & SANTINI, M. (1972) A cutaneous projection to the pyriform cortex in the cat. *Acta Physiol. Scand. 84*, 5–6A.

DOUGLAS, R. J. (1967) The hippocampus and behaviour. *Psychol. Bull. 67*, 416–442

DOUGLAS, R. M. (1977) Long lasting synaptic potentiation in the rat dentate gyrus following brief high frequency stimulation. *Brain Res. 126*, 361–365

GRANTYN, A. & GRANTYN, R. (1972) Postsynaptic responses of hippocampal neurons to mesencephalic stimulation: hyperpolarizing potentials. *Brain Res. 45*, 87

GRANTYN, R., MARGNELLI, M., MANCIA, M. & GRANTYN, A. (1973) Postsynaptic potentials in the mesencephalic and pontomedullary reticular regions underlying descending limbic influences. *Brain Res. 56*, 107–121

GRASTYÁN, E. (1959) The hippocampus and higher nervous activity, in *The Central Nervous System and Behaviour* (Brazier, M. A. B., ed.) *(Transactions of the 2nd Conference)*, pp. 119–206, Josiah Macy, Jr Foundation, New York

GRASTYÁN, E. & VERECZKEY, L. (1974) Effects of spatial separation of the conditioned signal from the reinforcement. (A demonstration of the conditioned character of the orienting response or the orientational character of conditioning). *Behav. Biol. 10*, 121–146

GRASTYÁN, E., LISSÁK, K., MADARÁSZ, I. & DONHOFFER, H. (1959) Hippocampal electrical activity during the development of conditioned reflexes. *Electroencephalogr. Clin. Neurophysiol. 11*, 409–430

GRASTYÁN, E., KARMOS, G., VERECZKEY, L. & KELLÉNYI, L. (1966) The hippocampal electrical correlates of the homeostatic regulation of motivation. *Electroencephalogr. Clin. Neurophysiol. 21*, 34–53

HEATH, R. G. (1964) Pleasure response of human subjects to direct stimulation of the brain. Physiologic and psychodynamic considerations, in *The Role of Pleasure in Behavior* (Heath, R. G., ed.), Harper & Row, New York

HERKENHAM, M. A. (1976) A thalamo-hippocampal connection in the rat. *Abstr. Soc. Neurosci. 2*, 389

HERKENHAM, M. A. & NAUTA, W. J. H. (1977) Afferent connections of the habenula nuclei in the rat. A horseradish peroxidase study, with a special note on the fiber-of-passage problem. *J. Comp. Neurol. 173*, 123–146

HESSE, G. W. & TEYLER, T. J. (1976) Reversible loss of hippocampal long term potentiation following electroconvulsive seizures. *Nature (Lond.) 264*, 562–564

JONES, E. G. & POWELL, T. P. S. (1970) An anatomical study of converging sensory pathways within the cerebral cortex of the monkey. *Brain 93*, 793–826

LAWRENCE, D. G. & KUYPERS, H. G. J. M. (1968) The functional organization of the motor system in the monkey. II. The effects of lesions of the descending brain-stem pathways. *Brain 91*, 15–33

LIVESEY, P. J. & BAYLISS, J. (1975) The effects of electrical (blocking) stimulation to the dentate of the rat on learning of a simultaneous brightness discrimination and reversal. *Neuropsychologia 13*, 395–407

LIVESEY, P. J. & MEYER, P. (1975) Functional differentiation in the dorsal hippocampus with local electrical stimulation during learning by rats. *Neuropsychologia 13*, 431–438

LIVESEY, P. J. & WEARNE, G. (1973) The effects of electrical (blocking) stimulation to the dorsal hippocampus of the rat on learning of a simultaneous brightness discrimination. *Neuropsychologia 11*, 75–84

MACKINTOSH, N. J. (1973) Stimulus selection; learning to ignore stimuli that predict no change in reinforcement, in *Constraints on Learning* (Hinde, R. A. & Stevenson-Hinde, J., eds.), Academic Press, New York & London

MOLNÁR, P. (1973) The hippocampus and the neural organization of mesodiencephalic motivational functions, in *Recent Developments of Neurobiology in Hungary*, vol. IV (Lissák, K., ed.), pp. 93–173, Akadémiai Kiadó. Budapest

O'KEEFE, J. (1976) Place units in the hippocampus of the freely moving rat. *Exp. Neurol. 51*, 78–109

O'KEEFE, J. & NADEL, L. (1978) *The Hippocampus as a Cognitive Map*, Oxford University Press, Oxford, in press

OLDS, J. A. (1956) A preliminary mapping of electrical reinforcing effects in the rat brain. *J. Comp. Physiol. Psychol. 49*, 281

OLDS, J. (1967) The limbic system and behavioral reinforcement. *Prog. Brain Res. 27*, 144–164

OLDS, J. A. (1969) The central nervous system and the reinforcement of behavior. *Am. Psychol. 24*, 114–132

OLDS, J. & OLDS, M. E. (1961) Interference and learning in palaeocortical systems, in *Brain Mechanisms and Learning* (Delafresnaye, J. F., ed.), pp. 153–188, Blackwell, Oxford

OLDS, J. & HIRANO, J. (1969) Conditioned responses of hippocampal and other neurons. *Electroencephalogr. Clin. Neurophysiol. 26*, 159–166

OLDS, J., MINK, W. D. & BEST, P. J. (1969) Single unit patterns during anticipatory behavior. *Electroencephalogr. Clin. Neurophysiol. 26*, 144–158

RANCK, J. B., Jr (1973) Studies of single neurons in dorsal hippocampal formation and septum of unrestrained rats. I. Behavioral correlates and firing repertoires. *Exp. Neurol. 41*, 461–531

RAWLINS, J. N. P. (1977) *Behavioural and Physiological Correlates of Limbic System Activity*. Unpublished D. Phil. Thesis, Oxford University

RAWLINS, J. N. P. & GREEN, K. F. (1977) Lamellar organization in the rat hippocampus. *Exp. Brain Res. 28*, 335–344

SEGAL, M. (1973) Dissecting a short term memory circuit in the rat brain: changes in entorhinal unit activity and responsiveness of hippocampal units. *Brain Res. 64*, 281–292

SHIPLEY, M. T. & SØRENSEN, K. E. (1975) On the laminar organization of the anterior thalamus projections to the presubiculum in the guinea pig. *Brain Res. 86*, 473–477

SOKOLOV, YE. N. (1960) Neuronal models and the orienting reflex, in *The Central Nervous System and Behavior* (Brazier, M. A. B., ed.) *(Transactions of the 3rd Conference)*, pp. 187–276, Josiah Macy, Jr Foundation, New York

SPENCE, K. W. (1966) Cognitive and drive factors in the extinction of the conditioned eye blink in human subjects. *Psychol. Rev. 73*, 445–458

STAFEKHINA, V. S. & VINOGRADOVA, O. S. (1976) Sensory characteristics of the hippocampal cortical input. Comparison between lateral and medial entorhinal cortex. *Zhurnal Vysshei Nervnoi Deyatel'nosti 226*, 1074–1081 (in Russian)

THOMPSON, R. F. (1976) The search for the engram. *Am. Psychol. 31*, 209–227
URSIN, R., URSIN, H. & OLDS, J. (1966) Self stimulation of the hippocampus in rats. *J. Comp. Physiol. Psychol. 61*, 353–359
VAN HOESEN, G. W. & PANDYA, D. N. (1975) Some connections of the entorhinal (area 28) and perirhinal (area 35) cortices of the Rhesus monkey. *Brain Res. 95*, 1–59
VINOGRADOVA, O. S. (1975) Functional organization of the limbic system in the process of registration of information: facts and hypotheses, in *The Hippocampus*, vol. 2: *Neurophysiology and Behavior* (Isaacson, R. L. & Pribram, K. H., eds.), pp. 3–70, Plenum Press, New York

Index of contributors

Entries in **bold** *type indicate papers; other entries refer to discussion contributions*

Andersen, P. 21, 22, 45, 46, 47, 80, 85, **87**, 102, 103, 104, 105, 106, 107, 123, 124, 125, 138, 139, 141, 195, 223, 301, 311, 312, 314, 315, 316, 318, 319, 321, 404, 405, 409, 411, 422, 424
Azmitia, E. 80, 174, 260, 261, 262, 264, 265, 319, 321, 371
Björklund, A. 46, 83, 104, 127, 128, 129
Black, A. H. 136, **179**, 194, 196, 268, 300, 301, 302, 318, 324, 345
Brazhnik, E. S. **145**
Chronister, R. B. **109**
Dalland, T. **351**
DeFrance, J. F. **109**, 122, 123, 124, 125, 141, 222, 372, 411
Ellertsen, B. **351**
Feldon, J. **275**
Gaffan, D. 172, 194, 303, 304, 305, 395, 396, 399, 403, 404, 405, 409, 410
Gall, C. **5**
Gray, J. A. **3**, 20, 44, 45, 85, 102, 105, 106, 122, 124, 125, 128, 136, 138, 141, 175, 194, 197, 226, 266, 267, **275**, 300, 301, 302, 303, 304, 305, 306, 315, 316, 318, 320, 323, 369, 371, 394, 405, 407, 408, 418, 422, 423, 425
Grossman, S. P. 23, 102, 132, 134, 192, 221, **227**, 260, 261, 262, 264, 265, 266, 267, 269, 270, 271, 272, 304, 344, 396, 417
Herrmann, T. **351**
Johnsen, T. B. **351**
Koella, W. P. 33, 103, 222, 323
Kramis, R. **199**
Livesey, P. J. 346, **351**, 370, 414
Lynch, G. S. **5**, 20, 21, 22, 23, 24, 83, 85, 104, 105, 112, 124, 175, 223, 224, 316, 317, 402, 408
McNaughton, N. **275**
Marchand, J. E. **109**
Molnár, P. 103, 104, 136, 193, 397, 423
O'Keefe, J. 173, **179**, 192, 193, 194, 195, 196, 197, 223, 226, 270, 271, 314, 410, 312
Olton, D. S. 317, **327**, 343, 344, 371, 419
Owen, S. **275**
Ranck, J. B. Jr 22, 105, 175, 195, 309, 311, 312, 313, 314, 315, 316, 319, 411
Rawlins, J. N. P. 46, 106, 139, 193, 266, **275**, 370, 401, 402, 407
Robinson, T. E. 128, 197, **199**, 221, 222
Rose, G. **5**
Segal, M. 84, 85, 102, 122, 123, 128, 129, 130, 134, 136, 140, 174, 193, 225, 315, 410
Srebro, B. 122, 139, 140, 141, 302, 424
Stanley, J. C. **109**
Storm-Mathisen, J. **49**, 82, 83, 84, 85, 104, 106, 123, 128, 140
Swanson, L. W. **25**, 44, 45, 46, 47, 84, 122, 128, 129, 138, 139, 400, 420, 422
Ursin, H. 21, 45, 106, 173, 267, 315, 319, **351**, 369, 370, 371, 372, 409
Vanderwolf, C. H. 134, 136, 173, **199**, 222, 223, 224, 225, 226, 303, 319, 322, 324, 395, 400, 419, 425
Vinogradova, O. S. 23, 47, 106, 140, 141, **145**, 172, 173, 174, 175, 197, 224, 261, 311, 314, 315, 322, 323, 348, 411, 421
Wahl, H. **351**
Weiskrantz, L. **1**, 21, 22, 44, 80, 85, 102, 106, 125, 128, 136, 138, 171, 172, 192, 194, 196, 225, 226, 260, 265, 267, 269, 272, 302, 303, 306, 315, 316, 317, 318, 321, 322, 343, 346, 370, 372, **373**, 394, 395, 306, 307, 400, 401, 402, 403, 404, 405, 408, 409, 411, 412, 414, 419, 423, 426
Winocur, G. 20, 172, 196, 225, 269, 305, 344, 388, 404, 419
Zaidi, Z. **351**
Zimmer, J. 47, 82, 83, 139, 140, 175, 423

Indexes compiled by William Hill

Subject index

access
  397
acetylcholine
  affecting GMP levels 116
  effect on pyramidal cells 131
  excitatory effects 114
  facilitation by 119
acetylcholine as transmitter
  52–54, 113, 130, 140, 208
acetylcholinesterase
  52
  localization 6–13, 15,
  52–54, 113, 130, 140, 208
active avoidance
  septal lesions and 230
activity, electrical
  see also theta activity, etc.
  during waking behaviour
  199–226
  low voltage fast waves
  209
  slow waves 204
adaptation
  409
afferents
  5, 425
  acetylcholine synthesizing
  54
  cholinergic 52
  connections 34
  deprivation 160
  origin 8
  potentials 21, 22
  spatial organization 165
  termination 113
afferent fibres
  impulse conduction 93

afferent volleys
  physiological role 100
  recording 92
aggression
  septal lesions and 229,
  230, 236, 237, 239, 253,
  272
alveus
  138, 141
amino acids
  in mamillary body 70
  in septum 70
  in stratum radiatum 67
  receptors 84
  transmitter function 83
γ-aminobutyrate (GABA) as transmitter
  59, 66, 70
  distribution 80
  effect of trauma 68
  membrane transfer 61
  synthesis 62
  uptake 71, 72
Ammon's horn
  30
  connections 31, 32, 47
  fibres 36
  inputs 38, 45, 122
  projections 46
amnesia
  295, 296, 373, 374, 388,
  395
  clinical description 375
  cued recall 377, 381,
  388–394, 402
  learning and 377, 382,
  384

  retrograde 343
  short-term memory in
  375, 395, 401
  topographical 382
amphetamine
  225
amygdala
  behaviour and 246, 249,
  262, 267
  connections 39
  input from 35
  stimulation 341
anaesthesia
  122
  low voltage fast activity
  and 211
anatomy
  3, 5–24, 237, 415, 419
  function and 5, 17, 420
  nomenclature 30, 234,
  321
  of projections 25–48
angular bundle
  45
anterograde degeneration method
  6, 13
anticholinergic drugs
  effect on behaviour 241,
  252
antimuscarinic drugs
  theta activity and 203
anxiety
  276, 286, 289, 306, 417
approach/avoidance conflict
  353, 354, 356, 357,
  371

431

approach-consummate cell
310, 416
aspartate as transmitter
62, 83, 85
  binding to receptors 84
  calcium release 69
  localization 63
  uptake 71, 72
atropine
322
  behaviour and 240
  effect on low voltage fast activity 210
  effect on theta activity 203, 212, 214, 223
avoidance behaviour
  effect of drugs 240
  time course of septal lesions 238
avoidance conditioning
383
avoidance learning
353, 354, 356, 360, 364
avoidance response
  pathways 253
axons
  cellular targets 15
  distribution 6
  uptake 46

barbiturates
276
basket cells
  see interneurons
behaviour
2, 3, 146, 409
  amygdala and 246, 249, 262, 267
  anatomy and 234, 419
  anticholinergic drugs and 136, 241, 252
  atropine and 240
  bilateral lesions affecting 329
  changes after neurotoxic drugs 82
  cholinolytic drugs and 240, 242
  cholinomimetic drugs and 240, 242
  cognitive maps and 414
  effect of crossed lesions 332

effect of fibre transection 246
effect of reserpine 213
electrical activity and 199–226
entorhinal afferents and 18, 20
entorhinal lesions and 252, 268, 331, 332
extra-septal pathways and 243
fimbrial lesions and 305, 331, 332, 336, 338
flexibility 337
fornix lesions and 245, 246, 268, 286, 301, 307, 331, 332, 338
habenula and 246, 248, 250
hippocampal lesions and 244, 306, 417
hypothalamus and 246, 262
large amplitude irregular activity and 202
monitoring 397
nicotine affecting 132, 134
partial septal lesions and 235
scopolamine affecting 240
septal lesions and 228, 247, 250, 260, 264, 266, 271, 303, 369
slow wave activity and 200, 209, 322
spatial 327–349, 412–417
stria medullaris and 248, 250, 268
stria terminalis and 248
subfield lesions and 336
theta activity and 192, 200, 202, 289
theta cells and 310
tranquillizers and 275, 276, 289, 292
behavioural correlates
146
behavioural inhibition system
276, 294, 297
behavioural responses to non-reward
275–307

blind-sight
399, 401
brainstem
  inputs from 36
  water intake after lesions 245
burst cells
208
bursting pacemaker
166

calcium
  transmitter release and 69
carbachol
240
catecholamine inputs
127
catecholamines as transmitters
55, 127–130, 292
cells
  see also theta cells, complex spike cells, etc.
  synaptic current injected into 103
  triggers for firing 315
  types 311, 315, 318
cell membrane
  excitability 89, 97
  potential 315
  resistance 98
cerebral cortex
34
chlordiazepoxide
276
chlorpromazine
216
cholecystokinin
58
choline acetyltransferase
53
cholinergic mechanisms
109–126
cholinergic neurons
140
cholinergic projections
212
cholinolytic drugs
  behaviour and 240, 242
cholinomimetic drugs
  behaviour and 240, 242
cingulate gyrus
34

inputs 38
projections 39, 45
coeruleo-hippocampal pathway
57
cognitive map
412–417, 418
behaviour and 414
building 413
lesion studies 187
problem-solving in 181
sensory inputs into
179–198
single-unit study 183–187, 196
theory 180
commissural fibres
84
complex spike cells
173, 197, 309–319
conditioned avoidance response
228, 230
drugs and 241
mechanism 236
septal lesions and 238, 247, 252
conditioned emotional response
231
connections
31
afferent 34
efferent 32
corpus callosum lesions
400
cortical afferents
156
cortical input
156
counter-conditioning
278, 285, 293, 294
cued recall in amnesia
388–394, 402

degeneration of hippocampus
173
dendrites and glutamate
69
dentate gyrus
23, 30
afferent projection fibres
18

as source of fast activity 200
connections 36
evoked response in 17
signals in 167
stimulation of 346, 347
theta activity and 203
dentate gyrus granule cell agenesis
301
descending hippocampal influences
167
diagonal bands
9, 12
diencephalic sources of input
35
discharge patterns
modification 21
disconnection
336
discriminative stimulus
189
discriminative training
190, 194
dopamine
129
effect on RSA and LVFA 216, 222
terminals 128
dopamine-β-hydroxylase immunohistochemistry
36, 56
dopaminergic blocking drugs
216
dopaminergic innervation
127
dopaminergic projections
130
dorsal ascending noradrenergic bundle
290–294

EEG activity
147
septal unit discharges and 146
sleep 221
eight-arm maze
345
electrical activity
see under activity

emotions
416
enkephalin
58
entorhinal afferents
66
behaviour and 18, 20
entorhinal area
30
connections 34, 35, 46–47
input from 167, 422
localization in 421
entorhinal area lesions
behaviour and 252, 268
septal behaviour and 331, 332
entorhinal cortex
46–47
efferent input 17
entorhinal-hippocampal connections
6
entorhinal projections
18
environmental stimuli
316, 318
epilepsy
85
exploratory behaviour
413, 425
extinction
277–284, 304, 305
extracellular postsynaptic potential
88

facilitation of synaptic transmission
87–108
afferent volley 92
changes in EPSP amplitude 96
changes in excitability of cell 97
effect of probability of discharge 96
mechanism 103
memory and 106, 407–412
nature of synaptic response 99
pattern of change 94, 100

facilitation of synaptic transmission, *continued*
  physiological role 100
  potassium 119
  specificity of induced charges 95
  specificity of results 99
  time course 114
familiarity discrimination 382
fear 358
feedback to septum 24
fibre system transections 246, 265
field potentials 21, 22
fimbrial lesions
  behaviour and 301, 305, 331, 336, 338
  partial reinforcement extinction effect and 284
fixed action patterns 294
flight 358
food intake
  brainstem lesions and 245
  septal lesions and 234, 271 237, 251, 253, 263, 264,
forebrain 27
  function 322
fornix lesions
  behaviour and 245, 246, 268, 286, 300, 301, 307
  memory and 396, 399, 405, 410
  partial reinforcement extinction effect and 284
  problem solving and 187
  spatial behaviour and 331, 332, 338
freezing 202, 358, 359, 360, 362
frequency potentiation 87 *et seq.*, 102–107, 109–120, 124
  memory and 106, 407–412
  function of system 2, 3, 37, 40, 260, 294–298, 417–427
  anatomy and 5, 17, 415, 420
  inhibitory role 294

GMP as second messenger 116, 123
gallamine 136
gastrin-like peptide 58
glutamate as transmitter 62, 83, 85
  binding to receptors 84
  calcium release 69
  dendrite sensitivity to 69
  localization 63
  membrane transport 65
  uptake 66, 72, 82
glutamic acid 103
glutamic acid decarboxylase 59, 71
granule cells 5–6
  agenesis 301
  discharge 23
  feedback from 23
  responses 89

habenula
  behaviour and 246, 248, 250
  projections 39
habenular nuclei input 37, 38
habituation 2, 151, 160, 162, 172, 174, 302
  inputs and 172–173
  mechanism 173, 176
  perforant path and 152–156
haloperidol 216
hippocampal lesions
  behaviour and 244
hippocampal pathology, comparative 373–406
hippocampus
  connections 31, 32
  development 26
  pathways 139
  relation to septum 26
histamine
  as transmitter 57
  depressing neurons 58
histidine decarboxylase 57–58
5-hydroxytryptamine
  as transmitter 54–55, 80, 82
  behaviour and 263
  effect on RSA and LVFA 215
  neurotoxic drugs affecting 81
hyperdipsia
  recovery from 263
  septal lesions and 233, 237, 243, 245, 250, 253, 264
hypothalamus 45
  behaviour and 246, 262
hypothesis preservation 346

information processing 40, 324, 415, 418, 421
inhibitory basket complexes 21
inhibitory function 4, 294–295, 362
inhibitory responses
  abolition 174
  in place cells 186
interneurons 21, 137, 146
  AChE-positive 140
  afferentation 136
  input to 15–17, 22
  non-response to acetylcholine 131
  self-reexciting in 104
  septal innervation 16, 136
  septal projections and 15, 21
  theta cells as 311, 313
interpeduncular nuclei 37
intracellular postsynaptic potential 88

SUBJECT INDEX

isolated hippocampus
 sensory reactions   164, 174

Korsakoff syndrome
 44, 374, 375, 380, 388, 390, 394

large amplitude irregular activity
 behaviour and   202
learning
 109, 344
 amnesia and   377, 382, 384
 location of discriminative stimulus and   189
 mechanism   411
 memory and   327, 409
locus coeruleus
 inhibitory responses and   174
 input from   127–128
 noradrenaline projections   214
locus coeruleus lesions
 transmitters and   56
low voltage fast activity (LVFA)
 effect of morphine   218, 222
 effect of serotonin   215
 mechanism   212
 reticular formation affecting   218
 types   209

mamillary body
 amino acids in   70
 connections   34
 inputs   37, 44
 memory and   394
 transmitters and   71
medial habenular nucleus
 35
medial septal lesions
 behaviour and   286
medial septal nucleus
 AChE-positive cells in   9, 12

medial septal stimulation
 123
 potentiation from   109
medulla oblongata
 127
memory
 295, 327, 344, 348, 388
 see also amnesia
 anatomy   394
 consolidation   376, 380
 cued recall   377, 381, 388–394, 402
 frequency potentiation and   106, 407–412
 hippocampal lesions and   393
 learning and   409
 lesions of fornix and   396, 399, 405, 410
 nature of   106, 402
 normal   393
 recognition hypothesis   381–382, 399, 404
 role of hippocampus   391
 spatial   328, 340, 343, 380, 401, 403
 topographic   400
mismatch cells
 413, 416
monoaminergic inputs
 127
morphine
 216, 222
motor activity
 3, 210, 295
movement
 theta activity and   202, 225, 296, 425
muscarinic receptors
 130

neocortical low voltage slow waves
 209
neurons
 8, 145–177
 see also interneurons, etc.
 activation   103
 CA3 and CA1   148
 characteristics   160, 164
 degeneration   158
 effect of afferent lesions   162–164

 effect of perforant path transection   152, 175
 effect of reticular formation stimulation   155, 162
 effect of septo-hippocampal disconnection   150, 160
 GABA in   59
 histamine depressing   58
 instability   154
 peptide-containing   58
 reaction to stimuli   160
 sprouting of axons   168
 tonic reactions   156
 types   309
neuronal plasticity
 87, 129
neurotoxic drugs
 81, 82
nicotine affecting behaviour
 132, 134
nicotinic transmission
 130
non-spatial tasks
 196, 414, 415
noradrenaline
 blockade   292
 depletion   214, 222
noradrenaline as transmitter
 55, 56
noradrenergic input
 127–128, 214, 290–294
novelty versus inhibition
 172

orbitofrontal lesions
 267
orientation
 302, 348

partial reinforcement extinction effect
 277, 292, 300, 302, 303
 fimbrial lesions and   284
 fornix lesions and   284
 septal lesions and   278
passive avoidance behaviour
 241, 242, 245, 269, 272
pathology, comparative
 373–406
pathways
 138

pathways, *continued*
  behaviour and 243
  mediating punishment
    242, 244, 252
  producing theta activity
    205
pathway lesions
  53
peptides as transmitters
  58
perforant path
  habituation and 152–156,
    174
  stimulation 407–408
  transection 152, 175
perirhinal cortex
  47
perphenazine
  216
phenothiazines
  225
physostigmine
  115, 159
pimozide
  216
place cells
  185, 192, 193, 196, 310
place hypothesis
  180, 412
  problem solving by 182
post-commissural fornix
  47
postsynaptic potentials
  88 *et seq.*
potassium
  103, 119
potentiation, frequency
  87 *et seq.*, 102–120
  from medial septal
    stimulation 109
  grading 124
  loss after electroshock
    411
  mechanism 105, 112
  memory and 407–412
  nucleotides and phospho-
    proteins in 124
  protein synthesis and 105
  role 125
potentiation, post-tetanic
  111, 112
potentiation, short-term
  109

potentiation, tetanic
  111, 119
pre- and parasubiculum
  33, 34
prefrontal cortex
  47
presynaptic afferent volley
  87, 90
  effect of glutamic acid
    103
problem solving
  180
  by place hypothesis 182
  by stimulus–response
    hypotheses 183
projections
  5–24, 44, 423
  activation 18
  anatomy 17, 25–48, 264
  cholinergic nature of 212,
    285
  cortical 45
  differences in 47
  distributions and targets
    13
  from raphe 80
  noradrenaline 214
  potentiation of synapses
    102
  septal into hippocampus
    6, 17, 235
  termination 21
  topographic considerations
    26
  transmitters and 70
protein synthesis
  105
punishment
  effects of 270
  inhibition and 252
  response to 231
  septal lesions and 266
pyramidal cells
  137, 314
  acetylcholine acting on
    131
  excitability 125
  output of CA1 and CA3
    138
  sensitivity to transmitters
    104
pyriform cortex
  424

radial arm maze procedure
  328
rage
  see aggression
raphe
  projections from 80
raphe lesions
  effects of 263
  transmitters and 54
reinforcement
  197, 232
  disinhibition of 235
Renshaw cells
  137
reserpine
  213, 222
residential maze test
  352, 354, 355, 357
responses
  distribution of 17
  non-rewarded 233
response enhancements
  112
response inhibition
  228
reticular formation
  cholinergic input from
    205–211, 218
  habituation in 168
  inducing theta bursts 165,
    204
  stimulation 155, 162, 224
  stimulation producing low
    voltage fast activity
    210
rhythmic activity
  158
rhythmical slow activity
  see theta activity

Schaffer's collaterals
  138, 139, 158, 173
scopolamine
  203, 207, 240
sensory input
  lesion studies 187
  non-spatial 196
  response to 192
  single-unit study 183–187,
    196
  to hippocampal cognitive
    map 179–198

SUBJECT INDEX

sensory information
  visceral 39, 40
septal input
  122, 141
  function 166–167
septal lesions
  228, 359
  aggression and 236, 237, 239, 253, 272
  anatomical considerations 234
  behaviour and 228, 247, 250, 252, 260, 264, 266, 271, 303, 331, 369
  behaviour and anatomy 234
  disinhibitory effect 248, 251, 260, 269, 272
  food intake and 234, 237, 253, 263, 264, 271
  hyperdipsia and 233, 237, 243, 245, 253, 264
  metabolic consequences 233
  partial 235
  partial reinforcement extinction effect and 278
  pharmacology 239
  recovery 237, 262
  theta activity and 281
  time course 238, 261
septal nuclei, projections
  44, 141
septal stimulation
  209
septal syndrome
  227–273
  factor analysis 356
  inhibition factors 371
  localization in 363
  multivariate analysis 351–372
septofimbrial nucleus
  33
septofimbrial nucleus
  in spatial behaviour 327–349
septo-hippocampal lesions
  bilateral 329
  crossed 332
septum
  connections with hippocampus 31, 141
  development 26
  divisions 28
  efferent connections to hippocampus 32, 141
  feedback to 24
  inputs from 34
  pathways to hippocampus 138
  relation to hippocampus 26
  role 17, 227 et seq., 424
  sensory response to 168
  ventral connections 265
serotonin
  see 5-hydroxytryptamine
sleep
  221, 311
  electrical activity during 199
  paradoxical (active) 204, 221
  theta activity and 204
space units
  192, 412
spatial analysis
  295, 296
spatial behaviour
  419
  septo-hippocampal connections and 327–349
spatial functions
  380
spatial memory
  340, 343, 380, 401, 403
spatial orientation
  348
spatial units
  326, 317, 410
species differences
  45, 46, 122, 140, 297, 318
  in pathology 373–406
stimulus-response hypothesis
  problem solving by 183
stratum lucidum
  63
stratum oriens
  91, 122, 141
stratum radiatum
  67, 91, 141
stria medullaris
  behaviour and 248, 250, 268
stria terminalis
  39
  behaviour and 248, 268
  function 236
stria terminalis lesions
  267
subfields, lesions
  336
subicular complex
  30
  connections 31, 32, 33, 39, 44
  glutamate and aspartate in 63
  inputs 34, 38
  projections 46
subiculum lesions
  transmitters and 68
substance P
  58
supracallosal serotonin axons
  82
synaptic transmission, facilitation
  87
  changes in EPSP amplitude 96
  mechanism 103
  nature of synaptic response 99
  pattern of change 94, 100
  physiological role 100
  probability of discharge and 96
  specificity of changes 95, 99
synchronization
  323

thalamic nuclei
  input from 37
theophylline
  effects 117, 118
  GMP and 117
theta activity
  136, 153, 171, 175, 197, 311, 423, 425
  amphetamine affecting 225
  atropine-resistant 203, 212, 218, 223, 318
  atropine-sensitive 203–209

theta activity, *continued*
  behaviour and   192, 200, 289, 425
  cause   311–312, 226
  control   285
  dentate gyrus and   203
  dopamine and   216
  effect of anti-muscarinic drugs   203, 207
  effect of atropine   214, 218
  effect of reserpine   222
  effect of serotonin   215
  frequency train   224
  functional significance   321
  low voltage fast activity and   210
  mechanism   204, 311–312, 319–321
  motor activity and   295
  movement and   131, 197, 226, 296, 425
  orientation and   294
  reticular formation stimulation and   204–209, 218, 224
  septal driving and   319
  septal lesions and   281
  significance of frequency   288
  sleep and   204, 222
  tranquillizers and   290
  types   200, 293
  walking and   202, 225
theta bursts
  165
theta cells
  131, 173, 197, 309–319
  as interneurons   311, 313
  behaviour and   310
  firing   310
theta rhythms
  146
  induction   159
tranquillizers
  306
  action   290
  behaviour and   275, 289, 292
  effect on septo-hippocampal system   276, 289, 292
  theta activity and   289, 290
  transmitters and   292
transmission, synaptic
  facilitation of   87–108
transmitters
  see also under substances concerned
  identification of   84
  localization of   49–86
  markers for   50, 71, 72
  raphe lesions and   54
  tranquillizers and   292
  trauma and   67, 68
d-tubocurarine
  130, 131, 136
  visual responses   316

walking, and theta
  131, 197, 202, 209, 225
water intake
  see also hyperdipsia
  septal lesions and   233
'wave phenomenon'
  154
Wernicke's disease
  374, 394
X-irradiation
  301